AF832604

Eucalyptus Kraft Pulp Refining

Vail Manfredi

Eucalyptus Kraft Pulp Refining

Springer

Vail Manfredi
Retired engineer from the Brazilian pulp and paper industry
Jundiaí, São Paulo, Brazil

ISBN 978-3-031-47284-8 ISBN 978-3-031-47285-5 (eBook)
https://doi.org/10.1007/978-3-031-47285-5

© The Editor(s) (if applicable) and The Author(s), under exclusive license to Springer Nature Switzerland AG 2024
This work is subject to copyright. All rights are solely and exclusively licensed by the Publisher, whether the whole or part of the material is concerned, specifically the rights of translation, reprinting, reuse of illustrations, recitation, broadcasting, reproduction on microfilms or in any other physical way, and transmission or information storage and retrieval, electronic adaptation, computer software, or by similar or dissimilar methodology now known or hereafter developed.
The use of general descriptive names, registered names, trademarks, service marks, etc. in this publication does not imply, even in the absence of a specific statement, that such names are exempt from the relevant protective laws and regulations and therefore free for general use.
The publisher, the authors, and the editors are safe to assume that the advice and information in this book are believed to be true and accurate at the date of publication. Neither the publisher nor the authors or the editors give a warranty, expressed or implied, with respect to the material contained herein or for any errors or omissions that may have been made. The publisher remains neutral with regard to jurisdictional claims in published maps and institutional affiliations.

This Springer imprint is published by the registered company Springer Nature Switzerland AG
The registered company address is: Gewerbestrasse 11, 6330 Cham, Switzerland

Paper in this product is recyclable.

This book is dedicated to Teresa, my life's accomplice, to our sons, and grandsons.

Preface

Pulp refining is one of the most fundamental operations in paper manufacturing. Although this fact is fully recognized by the great majority of papermakers, the ways to evaluate and to optimize quality and performance in pulp refining are poor in most paper mills. This situation is mainly due to the weak understanding of the effects of refining variables on fibers and the corresponding impacts on paper properties, when refiners are in operation. The knowledge of refining machinery variables for better results on the pulp quality developments is also another point to deserve improvements. Even the most important operation and quality parameters to evaluate refining behavior are not fully understood, not only by papermakers, but also by researchers on this field. Refining control is generally based on applied/consumed energy and pulp freeness development: something recommended just for beginners and not compatible to the overall demands for obtaining the most appropriated performances in paper machine runnability and paper sheet quality. The lack of technical knowledge and orientation to corrective actions by technicians both do impact costs, paper machine performance, and final product quality.

These fragilities in pulp refining knowledge by people working in refining rooms over the world surely are to be reduced by reading this amazing book has been written thanks to Professor Vail Manfredi's expertise. Master Manfredi is one of the most renowned experts on the theories and practical applications of pulp refining for papermaking purposes.

I know Vail Manfredi since the early days of his academic studies with the purpose to become forest engineer with a major in pulp and paper science and technology. During 1975/1976, I had the privilege to be his teacher and to advise him as undergraduate student and also as trainee on his initial practical studies with pulp and paper in the P&P laboratories at the School of Agriculture "Luiz de Queiroz" from the University of Sao Paulo/Brazil. Our first published technical article in partnership between we both and Professor Luiz Ernesto Barrichelo and Regina Fazanaro (another trainee student) was related to measure fiber population and fiber coarseness using very simple and immediate laboratory techniques. The title of that article, translated to English, is: "Quantitative Analysis of Cellulosic Fibers," published through the Brazilian magazine O Papel, in September 1976. Since those early times, I maintain in my mind the figure of Vail Manfredi as a dedicated, keen, and responsible person and an excellent professional, what has been proved along his career. After his graduation in 1978, Vail was also my graduate student, when working for his Master of Science degree at the same university. He has always shown to have a great ability, capacity, and understanding to face practical and theoretical issues. His master thesis was oriented to evaluate the changes of wood quality and kraft pulping ability along the stem heights of *Eucalyptus* planted trees.

From the university to the industry was an easy jump. Vail Manfredi had a fast growth in his career by developing R&D innovations in some Brazilian eucalyptus pulp/paper manufacturing companies (Ripasa, Aracruz, Bahia Sul, and Suzano). At Aracruz Celulose (from 1983 till 1992), he had his first challenge with pulp refining, thanks to the project and operation of a pilot plant to evaluate and to optimize *Eucalyptus* bleached kraft pulp refining in comparison to other pulp fibers. This project had also a strong marketing orientation, since the aim was to help *Eucalyptus* pulp users to take the most from this recent product made available in the world pulp markets. When Vail changed jobs from Aracruz Celulose (market pulp mill) to Bahia Sul (market pulp and printing and writing paper mill) in 1992, he had as initial tasks some similar challenges as before in Aracruz: responsibility for projecting the technical area for R&D studies, including the development of the design and operational issues for another pulp refining pilot plant. In both cases, the pilot plants had been built to be very close "to the real life" and to represent pulp refining in industrial operations. These studies were so important, that pulp refining has become a passion technical issue on Vail's professional life.

Vail Manfredi has also an admirable personal profile for sharing knowledge through articles, speeches, courses, and participation in technical associations (TAPPI/USA as a member of the IRMC – TAPPI International Research Management Committee; ABTCP/Brazil, etc.). I feel proud on this because his behavior is very similar to mine, and for this reason, I keep a happy and hidden feeling that my initial guidance in his career has been approved and taken as a simple example.

This present book on pulp refining may be considered as his masterpiece, crowning his professional career. The book has been written in his retirement period, what represented available time for reading, studying, thinking, and making reflections and comparisons for building the text.

As a researcher and professor in many courses on pulp refining, it is possible to observe that Professor Vail Manfredi had as purpose to guide the book readers through the history of paper and this essential product manufacturing process, finally reaching, with a lot of details, the process of pulp refining: and the consequences of refining in papermaking. The texts are very complete, associating the theoretical points of how refining happens with operational conditions, pulp fiber changes, and paper quality requirements. He pays attention to the different concepts and principles of the refining operation, trying to get the most from the different theories, basing his writing on deep thinking and interconnected reflections.

As a talented forest engineer with a rich career on pulp and paper technology, his book is written searching for associations among changes in pulp anatomical elements, refining variable operations, machinery devices, and the corresponding effects on paper quality. It is my understanding that the chapters about refining variables control, optimizations, and measures to evaluate refining performance were outstanding and very useful to technical people. Also, students and young engineers, devoted to pulp and paper manufacturing, are to have a great chance to upgrade their knowledge by a careful reading of all book chapters, since the book searches for connections from fibrous raw materials to final paper products. The young people, beginners in the P&P sector, are invited to reflect and to imagine how someone can be able to significantly improve his knowledge by reading the available technical literature. I was invited to follow Vail's writings, and I'm quite sure that he has done complete reading of over 1400 technical references obtained from the world technical and scientific literature. From the range of age for these technical materials, I'm also sure that this collection of knowledge has been gradually accumulated along his close to 50 years of dedication to the pulp and paper science and technology.

For all these points (and many others not mentioned by my own forgetfulness), I recommend this book to all those interested on learning about pulp refining and its connections with paper manufacture, behavior, and performance. I'm very confident that this book is to become an important part of the global history of papermaking technology.

Porto Alegre, Rio Grande do Sul, Brazil — Celso E. B. Foelkel

Acknowledgments

I thank the authors and associations below for permission, as a courtesy, to use photos or illustrative figures from their publications in the text of this book.

Associação Brasileira Técnica de Celulose e Papel – ABTCP – Brazil
Dr. Celso E. B. Foelkel – Eucalyptus On-Line – Brazil
Dr. Herbert Sixta – Aalto University, Finland
Dr. Huan Liu – Shaanxi University of Science & Technology – China
Dra. Patrícia Kaji Yasumura – Instituto de Pesquisas Tecnológicas – Brazil
Pulp and Paper Technical Association of Canada – PAPTAC Canada
REGMED Indústria Técnica de Precisão Ltda – Brazil

Special thanks to
Dr. Celso E.B. Foelkel
for the conviction, encouragement, and support given in the elaboration of this book.

Contents

Chapter 1
Introduction

Pulp refining is defined as the mechanical treatment applied to fibers in suspension to develop the desired properties in the paper. Since the most remote times (T'sai Lung, 105 AD), the treatment is carried out through successive cycles of mechanical impacts on the fibrous materials.

The treatment is done on fibers in suspension, promoting hydration of the fiber cell wall leading to their delamination, making it porous, flexible, and less resistant, or, depending on the amount of applied energy, leading to its complete destruction causing fiber rupture (reduction in the average fiber length).

The process is a stress fatigue one in which the fibers are subjected to compressive and shear forces that reduce the cell wall intrinsic resistance. The fibers tend to collapse becoming more flexible; deformations of the fibers and fines are generated which increase the specific surface available for the connections between fibers. The process is characterized by high energy consumption.

In the production of most commercial papers, after the selection of the pulp to be used, refining is the main determinant of the quality of the finished product, as it is the only unit operation that irreversibly modifies the morphology of the fibers.

The process, however, has some unwanted side effects, such as the reduction of both specific volume (bulk) and opacity of the paper, in addition to increasing the difficulty of removing water in the areas of paper formation and pressing in paper machines.

Nowadays, with paper machines operating at speeds of up to 2000 m/min, it is essential that refining is properly conducted so that productivity is not affected, and manufacturing costs are not exceeded beyond what was planned.

In this book, we present a brief history of papermaking followed by comments regarding wood as a source of fibers, including its chemical and anatomical characteristics and, also, the influence of these aspects on the quality of the pulp produced as well as the effects of the pulping process, mainly chemical process, on pulp quality, in addition to addressing how these wood characteristics influence both the refining process and the quality of the final paper.

© The Author(s), under exclusive license to Springer Nature Switzerland AG 2024

V. Manfredi, *Eucalyptus Kraft Pulp Refining*,
https://doi.org/10.1007/978-3-031-47285-5_1

For the refining treatment, in addition to presenting its history and development of the equipment involved, we will approach the evolution of refining models, the so-called "refining theories," developed with the aim of expanding the understanding of the whole process, especially how the treatment influences the characteristics of the fibers and how they influence the economy of the process and the quality of the final products.

We also present a broad discussion, based on experimental results, on the contribution of the main operating refining variables, as well as the main strategies that can be used industrially to optimize the operating results. From this evaluation, the parameter that complements the specific edge load theory was identified. This parameter is related to the retention time of the fiber flocs inside the refiner.

This evaluation also confirmed that refining cannot be adequately characterized using only two operating parameters, as is traditionally done. Four operating parameters are required: refining energy, refining intensity, consistency, and the pulp to be refined.

A specific chapter discussing the refining process optimization is also presented. In this chapter, a new proposal for disc configuration, which should allow considerable gains in refining energy, is discussed. The suggestion is the adoption of channels with variable width in the refining discs.

We hope, in this way, to contribute to improve the understanding of the refining process and, also, help students and technicians working in this extremely important topic in the paper manufacture.

Chapter 2
The Beautifull History of Paper

Before the invention of paper, various types of materials were used as a means of recording written information. Different cultures have used clay, wood, bark, leaves, stone, metal, papyrus, parchment, and cloth as recording media at one time or another. As a curiosity, around the fifth century, in Central America the Aztec and Mayan peoples also developed their appropriate and durable writing supports, which were made from birch (*Betula alba* L. and *Betula nigra* L.) and from the bark of the fig tree (*Ficus* sp.), being called huun and amate, respectively [1–3].

In antiquity, the Egyptians developed a way of using papyrus, soaking it with water and kneading it until it formed a parchment shape, with a thickness like a fabric. The oldest papyrus scrolls, on which we have something written, are over 5000 years old. Papyrus helped to shape and strengthen Egyptian society and was the support of writing in current use until the first centuries of the Christian era throughout Europe, Asian regions, and, of course, Africa, where it originated. Without the papyrus, the course of Mediterranean history and literature would have been quite different. The same is true for the effect on European history [1, 3].

The word paper originates from the Latin "papyrus," an aquatic plant that exists in the delta of the Nile River and is considered sacred because its flower resembles the rays of the sun, the maximum deity of the ancient Egyptian people [2].

Although there is evidence that paper, as we know it today, has an earlier origin, the first known record of a papermaking process dates back 105 AD. Tsai Lun, an official at the court of the Chinese Emperor Ho Ti, directed some workers to crush mulberry bark together with pieces of bamboo and old fishing nets. The crushed raw materials were dipped in water and boiled for a long period of time, being reduced to a pulp that was beaten and dispersed in vats with water. From these vats, samples were collected on a sieve, with a bamboo frame and a stretched cloth, and placed to dry in the sun in a natural drying process [1–11].

Thus, the first sheet of paper was born, whose basic principle of production remains almost unchanged after almost two thousand years of its invention.

© The Author(s), under exclusive license to Springer Nature Switzerland AG 2024
V. Manfredi, *Eucalyptus Kraft Pulp Refining*,
https://doi.org/10.1007/978-3-031-47285-5_2

In addition to being easier to produce than parchment, paper had other advantages such as less possibility of breakage and better ink absorption, making it difficult to erase or alter what was written, important aspects when it came to preserving official documents and records. Following the commercial routes of the great caravans, the use of paper extended through the four corners of the Chinese Empire. Paper became the preferred vehicle for maps, battle plans, and astronomical charts, as well as for books, architectural plans, and scientific, musical, religious, and mathematical notes, which facilitated experimentation in each discipline [2, 3, 6] (Fig. 2.1).

In China, paper was essential for the evolution of woodblock printing, and in the Tang dynasty (618–906 AD), thousands of religious and secular books were produced [6]. The first reported use of paper for toilet purposes also comes from China in the sixth century, where it was made from rice straw [3].

The Chinese kept the secret of papermaking for many centuries until, in the sixth century, their invention was taken to Japan by Buddhist monks. The Japanese first used paper only for records and official documentation, but with the rise of Buddhism, the demand for paper grew rapidly [2, 3, 11].

In the eighth century, the papermaking technique arrived in Korea, when the country was invaded by the Chinese. The pulp used to make paper was prepared by grinding fibers from hemp, rattan, blackberry, bamboo, rice straw, and even seaweed [3, 10, 11].

Fig. 2.1 Production of manual paper sheet in the ancient China [16]

Over time new raw materials were introduced, such as lime, to help with the defibration, ramie, silk, and textile fibers from rags and old clothes, an excellent raw material because due to the processing already received for fabrics, they could be easily prepared, with little mechanical treatment in pestles. Pressing the wet paper before sun drying as well as finishing techniques, such as coating the paper surfaces with a thin layer of a wheat starch dissolved in boiling water, were introduced [2, 6–11].

Taught by Chinese papermakers, Tibetans began to make their own paper in place of their traditional writing materials, a large palm leaf. Even though they adopted paper, Tibetan books traditionally still reflect the long, narrow format of the original palm-leaf books [3].

Through trade routes, the papermaking process was introduced to Central Asia, Persia, and India. In 751, during the Battle of Talas, when the Arab army attacked the city of Samarkand, an important outpost of Chinese trade caravans in Central Asia, prisoners were taken who knew the techniques of papermaking and exchanged secrecy for freedom. Samarkand quickly became known for the excellent quality paper it produced, mainly due to the flax and hemp used as raw materials and the quality of the water used in the manufacturing process [2–4, 10–12, 28].

In Baghdad, paper production began in 795, with the first factory already using hydraulic energy to crush the raw materials used [2–4, 12, 28]. In the Arab world, paper was used not only for books and documents, but also as wrapping material and napkins [3].

From then on, the diffusion of knowledge about paper production followed the Muslim expansion along the North African coast and from Alexandria, Tripoli, and Tunisia. Furthermore, in the eleventh century, trade routes involving Italian ports that maintained active trade relations with the Arab world and land routes from Spain to France were vehicles that disseminated papermaking techniques throughout Europe [2, 10, 12, 14, 28].

Outside of Arab, Japanese, and Chinese jurisdictions, paper began to be manufactured in Italy, in Fabriano, in 1276, where the identification of paper was also created using watermarks. In France, manufacturing began in the city of Troyes, in 1348, and in Germany, in 1390, in the city of Nuremberg. In England, the first record of papermaking occurred in 1494 [2, 3, 10].

Western Europeans were initially suspicious of the paper. The Christian world probably thought of this as a manifestation of Muslim culture and rejected it. In fact, in 1221, Holy Roman Emperor Frederick II declared all official documents written on paper invalid. This decree also helped protect the interests of powerful European landowners who monopolized the parchment markets. It was not until the introduction of printing in the fifteenth century that Western Europeans would fully embrace paper [3].

The raw material chosen by European manufacturers was, most of the time, cotton or linen fiber rags that were separated, cleaned, and heated in an alkaline solution, first in an open vat and later under steam pressure. The rags were then washed and macerated until they formed a pulp that was used in the manufacture of paper [3, 11].

Most paper mills at the time used mills powered by water wheels for this task. The flow of water would turn a wheel and its energy would be transferred to a hammer mill that would break apart the rags [3].

In Europe, the main medium for writing during most of the Middle Ages was parchment. Even though it is much more resistant than paper, as it is animal skin, it was gradually replaced by paper because it is also more expensive. But writing by hand was a slow and laborious process, both to produce the original of a work and to reproduce it.

The Chinese had already invented xylography, a technique in which graphic signs were carved in relief on wooden boards and applied to the paper as if they were a stamp. This process reached Europe and, in 1450, was perfected by the German Johann Gutenberg, who created movable types made of metal, which could be regrouped to print different texts, better adapting to the reduced European alphabet in relation to the thousands of Chinese characters.

The texts considered important were, until then, handcrafted by copyist monks through long and laborious processes, they were produced much faster, and the dissemination of information became much faster and accessible to all, which required increasing amounts of paper [2, 9, 11–14]. The papermakers faced two major problems: the supply of adequate fibrous material and the tedious artisanal papermaking process. Solving these problems took centuries, but when it did, it profoundly impacted the industry's technology development [7-10, 13].

In Europe, the use of white cloth rags, mainly those made of linen and hemp, gave rise to what became known as rag paper, and for several centuries these inputs were the fundamental raw materials for the paper production in the West. With the increase in the demand for paper, the great demand for this raw material created an international trade in rags. Great Britain, in addition to having its own stocks of rags, receives them from Japan, South America, and the Canaries. But its main suppliers were Germany, France, Russia, Italy, Egypt, Turkey, Belgium, and Holland [10].

The rags were separated, torn, and then crushed using a pestle system with hammers, moved initially manually activated and then by the water wheels, forming a paste that was placed in a vat with water, allowing the introduction of a screen capable of retaining a layer of fibers, and allowing the water to drain. After removing the frame from the vat, the water was pressed out and the remaining pulp was left to dry. The frame could not be reused until the previous sheet of paper was removed from it [10–12].

It was a completely handmade and time-consuming process in which a good worker could not make more than 750 sheets a day, which explains the high value of all literary pieces of the time and their little use. The process remained artisanal and time-consuming until the nineteenth century when, in December 1798, when the French mechanical engineer Louis-Nicolas Robert invented the first papermaking machine (Fig. 2.2) [2, 11, 12].

The machine could produce wet paper 12–15 m long by about 40 cm wide. It consisted of a movable screen that received a continuous flow of suspended fibers

Fig. 2.2 Diagram of Didot's paper machine [5]

and delivered an entire sheet of damp paper to a pair of pressure rollers. When the continuous strip came out of the machine, it was manually hung over a series of cables or bars to dry. This sheet was cut into sheets [2, 12].

Two English industrialists, Henry de Fourdriner and his brother, perfected the invention by transforming it into a continuous production machine, known as the Fourdrinier Machine [2, 7, 8, 11–14]. The machine basically consisted of an endless screen moving on a horizontal table on which the suspended fibers were applied at one end through an inlet box. Drainage occurred when the forming fabric was moved to the other end, where the wet sheet was removed after pressing. Continuous paper drying came later.

This was probably the most important invention in the history of papermaking, having a great impact on the development of modern civilizations.

With the invention of the printing press, allowing linotype printing on paper and the industrialization of paper manufacturing, the dissemination of information became much faster and accessible to all. The books, newspapers, and magazines appeared, in an increasingly democratized process in which generalized culture, science, faith, terror, and printed advertising have since spread [2, 4, 14].

America's first paper mill was founded in 1690 by William Rittenhouse near Germantown, Pennsylvania. The first newspaper in the North American colonies was the Boston, published in 1705; the second was the Boston Gazette, first published in 1719. The third, also dated 1719, was Bradford's Mercury, published by Andrew Bradford, son of the printer William Bradford. To supply paper for the New York Gazette, William Bradford opened a paper mill in New Jersey around 1726 [11].

When paper production grew in volume, the availability of raw materials became a serious problem. At the time, rag overcoats were used, but the clothes available were not enough to meet demand, as there was strong competition with the textile industry in addition to export bans by some sovereigns. Rags became scarce and the search for alternative and abundant raw materials began [2, 4, 11].

In 1719, Reamur suggested using wood instead of rags. American papermakers began experimenting with alternative raw materials as early as the 1790s by testing local fiber sources such as bark, straw, and corn stalks. In search of something to replace the rags, Mathias Koops edited, in 1800, a book printed on straw paper [2, 11].

In Brazil, currently the world's largest producer of eucalyptus pulp, the first paper mill emerged with the arrival of the Portuguese royal family. Located in Andaraí Pequeno, in the city of Rio de Janeiro, it was founded in 1808 by Henrique Nunes Cardoso and Joaquim José da Silva, Portuguese industrialists transferred to Brazil [2]. In 1809, Friar José Maurício da Conceição Velloso was then allowed to open a paper mill embira, a shrub from which a very resistant fiber is extracted, as raw material [2, 10].

In 1837, André Gaillard opened a paper factory in Rio de Janeiro with machines imported from France, and, in 1841, another factory was opened in Engenho Velho by the sculptor Zeferino Ferraz. Both factories produced low-quality paper, which was used only for packages and packaging [2, 4].

In the mid-nineteenth century, the development of coffee culture brought great progress to the then Province of São Paulo and, with the arrival of European immigrants, it began to experience a great industrial development that generated several enterprises. One of them, idealized by the Baron of Piracicaba, in the city of Itú, intended to create conditions for the installation of industries taking advantage of the hydraulic energy possible in the region due to the existence of the waterfall on the Tietê river. In this place, in 1889, the company Melchert & Cia started the construction of the Salto Paper Factory, which operates until today, duly modernized, producing special papers, being one of the few factories in the world that manufactures money paper [4].

In 1844, Friedrich G. Keller made fiber pulp, using wood by the griding process, but still adding rags to the mixture. The pulp was obtained by rubbing pieces of wood against a rotary grinder in the presence of running water (Fig. 2.3).

Heinrich Voelter bought Keller's invention in 1847 and, in 1848, hired Johann Matthäus Voith to build the first wooden grinder, allowing, from 1850, the installation of the first machines to obtain wood fiber for paper production.

The fibers were separated and transformed into what came to be known as "mechanical pulp." In 1852, the Voith spindle grinder appeared and, soon after, in 1859, the Voith refiner was introduced to improve the quality of the paste obtained by shredding the tailings [2, 14].

In 1854, a process for producing cellulosic pulp was discovered in England through treatment with chemical products, giving rise to the first "chemical pulp." Seeking to separate the cellulose fibers from the lignin, several processes were discovered: such as the soda process, the sulfite process, and the sulfate process ("kraft process").

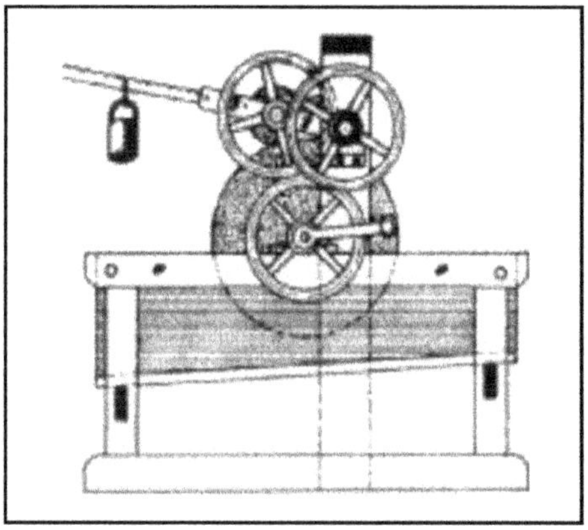

Fig. 2.3 Equipment to obtain pulp from wood developed by Keller [5]

From the second half of the nineteenth century onward, wood began to replace rags in papermaking in Europe, and is currently the main fibrous raw material used in the production of a large types of paper and paperboard [2, 8-10, 13, 16–21].

The use of wood as a raw material, although relatively recent, was one of the most important facts in its history, having an effect comparable only to the development of continuous production in paper machines [7–9, 23].

Although technology has significantly changed, the basic operations remain the same. Papermaking involves a five-step process in which the fibrous material is broken down and the suspended fibers are drained through a screen to form a fiber sheet. The sheet is pressed, dehydrated, and modified according to their final use [3].

Chapter 3
The Wood Pulping Processes

The purpose of the pulping process is to release the constituent fibers of the wood, which can be done either mechanically or chemically. These two processes have considerably different yields and fibers are produced with different inherent properties. In general, chemically produced fibers have a longer average length, are less damaged, are more flexible, and are more hydrophilic than mechanical pulps [24–27].

An efficient pulping process depends on the uniformity of the wood's characteristics and must extract the maximum benefits that wood can offer. In these two cases, both the production cost and the quality of the pulp obtained are related to the characteristics of the used wood [29–31].

However, wood is a naturally variable material, and, to maintain relative uniformity, several authors [18, 20, 31] have proposed the classification of woods be used according to their characteristics, looking to optimize both pulping and the quality of the resulting pulps. This procedure also opens up opportunities to develop pulps with different characteristics for specific uses. Thus, there must be a close connection between the forest and the pulping process. It is worth remembering that it is possible to produce bad pulp from good wood, but the reverse is not true [23].

3.1 Mechanical Pulping

Mechanical pulps are produced by the separation of wood fibers by abrasion (shearing), in the presence of water, from the application of mechanical energy in a friction device.

The processes are characterized by high yield (90–98%) and energy consumption (1000 kWh/t). Yield is high because only a few carbohydrates and extractives are dissolved during the process [34, 267].

© The Author(s), under exclusive license to Springer Nature Switzerland AG 2024

V. Manfredi, *Eucalyptus Kraft Pulp Refining*,
https://doi.org/10.1007/978-3-031-47285-5_3

There are two basic types of equipment used: the "grinders," which evolved from the initial equipment using millstones, and the refiners.

Friction damages the fibers producing a wide range of particle sizes. However, the extent of these damages can be partially controlled by altering the elasticity of the wood through heat pretreatments associated or not with chemicals [27, 34]. Figure 3.1 illustrates the various types of mechanical pulping processes available for using wood as a raw material. The process is called "refining" when friction occurs in refiners where high rectangular bars in the plates compress and cut the woody material, causing the fibers to separate [34].

The mechanical pulping processes per refiner currently in use are: thermomechanical pulping (TMP), chemo-thermomechanical pulping (CTMP), and high-temperature thermomechanical pulping, more commonly referred to as the medium density fiber (MDF) process [27].

For the TMP and CTMP processes, fiber separation is mainly achieved at low temperature (100–140 °C) by physical crushing and rupture of the S1/S2 layers interfaces of the cell wall (TMP process). Additionally, the temperature can be reduced if the chips are pretreated in a solution of sodium hydroxide followed by preheating with steam. The lignin becomes sulfonated, and the process is called chemo-thermo-mechanical (CTMP). These fibers contain microfibrillar structures from the secondary layers of the cell wall and some remaining middle lamella

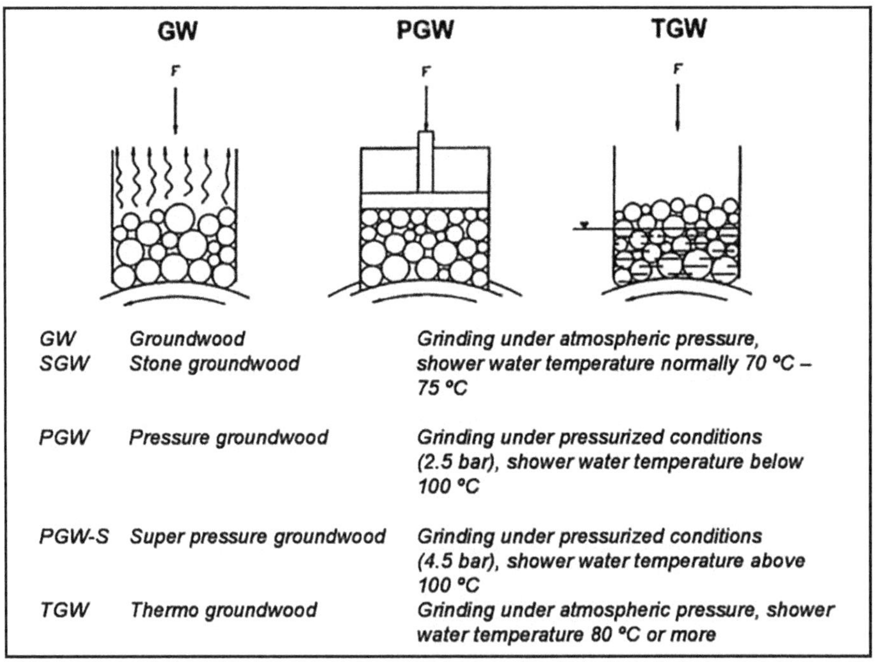

Fig. 3.1 Principles of the griding procedures [5]

fractions. This heterogeneity confirms that physical disruption occurs through several structural layers of the cell wall [27, 35, 267]. In contrast, the high-temperature MDF process (165–180 °C) causes fiber separation within the middle lamella and therefore results in hydrophobic fibers with a lignin coating in a smooth film [27].

In general, fibers from mechanical pulps are more rigid and less collapsible than fibers from chemical pulps, as they retain substantially higher levels of lignin. However, they present large fine content which are important as they provide optical properties to the pulp, and, if they are present in large proportions, they can contribute to the mechanical strength as the hemicelluloses and the specific surface of the pulp will be high [36, 267].

For papermakers, the advantages of mechanical fibers include their high yield, retaining more than 90% of the solid wood material and producing pulps with high specific volume and rigidity, being widely used in the inner layers of paperboard.

Disadvantages include a high need for electrical energy and limited strength properties due to the reduction in the average length, in addition to the presence of hydrophobic materials on the fiber surfaces, that interfere with the development of hydrogen bonds between fibers when the paper is dried [34].

3.2 Chemical Pulping

Kraft pulping or sulfate pulping is the nowadays a most used process. Other chemical processes are sulfite and organosolv.

Kraft pulp is obtained by treating the chips with a mixture of sodium hydroxide and sodium sulfide. In the sulfite pulp, various sulfur acid salts are used, whereas in the organosolv process, alcohol is the main reagent.

In wood, the lignin present in the middle layer between the fibers has the function of uniting them, maintaining the structure of the plant's tissues. In chemical pulping, the objective is to remove lignin, in particular from the middle lamella between the fibers, to enable the fibers' individualization [34, 37, 267].

In kraft pulping, wood chips are exposed to an aqueous solution of sodium hydroxide and sodium sulfide at a temperature of about 170 °C. The process is a low selectivity one and carbohydrates are degraded by acetylation ("peeling") at the beginning of cooking, by alkaline hydrolysis at temperatures above 150 °C [34, 38, 265, 267].

About half of the wood's constituent material is removed during the process. Therefore, a certain residual amount of lignin, estimated by the kappa number of the pulp, will remain in the fibers [267]. Part of the dissolved material may be reabsorbed by the fiber surface at the end of cooking and during acid stages of bleaching.

The bleaching of kraft pulp, which is done by means of multi-stage sequences using reagents such as chlorine dioxide, sodium hydroxide, oxygen, ozone, and hydrogen peroxide, generally with at least one alkaline extraction step and several washing operations, results in a purification of the fibers [34, 351].

Compared with high-yield pulp fibers, those obtained by chemical pulping show a more hydrophilic surface, are more flexible, and have a greater tendency to flocculate conforming better to each other when forming the paper. In this way, the paper produced will have lower specific volume, small opacity, and porosity, but with higher levels of strength and better formation [26, 27, 34].

Chemical pulps are often mechanically refined to delaminate the cell wall layers releasing the cellulose microfibrils to the fiber surface [27].

3.3 The Main Uses of Each Type of Pulp

The following tables present the most common commercial pulps and their main contributions in the paper production or the quality of the finished product (Table 3.1) and the main contributions of each type of pulp on both papermaking process and paper quality (Table 3.2) [41, 42, 57, 343].

Table 3.1 Most common uses for chemical, semi-chemical, and mechanical pulps

Paper or paperboard	Pulp	Effect on production processes or product quality
Wood-free print and write papers	Long fiber kraft	Reinforcement for machine runnability
	Short fiber kraft	Opacity and formation – normally the mainly component
	Short fiber sulfite	Reinforcement for machine runnability
		Filler
Wood containing print and write papers	Long fiber kraft	Reinforcement for machine runnability
	Bleached mechanical pulp	Opacity and printability
Newsprint papers	Long fiber kraft	Reinforcement for machine runnability
	Mechanical pulp	Bleached in the case of dark woods or to improve the quality of the paper – it is the main component
Tissue paper	Chemical pulps	Softness and water absorption
	CTMP[a]	Water absorption and specific volume
Paper packing	Long fiber kraft	Physical strengths – tensile and delamination
Cardboard boxes	Chemical pulps in general	Appearance, printability, surface strength, and internal strength
	CTMP[a], TMP[b]	Specific volume and rigidity

[a]Chemical-thermo-mechanical pulps
[b]Thermo-mechanical pulps

Table 3.2 Contribution of each type of pulp to the process or product quality

Long-fiber pulps	Short-fiber pulps	Mechanical pulps	Fillers	Additives
Strength	Formation	Opacity	Opacity	Retention
	Opacity	Specific volume		Hydrophobicity
Machine runnability	Surface smoothness	Internal structure and porosity for printing		Strength
				Brightness

Chapter 4
The Modern Paper

Paper can present many different properties' combinations characterizing its quality for the final use. The wide range of paper types available on the market and the wide variation in their uses confirm this; each type of paper has its typical characteristics set in order to meet consumer demand [41].

To ensure the desired specifications, the type of pulp and the additives that will be used must first be selected, in addition to the method of pulp slurry preparing, which varies according to the type of fiber selected [411, 555].

For all types of papers, the quality must be uniform which means that the paper's structure and surface chemistry are homogeneous on both small and large scales. Such homogeneity, nowadays, is the result of the modern paper machine designs, the use of more adequate pulps, and the improvement in the operating control processes [41].

It is evident that the use of fibers having characteristics closer to the desired ones is an important factor for competitiveness. Thus, fibers that require lower complexity process and with lower levels of energy application should have priority when selecting the pulp to be used commercially.

4.1 Desired Paper Properties

4.1.1 Low Production Cost and Machine Runnability

Regardless of the type of paper being produced, the papermaker has basic needs to be met, such as maintaining high productivity, with high operational efficiency, low production cost, and uniform quality in the finished product.

As operating efficiency, it is possible to relate few breakdowns, few operating problems, little generation of scrap, few stops, that is, few problems… including in the conversion equipment [47, 49, 57, 67, 115].

© The Author(s), under exclusive license to Springer Nature Switzerland AG 2024

V. Manfredi, *Eucalyptus Kraft Pulp Refining*,
https://doi.org/10.1007/978-3-031-47285-5_4

When the papermaker demands a uniform pulp, he is not asking for a specific property but for a series of directly related properties with a direct impact on the machine behavior and on the quality of the finished product. It also considers the suitability of the fibrous material to the constraints, or operating limits with which he has to work and, even so, allows it to produce the desired paper [41, 50–62, 665].

The quality criteria for fibrous material must include properties and criteria that are practically and economically measurable. The most parameters used in this evaluation have been:

- Strength properties of paper and individual fibers
- Ease of refining and drainage
- Whiteness and whiteness stability
- Environmental compatibility
- Process yield and paper production efficiency
- Recyclability, that is, fiber strength and optical properties with multiple cycles of use, which is the most difficult parameter to determine

The quality specifications for the pulp will be defined in order to ensure the best possible performance for the paper machine facilities. Among the most common restrictions in paper mills are the refining capacity, drainage and retention on the flat table, production speed, steam availability, wet or dry paper strength, and the finished product uniformity. The machine runnability is the main parameter to be considered because it is directly related to the operating production costs.

A good pulp, according to the papermaker's point of view, is one that allows adequate levels of drainage and retention in the forming section of the paper machine, attributing strength to the sheet along all the machine, mainly at the exit of the forming and pressing sections, and that meets to paper quality specifications as per customer demands [49, 57].

A good performance (runnability) is the result of a satisfactory performance in all production stages and not only of a particular subprocess. This behavior (machine performance) is conditioned by a series of factors combining structural aspects and mechanical properties of the paper, as well as the stress applied to the paper, as it runs through the paper machine [56, 60, 63–67, 69, 70, 406].

Most problems related to "machine running" (runnability) are caused by variations in tension in the pressing room, combined with wet paper moisture and basic. Along the paper machine, or in the paper converting machines, sheet tensions are created by speed variation in the sheet transfer from the pressing to the drying sections, or between rolls [70–74, 406]. Controlling the paper formation and fiber orientation contributes to optimizing both machine behavior and some paper properties [411, 60, 67, 76–79].

As fibers are the most abundant pulp component, their characteristics, especially their bonding and collapsing capacity, will be decisive factors, interfering with production costs and paper quality [8, 23, 32, 39, 41, 57, 80–108, 343, 798].

4.1.2 *Printing and Writing Papers*

In general, the paper must present a set of properties that guarantee its printability and machine performance during the printing and finishing processes.

For a graphic, the characteristics that a printing paper must have are: dimensional stability, absence of dust or imperfections, good contact with the printer's surface (smoothness), equal finish on both paper sides, compatible absorption for the type of ink to be used, color fidelity with high print gloss, high opacity, porous structure with good formation and high bulk (specific volume), neutral color, and, of course, adequate strength to ensure the printer runs without breakages [41, 51, 53, 56–58, 91, 109].

Although the properties required to guarantee a good printing surface may vary depending on the printing process, one of them is common to all: the paper must have good surface resistance, which allows for a uniform ink application and without the release of particles, which, when adhering to ink transfer surfaces, cause print failures or defects [50, 55, 56, 104, 665].

The printing and writing paper manufacturers prefer pulps that are easy to refine, that is, that have high strengths with low refining energy consumption because this reduces production costs [49, 57]. Theoretically, pulp fibers with a high specific surface, greater bonding capacities, and higher number of fibers per gram, such as eucalyptus ones, are the most suitable for use in these papers.

For printing and writing (P&W) papers production, higher hemicellulose contents in the fibers are desirable due to their positive effect on the refining and tensile strength of the pulp [115].

4.1.3 *Tissue Papers*

Absorbent papers correspond to one of the market segments with the highest growth rate and for which eucalyptus pulp has shown high acceptance.

There are four groups of papers classified as "tissue." They are toilet paper, napkins, paper towels, and tissues for various purposes. The properties required for tissue line papers in general are described below [52, 57, 58, 110–116, 621].

- Specific volume – The specific volume (bulk) is closely associated with the porosity of the paper. This is an important property of tissue paper as absorbency and structural softness correlate well with porosity and paper thickness.
- Poorly consolidated paper structure – Tissue papers require loose fibers in the paper structure and, to a certain extent, large number of bonds between the fibers is undesirable. The fibers cannot collapse because this would result in a flattening of the fibers and the surface of the paper, leading to increased tensile strength but reducing all tactile properties (feel of softness) due to the paper structure compaction.

- Liquid absorption (fast absorption and water holding capacity) – An important aspect related to absorption is the paper porosity. The microscopic spaces between the fibers form a capillary geometry, through which water passes, saturating the fiber surfaces. This property is more pronounced in papers produced with short fiber pulps, such as the eucalyptus one, due to its high specific surface,
- High surface and structural smoothness – Sensation when touching the surface of the paper in the act of smoothing or crumpling the paper, leading to the perception of a smooth, fluff, and soft paper. This property is more pronounced in papers produced with short fiber pulps due to its high fibrous population and formation of smaller flocs.
- Designs caused by dry creping and embossing (these designs improve the feeling of softness and provide better absorption and beauty to the sheet of paper).
- Paper's ability to retain anatomical pulp elements (fines and vessels) to avoid excessive dust generation during paper manufacturing or converting operations.
- Exact paper strength (wet and dry) to avoid machine breakages during paper production or conversion, and to ensure that the paper does not fall apart when the customer is using it. Wet resistance is an important property with different levels of requirement depending on the type of product. Paper towels are the ones that require greater resistance as they need to maintain their integrity when absorbing liquids. Toilet paper is generally the least resistant because it must disintegrate quickly in the flushing system. Napkins and handkerchiefs require intermediate strengths.
- Fast drainage in the paper machine.
- Low content of fines to avoid both the generation of dust, especially when drying the paper, and the accumulation of fines in the white water from the paper machine, which will be reflected as drainage losses and higher steam consumption in the drying process.

Absorbent papers require high specific volumes and porosity, with emphasis on uniform pore size and distribution, in addition to certain specific properties such as: rapid adsorption of liquids, high flexibility, structural softness (feeling of being soft or fluffy), and surface softness (soft to the touch). Higher water absorption rate and easy disintegration in water are also required [52, 57, 58, 110, 111, 114–117].

During production, tissue papers must not significantly lose strength when embossed, laminated, etc. The most modern papermaking technolgies involve structuring the paper looking for generate bulk softness and absorbency increases. This approach allows to obtain the highest quality products, combining a high specific volume and high softness [111].

Good quality papers are obtained by the appropriate combination between fiber characteristics, equipment and control systems of the paper production, and conversion processes [57, 113, 115, 118, 119].

If the fibers collapse, increasing their bonding capacity, there will be damage on papers' softness and porosity. Tissue paper manufacturers prefer pulps with high softness, high specific volume, and good absorption properties. The desired characteristics of the fibers include stiffness, predominance of short fibers, and a low content of hemicelluloses and extractives, to avoid dirt and incrustations in the drying rolls. Eucalyptus fibers, due to their favorable characteristics, have contributed to obtain such high standards of softness, specific volume, and machine runnability [49, 57, 113, 115, 120], being one of the most used in the production of this type of paper.

4.1.4 Coated Papers

The main properties required in coated papers are high smoothness, high surface resistance, ability to absorb coating inks, low porosity, and high blistering resistance (defects caused by the vapor bubbles formation in the internal paper structure during drying). For the base paper, high opacity and good formation are desired [53, 56].

High smoothness and low porosity are obtained from pulps with low fiber coarseness (thinner cell walls) in order to obtain greater collapse during drying. A higher proportion of short fibers also makes it possible as the voids inside the paper structure will be filled [397].

4.1.5 Special Papers

Among the papers considered as "special" ones, we have the auto-copying, thermo-copying, filters, resin absorption, labels, etc. In most cases, the eucalyptus fibers are used to improve paper formation, opacity, smoothness, dimensional stability, specific volume, and [57].

Papers for filters and for resins absorption have similar characteristics to absorbent papers in terms of porosity. Here again, uniformity in pore size and distribution is critical. In the case of papers destined for impregnation with resins, the need for resistance after water absorption is added as a way of ensuring adequate performance in the impregnating machines.

For papers used in the labels manufacture and for auto- and thermo-copying papers, the absorption capacity of the coating inks, the surface and blistering resistance, as well as the specific volume and formation of the base paper are important properties [56]. For these papers, except in cases where high mechanical strengths are required, eucalyptus fibers have been successfully used.

The manufacture of filter papers, decorative paper, and base paper for impregnation with resins demand the following properties [57]:

- Good formation
- Dry sheet strengths (tensile, tearing, bending)
- Internal paper strength and surface smoothness
- Paper surface strength
- Good porosity with adequate porous structure
- Good absorption of liquids, including inks and mordants
- Dimensional stability
- Good opacity

4.1.6 Paperboard

In the production of paperboard, depending on the properties required for the finished product, different levels of mixtures using high-yield pulps and chemical pulps are used.

Specific volume, compressibility, good rigidity, and bending resistance are the most important structural properties. The current challenge in the manufacture of paperboard for packaging is to meet the increasing strength requirements for often lower grammages. Stiffness is one of the critical properties, being dependent on the thickness and elasticity modulus of the layers forming the paperboard [54, 56, 61]. The strength properties required are ensured by the refining process [1404].

The most important properties in the pulps used in the packages manufacture require a low level of impurities, rigidity, mechanical resistance, brightness and brightness stability, and, in many cases, good optical and printing properties. In the case of food packaging, the absence of odor is added to the list of quality characteristics. In this case, bleached eucalyptus pulp is one of the most suitable.

Mechanical pulps have been used in internal paperboard layers. They attribute greater specific volume while increasing the paperboard stiffness, even for lower grammages. Chemical pulps, with higher elasticity, lower brightness reversion, and good printing characteristics, have been used mainly in superficial layers [61, 123].

4.2 The Most Important Paper Properties

A sheet of paper is a flat, porous three-dimensional structure formed by fibers and other solid particles, such as fillers and chemical additives, interlinked through intermolecular hydrogen bonds between the surfaces of adjacent fibers [124].

Thus, paper is a two-phase system (solid and voids) in which the pores within the fibers and the voids between the fibers are an important part of its structure. Its properties are mainly determined by the properties of the fibers, with a significant effect of the microstructure of the formed paper [63, 125–128].

The spatial fibers distribution in the paper structure depends on the fiber characteristics, the pulping process, and the equipment used in the papermaking facilities (Fig. 4.1).

The paper structure depends on the fiber collapsibility and on the wet fiber flexibility. These factors, as well as fiber swelling, which determine the density of the paper structure in wet pressing and drying, are determined by the fibers' physical-chemical characteristics and the additives used in their processing [129, 130–132] and can be functionally classified as presented in Table 4.1.

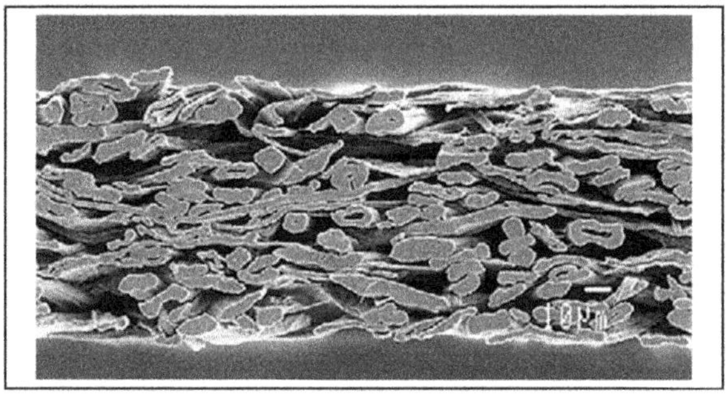

Fig. 4.1 Paper from eucalyptus pulp components in the consolidated paper structure [115]

Table 4.1 Paper properties classification according to the contribution of fiber-to-fiber bonding

Classification according to fiber-to-fiber bonding		Paper properties
Paper properties with no primary dependence on the fiber-to-fiber bonding	Structural properties	Dimensional stability
		Formation
		Porosity
		Specific volume (Bulk)
	Optical properties	Opacity/light scattering coefficient
	Surface properties	Smoothness
Paper properties with strong dependence on the fiber-to-fiber bonding	Plastic properties	Elongation
		Elasticity modulus
	Strength properties	Burst strength
		Tear strength
		Tensile strength
		Vessel picking
		Wet strength
		Internal strength
		Tensile energy absorption

4.2.1 Structural and Optical Paper Properties

4.2.1.1 Formation

The paper must have a highly uniform spatial distribution of fibers, absence of basis weight variations, and an efficient fixation of fine particles on fiber surfaces in a porous structure [34].

Formation is one of the most important properties of paper; it reflects the paper structure, such as, fibers flocs size, aggregation, and distribution, and the variation in grammage along the paper. In this way, formation contributes to the characterization of most of the paper's functional properties, such as its appearance, printability, and mechanical properties, and even its behavior in the machine (runnability), since all these properties are dependent on the paper weight variation [69, 78, 133–164, 343].

For some types of paper, such as, printing and writing ones, good formation is more important than high strength values. Furthermore, for papers where strength is important, the maximum values are only obtained when the sheet has a good formation, that is, only when the fibers are uniformly distributed in the paper structure.

The normalized formation index (NF) is often used to characterize paper formation and is derived from the measurement of mass variation and defined [163] as:

$$NF = \frac{sd}{\sqrt{BW}} \tag{4.1}$$

where:

NF formation index
BW mean basis weight, g/m^2
sd standard deviation basis weight, g/m^2

Fiber characteristics generally have a greater effect on paper formation and quality than operating variables related to processing equipment.

Fiber characteristics such as fiber length, diameter, coarseness, and flexibility are key properties in paper formation. Such properties can be controlled through the selection of the pulp to be used, the contents of each pulp in mixture, or during the refining. Higher the short fiber content in the mixture, easier a good paper formation is obtained. These fiber characteristics influence the pulp flocculation, which is the main reason for poor paper formation [8, 26, 80, 102, 136, 164–167, 170–178, 555].

4.2.1.2 Specific Volume (Bulk)

The specific volume (bulk) is probably one of its most important paper properties, being defined as the volume, in cubic centimeters, of one gram of paper. It is a measure of the compaction degree and structural organization of the paper being estimated from the fiber average dimensions. It relates to all structural, strength, and optical paper properties, being dependent on fiber characteristics as well as on process operating variables [92, 115, 129, 179–185, 555, 665].

An increase in the paper structure specific volume can be attributed to lower fiber flexibility due to lower fiber swelling. Other possibility is the change on fibers shape, curling, folds, and twists of the fibers that can impair the paper consolidation [62, 129, 186].

Specific volume interferes on several functional paper properties. A higher specific volume makes it possible to:

- Save pulp fibers, for the manufacture of papers of the same thickness, but meet the other quality requirements for the paper or, as an alternative, to produce papers with the desired characteristics, but with a lower apparent density [470].
- Produce toilet papers for facial or hygienic purposes with greater absorption capacity and a softer structure.
- Produce papers that require high porosity, such as those that require impregnation of ink or resin (decorative papers and industrial filters).

Specific volume has a significant economic impact on the paper production. It is the sheet density. To produce papers with the same thickness, pulps allowing a greater specific volume can be used to reduce the grammage impacting the production cost [62].

However, a large volume also has some disadvantages as more voluminous papers may have less fiber bonding capacity and weaker papers can be produced. One way to solve this problem is using high bulk pulps but with a higher hemicellulose content, such as those obtained from *Eucalyptus globulus*.

4.2.1.3 Dimensional Stability

Paper expands or shrinks with changes in moisture content. The change in paper dimensions due to relative humidity variation is a complex phenomenon influenced by several factors, basically, the anisotropy and fibers orientation in the paper structure, being strongly influenced by processes that involve water (hydro-expansion) or drying (shrinking) as well. The fiber deformation (curl, wrinkle, cockle, grainy edge) also contributes to this process [65, 188–191]

The ability to minimize or control such dimensional changes is particularly important for printing papers and for papers used in packaging [189, 191]. This property is excellent in paper sheets made from refined short eucalyptus fibers.

4.2.1.4 Porosity and Capillarity

The porosity, that is the inverse of the resistance to the passage of air through the paper, and the capillarity are dependent on the porous structure of the paper and are related to the size (volume) of the internal channels of the paper that are formed by the voids between the fibers.

The spatial distribution of these channels influences their performance in various applications. The retention of specific size particles and the adsorption of polar and non-polar fluids, such as water, resins, printing inks, and coatings, are dependent on the pore size and distribution [192, 193].

The resistance to the passage of air represents the difficulty that a given volume of air or steam has to pass through a sheet of paper [113, 193–196, 470, 555].

Blistering, which is the rupture of the internal structure of coated papers caused by the formation of steam bubbles during the drying process that complements the paper coating or the wet off-set printing [197], is related to porosity.

The factors that influence porosity are the fibers' anatomical characteristics, their degree of hydration, and their orientation in the internal paper structure. The compaction level of the paper structure is also important. Porosity and specific volume are interrelated [113, 187, 195, 198, 199].

Traditionally, the capillarity of paper is determined by the Klemm capillarity test, which reflects the ability of a fluid to penetrate its internal structure. Like porosity, capillarity is related to the size and distribution of pores in the internal structure of the paper.

It is important to say that there is a big difference between the Klemm water absorption test, which measures the capillary penetration in the paper sheet, and the pulp water retention value (WRV), which measures the adsorption and retention of water by micro-capillarity and ionic groups of the cell wall. They indicate different phenomena and cannot be confused. To be easily penetrated by capillary water, a sheet of paper must have been produced with low WRV pulp; otherwise, the fibers hydrate, swell, and close the paper structure, and capillarity is lost.

4.2.1.5 Liquid Absorption

The paper sheets produced with poorly refined eucalyptus fibers have a high capacity for absorbing and retaining water or micro-capillary inks. Therefore, eucalyptus fibers are widely used in the manufacture of sanitary and printing papers. In the case of decorative papers, resins also penetrate more easily into the paper structure.

Cellulose has hydroxyl groups that allow it to form hydrogen bonds with water. This is the main reason why paper absorbs water. Water can be accommodated in the fiber spaces, lumens and cavities, and pores of the swollen cell wall. The water absorption capacity of paper is generally in the range of 2–10 g/g [111, 115, 200].

The water absorption is a complex diffusion process, involving vapor phase diffusion pores, surface diffusion along the fibers, and bulk diffusion through the cellulosic material.

4.2.1.6 Structural Softness

The structural softness (soft feel) is one of the most desired properties in tis-sue papers. There are two types of "feeling" of softness, the one that corresponds to caressing the surface of the paper (surface smoothness or softness to the touch) and the one that is felt when folding or crumpling the paper with your own hands (structural smoothness).

Structural softness is related to the paper volume (thickness and bending stiffness) [111]. With refining, the paper develops strength, but its volume is reduced by the fiber collapsing, and the "feeling of softness" is lost. That is why it is said that "to be soft, paper must be made with little or no refining."

4.2.1.7 Light Scattering Coefficient (Opacity)

Opacity is the paper's ability to be penetrated by light. It is vital for printing and writing papers as well as decorative ones. Opaquer the paper is the greater its ability to hide what is under it.

The structural arrangement of the fibers, the fibers' surface area, and their collapsibility capacity define the paper opacity since the light scattering depends on the fiber-air interface area [42, 105, 127, 139, 153, 195, 201–207, 470].

This is because the refraction and reflection of light increase with the increase in the fiber-air interface area. Much of the light is deflected and does not pass through the entire thickness of the paper (Fig. 4.2).

The greater the fibrous population, the greater the internal paper surface structure, favoring light dispersion and increasing the opacity of the paper (light scattering coefficient) [105, 153, 195, 196, 202, 204–207, 211].

The light scattering coefficient also works as an index of the bonded area between the fibers because only the unbonded surfaces, including the fiber lumen, contribute to the light scattering. In this way, the variation of the paper light scattering coefficient can also be used as an indicator of the relative bonding area between the fibers (RBA) [127, 195, 202, 208, 213–215].

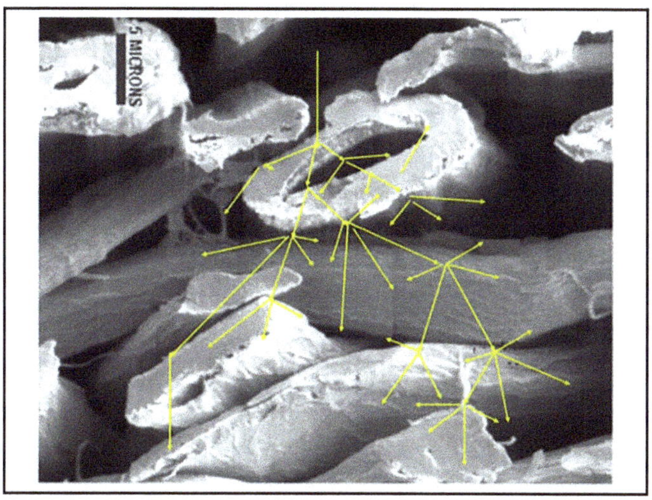

Fig. 4.2 Light scattering by paper structure. The light changes its direction at each fiber-air interface

4.2.2 Resistance Properties

Papers must resist the different types of forces that exist along the machines used in their manufacture and conversion [41, 50–58, 61, 102, 212, 665].

The strength properties are defined by the fiber type together with the pulping process and level of applied refining energy. These parameters determine the individual fiber strength and the number of connections between fiber and the connection area.

The strength of the paper is also determined by the fibers' orientation and distribution in the paper (formation) and by the type of treatment applied to the paper during pressing, drying, and calendering [60, 78, 102, 108, 126, 197, 215–217, 226–228, 343, 411, 747].

The strength properties are basically related to the collapsing ability of the fibers and their specific surface (area available for bonding between fibers). The increase in strength is mainly caused by a greater fiber flexibility and plasticity in denser paper produced from pressing and drying stages [229].

The strength of any material can be characterized by its Young's modulus as shown in Fig. 4.3. When subjected to a force, such as tension, compression, or even tearing, which are the most common for general paper, it will show some deformation under a tensile force, for example, the paper will initially deform by stretching to its yield point (Fig. 4.3a). If the tension is released before the yield point be reached, the paper will return to its original size.

If the applied tension exceeds the limit of elasticity, the paper begins to undergo deformations in an irreversible way, which corresponds to the plastic deformation of the paper, which will no longer return to its original size.

The area of the graph between the beginning of the application of tension and the limit of elasticity represents the modulus of elasticity of the paper (MOE) (Fig. 4.3b) which comprises the paper elastic deformation.

As the tension force continues to increase accentuating the paper plastic deformation, it may promote the paper rupture. The graph's area between the beginning of stress application and the breaking point corresponds to the work done by the sample, that is, the total stress energy absorbed by the specimen (Fig. 4.3c), which is equivalent to the maximum paper tensile strength level.

These stress-strain curves provide, for any material, accurate information about it as it considers both the structure and structure modifications. It can be used to evaluate both the material deformation and the strength of the material, as such, to follow the kinetics of rupture, determining the energy of the internal bonds and the time of stress relaxation [230].

The graphics illustrated in Fig. 4.3 are applied to any material including individual fibers and fiber flocs. This information is now presented because in Chap. 8, Sect. 8.3, the effects of refining on fibers will be discussed and the same basic concept will be applied. The specific elasticity modulus expresses the material capacity to resist when submitted to a deformation effort, by each increasing variation of

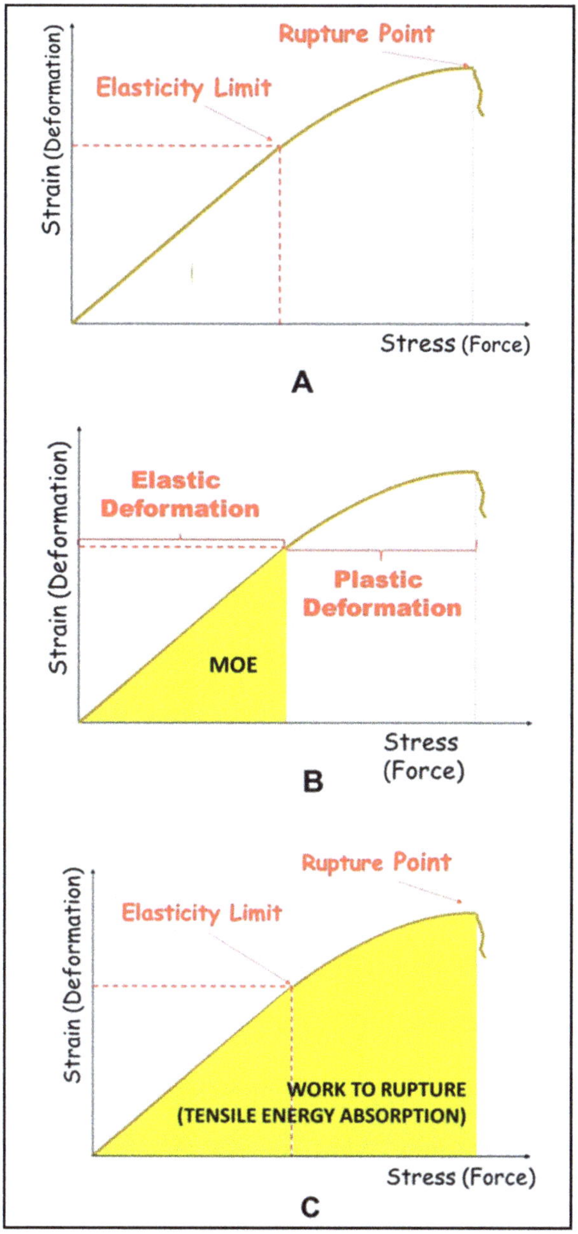

Fig. 4.3 Simplified view of Young's modulus material resistance

applied load per unit of specific mass of the material. Higher values of specific elasticity modulus indicate greater capacity of the material to absorb load without, however, suffering substantial deformations. This characterizes its stiffness potential [196].

In papers, the modulus of elasticity (MOE) is dependent on the parameters promoting the development of the fibers bonds, that is, the flexibility, the collapsing, and the conformability of the fibers, which are dependent on the fiber transverse modulus of elasticity. It is interesting to note that the parameters most influencing the MOE are also the main determinants of paper density [129, 147, 196, 199, 231–243, 1260].

The elasticity modulus and the fibers' structural distribution contribute significantly to determining the strength of the paper, which is governed by two factors related to the paper elasticity modulus:

- Combination of fiber-to-fiber bond breaking and individual fiber segments, with the bonding strength between fibers and the strength of individual fibers determining the relative importance of each type of rupture mechanism) of the paper.
- The structure of the failure paper, determined by its weight and formation, governs the path of rupture, with the lines of rupture generally following paths of lower basis weight. As the fiber flocs are stronger than the zones around them, the breakage occurs first in the zones around the flocs.

Thus, the strength paper properties depend on the elastic modulus of the fibers, which is directly related to the fiber's intrinsic strength and to their bonding capacity. This is the reason why, traditionally, the production of more resistant papers is attributed to long-fiber pulp, while short-fiber woods are recognized for producing papers with greater opacity, greater surface smoothness, and better paper formation.

As an example, the classical Page's equation [126, 245] for the paper tensile strength (Eq. 4.2) combines the intrinsic fiber strength measured as the zero-span tensile strength (ZS) with the strength given to the sheet by the inter-fiber bond (SB) in addition to the length and perimeter of the fibers, which are also related to the binding capacity between the fibers. The equation also considers the fiber's cross-sectional area and its density which are also related to the elasticity modulus.

$$\frac{1}{TS} = \frac{9}{8.ZS} + \frac{12.A.\rho_f.g}{SB.FP.l.RBA} \tag{4.2}$$

where:

TS tensile strength of the sheet (as breaking length)
ZS zero-span tensile strength, cm
A fiber cross-sectional area, cm^2
ρ_f fiber density, g/cm^3
g acceleration due to gravity, cm/s^2
SB shear bond strength per unit area, $dynes/cm^2 = 0.1$ Pa
FP fiber perimeter, cm
L average fiber length, cm
RBA relative bonded area

The specific energy of paper rupture corresponds to the energy consumed in the rupture of the bonds, in the fracture of the fibers, and in their plastic stretching. The energy consumed in breaking a fiber is much greater than that required for breaking a bond between fibers [218, 246–248].

The paper breaking strength, which depends on how the tear line propagates, is strongly related to its tensile strength and elongation [204, 218, 230, 249–251]. The fibers located along the rupture line are the ones that will control the tension. As the probability of a fiber crossing the tear line is directly proportional to its length, this becomes a decisive factor in controlling paper breakage [78, 126].

4.2.2.1 Wet Strength (Paper Machine Runnability)

The breaking strength of a given material corresponds to its ability to resist the propagation of a rupture line, a failure or breakage of its structure. During the paper production, its conversion, or printing, different types of forces are applied on it, which can generate flaws or points where its structure is damaged. Tear strength is therefore a key property for many types of papers as it relates directly to machine behavior ("runnability") [68, 121, 204, 251–258, 1409, 1413].

Machine behavior is a paper's fundamental property being during its production or during its conversion or printing, as these machines tension the paper in rolls.

Eucalyptus kraft pulps are used as raw materials in the manufacture of different types of paper. However, for each paper type, as also for each paper mill project, the pulp quality requirements may be different because these facilities, as a rule, have different productivity, quality, cost, and efficiency goals. This means that there is no universal pulp, that is, a pulp that can perform well everywhere [115].

Productivity means high paper machine speed, with fast drainage in wet end section, high dry content after wet presses, excellent paper consolidation, and no breaks along the machine. Quality implies the maximum percentage of paper under specification ranges, with minimum generation of scrap.

Efficient paper machine operation is the goal to be pursued by any papermaker. He wants his machine to run at the lowest possible operating cost, without problems, at the maximum possible speed, without breakdowns, without maintenance problems, and achieving the required product quality. The consequence of all this is that the specific unit cost is also optimized.

There is no doubt that a good pulp provides good operability in the paper machine and the desired quality in the finished product, regardless of whether this paper is a commodity (toilet paper, cut size paper, etc.) or a special paper [115].

The strength paper properties are strongly influenced by the paper moisture. The higher the dry content greater the paper strength is. This is the reason why it is so important have good drainage, good dewatering, and good wet sheet pressing to avoid paper breaks in the machines [54, 260, 343, 406, 588].

Then paper wet strength is very important. Many breaks occur in the pressing and surface finishing regions in sizing press or coating in the early stages of drying. Higher the wet strength lower the breaks are in these machine sections [56, 64, 261].

The increasing moisture (lower dry content) reduces the paper strength which is explained by the rupture of the hydrogen bonds that connect the fibers. Such bonds, in combination with the intrinsic fiber strength, are responsible for the dry paper strengths definition [54, 260, 343, 588].

When the graph in Fig. 4.3 refers to the relationship between the deformation (elongation) and the tensile strength, it is called a "failure envelope," which can be expressed for various paper moisture contents.

Thus, the "failure envelope" is the graphic expression of wet strength, and it relates to paper's break tendency during its manufacture [217, 218, 758, 262–264]. Figure 4.4 illustrates the effect of dry content (SC) on paper web strength by presenting typical stress-strain curves, also known as "failure envelope," of paper different papers' dry contents [216]. These curves are equivalent to those presented in Fig. 4.3 (Sect. 4.2.2) being used identically.

For almost all dried papers the fibers are tightly bound without any slipping, and the paper structure has a high modulus of elasticity and strength, while for wet papers the fibers bonding is loose and the paper shows a lower modulus of elasticity and greater strain to failure [264].

Higher wet strength values are, as a rule, determined for pulps with higher wet elongation values. This behavior is probably due to greater flexibility and greater number of bonds and micro-compressions in the wet fibers. On the other hand, when other fiber characteristics are held constant, wet strength tends to be linearly proportional to fiber length [67, 217, 262–267].

For the same pulp, higher tensile strength values mean lower elongations, and vice versa. Keeping constant one of these properties, higher values for the other property means greater resistance to wetting, that is, lower machine breakdowns (higher paper machine "runnability"), allowing production increasing due to higher machine speed [208, 217, 218, 249, 252, 254, 262, 263, 268, 270, 588, 758].

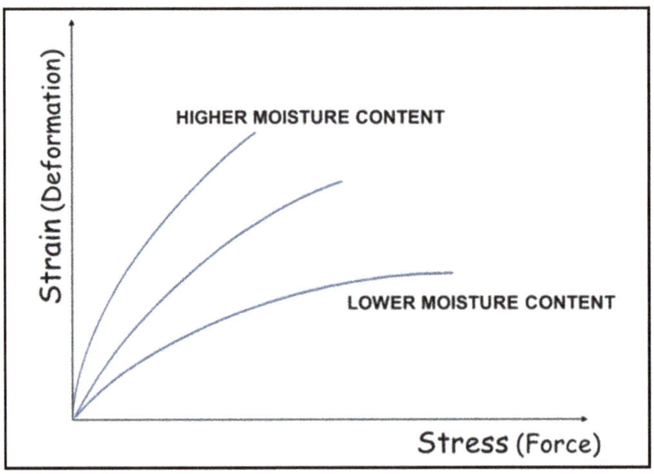

Fig. 4.4 Schematic paper failure envelope as a function of the paper moisture content

4.2.2.2 Elongation

Paper is a viscoelastic material that can show significant plastic deformation when tensioned. Strongly related to the mechanical strengths of the paper and its dimensional stability, the elongation corresponds to the percentage dimensional variation of the paper from the beginning of the tensile stress application until the moment of paper rupture [78, 258, 271].

The bonding capacity between the fibers is the main factor in determining elongation. For unrefined pulps, the fiber length exerts the main effect while for refined pulps the contribution of the bonding area between the fibers and the content of fibrillar fines increases in the determination of the result [242].

The elongation varies significantly between the pulping processes. In relation to the wood species, the elongation tends to be smaller when species with higher basic density are used [84]. This occurs because these fibers attribute a smaller bonding area between the fibers.

4.2.2.3 Tear Resistance

Traditional tear strength tests are the in-plane tear and the Elmendorf tear. The tear in the same plane is measured on a universal test equipment with a constant distance variation between the grips. Therefore, without application of forces perpendicular to the paper, just like on paper and converting machines.

The Elmendorf tear is the most used and corresponds to the test in different planes, that is, it measures the resistance in the shear or torsion mode, reflecting, in addition to the interactions in the connections between the fibers, the interactions in the transverse direction of the fibers.

Tear occurs under forces acting in a line perpendicular to the plane of the sheet and corresponds to the measure of energy required to propagate an out-of-plane tear line over a predetermined distance on a sheet of paper [30, 38, 196, 250, 272–276, 555].

The most widely accepted theory of tear strength was introduced by Van Den Akker in 1944 and was based on the observation that when paper is twisted, some fibers are broken while others are pulled out intact. Therefore, tear strength is determined entirely by the number of fibers extracted and the average work required to pull a fiber. [256].

The total energy required to propagate a break line is therefore the energy used to break a fiber, plus all the fibers in the break line that have broken, plus the removal energy needed to pull a fiber out of the line net, plus all fibers in the line of rupture that did not break [86, 256, 257, 272–274]. This can be expressed as

$$\text{Tear} = \left(N_R.E_R\right) + \left(N_P.E_P\right) \tag{4.3}$$

where:

N_R total number of fibers along the failure line that rupture
E_R average rupture energy
N_P total number of fibers along the failure line that pull out of the network
E_P average stripping energy associated with fiber pull-out [272]

The distance over which the force that acts pulling a fiber out of the paper structure is much greater than the distance over which the breaking force must act to break a fiber, so EP≫ER [274].

The breaking strength of the paper increases linearly with the increase of the bonding capacity between the fibers and decreases with the reduction of the intrinsic strength of the fibers. In this way, during refining and for the same type of fiber, the increase in the bonding capacity between the fibers compensates the reduction in the intrinsic strength of the fibers, increasing the tear strength up to the break-even point, *that means* the maximum value of tear strength from which the reduction in intrinsic strength prevails and the tear strength starts to show decreasing values [63, 78, 84, 147, 165, 199, 204, 208, 218, 242, 248–253, 266–268, 274–299, 470, 478].

Although commonly used to assess the probability of breakage of paper in process (machine runnability), the Elmendorf tear strength test does not simulate the behavior of the paper during its manufacturing or converting operations [45, 47, 250, 276, 476]. In these cases, the use of the Tensile Energy Absorption Index (TEA) (Fig. 4.3c) should be preferred.

4.2.2.4 Tensile Resistance

Tensile strength is a measure of the resistance of wet or dry paper to the tension produced by traction. It corresponds to the maximum stress developed before the sheet breaks and is traditionally associated with the durability of the paper in use subject to direct stress forces [477, 555].

Tensile strength is one of the most important quality parameters of writing and printing papers, where it indicates the probability of breakage during the converting and printing processes, but with less importance in the tissue paper segment [478, 555].

Measured by conventional methods, tensile strength is controlled by factors such as individual fiber strength and the strength of the bonds between fibers. These factors are related to the area of bond between the fibers (RBA), strength of the bonds between fibers, intrinsic strength of the fibers, paper formation, and structure of the paper. In unrefined pulp as well as in the initial refining levels, the tensile strength depends on the binding capacity and correlates well with the hemicellulose content in the pulp. Thus, an increase in the bonding area between fibers is a major source of improvement in tensile strength, while cell wall intrinsic strength becomes important for higher levels of refining [30, 42, 78, 83, 91, 102, 113, 127, 184, 208, 215, 218, 242, 249, 266, 272, 302–310, 312–315].

During the refining, the tensile strength reaches a maximum value and then decreases, evidencing the relationship with the structural modifications of the fibers, especially the continuous reduction of the cell wall intrinsic strength throughout the refining treatment [479].

Long fiber kraft pulps have a higher tensile index than short ones because they have greater intrinsic strength and greater average fiber length, which allows more contact points between the fibers [289, 314].

4.2.2.5 Tensile Energy Absorption (TEA): Work

TEA corresponds to the work performed by a paper sample until it breaks when subjected to tensile stresses (Fig. 4.3c). It indicates the paper's ability to absorb energy under conditions of increasing load applications [196].

This property is strongly dependent on the formation characteristics and the conformability potential of the fibers in the paper sheet, being directly related to the tensile strength and elongation [196, 249, 251].

4.2.2.6 Internal Strength of Paper

The internal strength of paper is the resistance perpendicular to its surface. This is an important property of paper as low strengths will result in paper delamination during coating or printing operations.

Internal strength is the product of the length of the bonds between fibers and the strength of the bonds between fibers. There are several denominations for this property, the best known being the "Scott-bond" strength, strength in the z-direction, strength to delamination, and strength perpendicular to the plane [42, 47, 78, 194, 215, 267, 473, 474].

In addition to the fiber-to-fiber bond breaking energy, the delamination energy includes the energy dissipated in the fibrous web of thick papers, while the tensile energy in the z-direction of the paper includes the bond breaking energy between the fibers, especially for papers made from chemical pulp with flexible fibers [42, 127, 215].

Increases in the content of long fibers and fibrillar fines, which increase the binding capacity between the fibers due to the increase in the contact points between the fibers, have a positive effect on both longitudinal and internal paper tensile strengths, being the most pronounced effect on internal resistance [63, 126, 473]

4.2.2.7 Burst Resistance

Burst resistance is defined as the pressure necessary to produce bursting of the material when uniformly increasing pressure is applied, transmitted by an elastic diaphragm of circular area equal to 92 mm^2. The burst resistance acts perpendicularly to the surface of the specimen; hence, the measurement of resistance is independent of the papermaking direction [196, 555].

The burst strength is highly dependent on the strength of the bonds between the fibers and the formation of the paper, being considered as a measure of the relationship between the tensile strength and the elongation. For the same tensile strength, papers with high elongation have greater burst strength than papers with lower elongation [84, 195, 242].

The burst strength increases with refining and paper weight and is significantly influenced by paper sizing [78, 134, 147,172, 195, 196, 281, 555].

4.2.2.8 Rewetting Resistance

To meet good stiffness properties in paper made from short fibers depends a lot on the characteristics of the fibers that promote the bond between the fibers [242].

Considering the final use of paper, its strength is of less interest when used only for writing and printing purposes. However, as paper has been increasingly used in other uses, its strength when re-moistened becomes important, especially when recent trends in grammage reduction and increased use of short fibers and recycled fibers are present as in the examples below [64, 588]:

- Tissues, paper towels, and similar products for drying or as absorbents
- Paper that comes into contact water when used, for example filter paper.
- Paper for packaging liquids or wet substances
- Packaging paper, sack paper, bags, paper board, and other types of paper for outdoor use

4.2.3 Surface Properties

4.2.3.1 Printability

Print quality is affected by paper properties such as formation, surface smoothness, gloss, and opacity. The paper strength requirements must match the demands on the ability to run on high-speed printers. However, the objective should be to produce a superior print surface within the constraints defined by the execution capability [377].

4.2.3.2 Picking and Linting Resistance

Surface resistance refers to the release of particles from the paper surface. In uncoated papers, the released particles correspond to fibers, parenchyma cells, and filler particles in addition to the vessel elements in the case of short fiber pulps. In coated papers, the coating particles released from the paper surface is also included in this group.

The detachment of particles from the paper surface generates several problems related to the formation of dust. These are serious problems in tissue paper machines and offset printers.

In printers, the material removed from the paper surface (linting) accumulates in the printing elements, requiring frequent stops for cleaning the equipment, that results operating time losses [320, 327, 338, 406, 424, 425, 471, 472].

The places previously occupied by the detached vessels in the printed paper become points without ink application characterizing the "vessel-picking" defect, which is closely related to the frequency and dimensions of the vessels present in the unrefined pulp. The paper manufacturing process and printing operating conditions also influence the vessel pullout intensity (Fig. 4.5).

As in the case of any short fiber pulp, eucalyptus pulp can produce papers with vessel picking problems during offset printing [41, 56, 57, 68, 81, 82, 84, 100, 121, 217, 316–339, 555]. Picking occurs mainly at surface locations with higher density points (flocs). Thus, reducing the average floc size is beneficial for the pull out and linting reduction [327].

In some cases, the vessels, although they do not detach from the paper surface, do not absorb the ink in a similar way to the rest of the paper surface leaving translucent spots on the printed surface (ink refusal). This occurrence may be indicative of low surface energy (poor paper surface wetting), which reduces ink transference to paper [187, 324, 329].

These translucent dots, known as "speckles," as well as "vessel picking," deteriorate the print quality.

Fig. 4.5 Photo of the surface of paper made with eucalyptus pulp showing vessels still intact on the surface [115]

4.2.3.3 Surface Smoothness

Roughness is the inverse of smoothness and corresponds to the volume of air, under standardized conditions of flow and pressure, that escapes between the smooth surface of the measuring head and the surface of the paper supported by a glass plate. The purpose is to numerically represent the existence of irregularities on the paper surface, with larger volumes of air escaping between the surfaces indicating greater roughness – or lower smoothness – of the paper surface. The most common equipment and methodologies for this determination are the Bekk, Bendtsen, Sheffield, and Parker Print Surf [187, 194, 480].

Smoother papers are ideal for printing and coating as they allow savings in terms of printing inks or coating formulas. Thus, adequate levels of smoothness should always be sought. For many paper uses, surface roughness is one of its most important characteristics. In several printing methods, it is the property with the greatest influence on the quality of the printed product being related to other properties of practical relevance, such as resistance to paper friction and glue adhesion [65, 179, 480, 481, 555].

Rigid fibers with thick cell walls tend to form papers with a rough surface and higher specific volume while the opposite occurs for short-fiber pulps, such as eucalyptus [63, 69, 90, 100, 101, 467, 468, 665, 1001].

The thickness of the paper is determined by its weight and controlled by the paper calendering, which leads to the reduction of its specific volume. Thus, if we want to reduce the paper roughness, the reduction of its specific volume is one of the alternatives [665].

4.2.3.4 Surface Softness

The "feeling of softness," or tactile softness, is one of the most desired properties in tissue papers. It refers to the perception of softness when stroking the surface of the paper (softness to the touch) and is related to the paper surface smoothness. Long fibers, due to their larger flocs, convey a feeling of roughness while short eucalyptus fibers, which form smaller flocs, provide an excellent feeling of softness and smoothness.

Chapter 5
Wood as Raw Material Source

The fibers used are of natural origin. The most common source is wood from which the fibers are extracted by chemical, mechanical, or chemo-mechanical processes. The wood used and the pulping process employed have a significant effect on the properties of the fibers produced [47].

In the beginning, long-fiber woods, such as pine, were the most used. However, with the development of chemical pulping processes and, motivated by its availability, manufacturers began to use short-fiber woods in their processes [2, 17, 21]. Among the short-fiber species, eucalyptus stands out, the subject of our book.

The eucalyptus pulp first appeared on an industrial scale in the early 1960s and was still considered a "novelty" until the 1970s. Eucalyptus pulp is the one that had the fastest growth cycle, becoming one of the main sources of fibers for the paper production [2].

The wood quality is the combination of its physical, chemical, and anatomical characteristics which allow better use of wood for a given purpose, that is, each use requires its own quality characteristics. In other words, the quality of the paper produced is closely related to the characteristics of the fibers used [23–25, 39, 56, 57, 79–81, 85–96, 100–103, 308, 341–345, 588, 798]. Such characteristics determine the possible combination of paper properties directly influencing its structure, properties, and production costs.

Among the main characteristics characterizing wood as a fiber source for papermaking, its basic density, anatomical characteristics, percentage of fibrous tissues, and its hemicellulose and extractive contents stand out [24, 31, 85, 87, 95, 99, 344, 347–350, 397, 352, 355, 468, 555, 650].

There are important variations in wood characteristics not only between trees from different families, genera, and even individuals of the same clone but also between species, origins, ages, climate and soil conditions, cultural practices, growing rate, etc.

© The Author(s), under exclusive license to Springer Nature Switzerland AG 2024

V. Manfredi, *Eucalyptus Kraft Pulp Refining*,
https://doi.org/10.1007/978-3-031-47285-5_5

Such variations result from a complex mechanism of interaction between the wood anatomical structure and the ecological and genetic factors that act on the seasonal activity of the secondary cambium and the morphogenesis of the xylem, being able to modify the wood structure, properties, and the performance of the conversion process [18, 20, 24 30–32, 40, 103, 288, 342, 344–347, 357–379].

5.1 Using the Eucalyptus Wood

The genus was named in 1788 by Charles-Louis L'Heritier de Brutelle, based on material (*Eucalyptus obliqua*) sampled from the south coast of Tasmania. The *Eucalyptus* genus belongs to the Myrtacea family and has more than 600 species and a growing number of hybrids. Except for a few species, the vast majority are native to Australia. *Eucalyptus urophylla,* for example, is native to the Sumba archipelago (latitudes 122–127 °E), in regions with an altitude range (100–1960 m) higher than that of the natural occurrence of most species [50 57, 81, 383–386].

The genus *Eucalyptus* presents a wide range of adaptation (sites), types of conduction (management), and varieties of use for wood, being considered as more efficient than other forest species for growth in sites of low fertility [50, 57, 215, 464, 609].

About 40 eucalyptus species are commercially important. Outside Australia, most eucalyptus plantations are based on a few species, all with good vegetative reproduction: *Eucalyptus grandis*, *E. saligna*, *E. globulus*, *E. tereticornis*, *E. urophylla*, *E. maculata*, *E. paniculata*, *E. camaldulensis*, *E. deglupta*, *E. exserta*, *E. citriodora*, *E. robusta*, and *E. viminal*is [81, 215, 385, 755].

The first reference to the use of eucalyptus wood in the chemical pulp production comes from Portugal, where bisulfite pulp was obtained by D. E. Berggvist, in 1906, from *Eucalyptus globulus*. However, it was only in 1919–20 that eucalyptus pulp began to be produced commercially, being exported to England [56, 81].

Spain was the first to use eucalyptus to produce mechanical pulp on a commercial scale. *Eucalyptus globulus* and *E. camaldulensis* were already in use around 1914–18. At the same time, studies were reported in France and the United States, involving the production of pulp from *Eucalyptus globulus*, *Eucalyptus cladocalyx*, and *Eucalyptus tereticornis* [56].

Brazil was one of the first countries to use eucalyptus to produce commercial pulp. In 1925, Cia. Paulista de Estradas de Ferro, the railroad responsible for the eucalyptus introduction in Brazil, sent wood from *Eucalyptus saligna* (15 years) and *E. tereticornis* (13 years) to the Forest Products Laboratory, in Madison, Wisconsin (USA) for studies aimed at producing paper pulp. Based on these results, the company Gordinho Braune & Cia, located in Jundiaí, São Paulo, began in 1927 the manufacture of various types of paper with sulfite cellulose from *Eucalyptus saligna* mixed with mechanical pulp from *Araucaria angustifolia* (Parana pine) [22].

From 1940, there was a massive growth in the use of eucalyptus for the pulp production in many countries, among them: Angola, Brazil, Portugal, Chile, Congo, India, Italy, Morocco, and Spain. Eucalyptus kraft pulp began to have a more marked presence in the international market from the 1960s onward. Due to its ability to improve formation, increase smoothness, and allow gains in paper opacity and uniformity, it became the most important short-fiber pulp, supplying the markets for absorbent, writing and printing, paperboard, and specialty papers [20, 52, 57, 58, 68, 82, 85, 107, 115, 121, 165, 184, 202, 209, 217, 295, 320, 333, 379, 388–403, 411, 555].

In Brazil, it was only between 1957 and 1958 that the massive industrial production of eucalyptus pulp took place, which was still considered a novelty until the 1970s. Today, eucalyptus wood is the most used raw material in the national production of pulp and paper [2, 4].

Eucalyptus wood has become the preferred choice in all regions with climatic conditions that allow the establishment of fast-growing plantations. Eucalyptus plantations have been used in several countries due to its rapid growth, less variability compared to natural forests, and the excellent fibers papermaking properties [20, 23, 32, 57, 81, 319, 333, 357, 402, 404, 407, 464, 465, 555].

Eucalyptus kraft pulp fibers provide a unique combination of formation, resistance, specific volume, softness, porosity, and opacity for the produced paper. Such characteristics associated with its high bleached yield, the relatively low demand for refining energy, and good wet resistance justify its excellent market positioning. This occurs because its fibers are short and fine, with low coarseness, but rigid and present in large numbers per unit mass when compared to other short-fiber pulps [49, 52, 56–58, 60, 68, 81, 85, 107, 115, 184, 202, 217, 295, 328, 333, 351, 392–411, 555, 868].

The most important and recognized qualities that eucalyptus fibers confer on paper quality are the following:

- Sheet formation
- Specific volume (Bulk)
- Porosity (resistance to air passage)
- Smoothness, surface softness, and structural softness
- Opacity and printability
- Adequate level of resistance both dry and wet
- Dimensional stability

There is yet another key factor for the success of eucalyptus fibers: their market price. Thanks to lower manufacturing costs and higher process yields, with less use of wood and chemicals and higher pulp production per invested amount, eucalyptus pulps can be offered at more competitive prices than softwood pulps. As papermakers continue to strive to improve their operations and the increased global use of eucalyptus pulp and other short fibers, the identified benefits (reduced refining energy consumption, better paper quality, and better paper machine operability) are of great importance [57, 68, 333].

5.2 Wood Characteristics

From the beginning, most of the fibers used to make paper came from plants such as cotton, flax, agricultural waste, or tree bark. To save costs, papermakers used rags. This situation remained until the middle of the nineteenth century, when wood pulp began to be used, currently being the main raw material used in papermaking [9, 34, 456].

Wood is a complex material with a highly hierarchical structure. The various wood cells in the living tree can be grouped according to their main functions which include sap transport, storage and support, and mechanical strength. Fibers are grouped in regular patterns to provide maximum mechanical performance in the most common loading modes, such as compression under the attached branch between the trunk and a branch in a tree [427, 462].

There are two main categories of trees: angiosperms, which are trees with flowers and fruits, and gymnosperms, which do not produce flowers and fruits [27].

Eucalyptus belongs to the angiosperms group which, in addition to flowers and fruits, have broad leaves which is why they are also called broadleaf or hardwood due to the higher wood density than gymnosperms. Pine trees are different from gymnosperms, which do not bear fruit and house their seeds in cones; hence, they are also known as conifers. Hardwoods and conifers have woods that differ in basic density and average fiber length. In hardwoods, such as eucalyptus, wood is made up of fibers with a shorter length and higher density. Hence, hardwoods are also known as hardwood or short-fiber woods, while pines are also known as softwood or long-fiber woods.

Hardwood and softwood fibers differ in several respects: their fiber length, cell wall thickness, number of fibers per gram of pulp, stiffness and chemistry, types of hemicellulose and lignin, cell wall structure and measurements, fibril angle, and types of cells are all different from softwood and hardwood.

Trees grow in height and diameter throughout their lives, with the newest tissues located on the outsides and tips of trunk, branches, and roots. The only section of these trees that contains the wood formed from seedling to its present age is the base of the tree. For this reason, the top of the tree is the youngest section while the base is the oldest [406]. The same tree contains tissues with different ages.

Primary growth, which corresponds to growth in height of a plant, produces leaves and non-woody material with the function of providing mechanical support and form for the plant. The secondary growth of the plant, which causes the increase in the trunk diameter, produces mainly woody material used both in the sap conduction and in the mechanical support [27, 461].

All the cells that make up these tissues originate from the meristematic secondary cambium differentiation. The secondary cambium is the embryonic tissue responsible for the growth in diameter of trees through the radial overlap of new tissues (Fig. 5.1). The cells of the meristematic cambium have thin and plastic walls formed by pectin, protein, non-cellulosic carbohydrates, and a small proportion of cellulose [57].

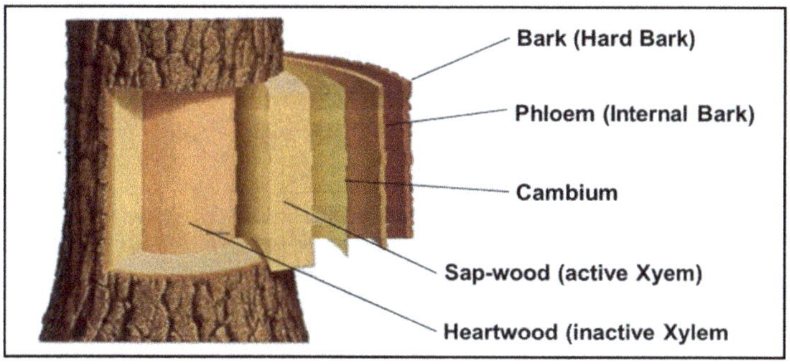

Fig. 5.1 Plant tissues in tree trunk [33]

The bark is external to the trunk. Beneath the bark is the vascular cambium, which is composed of a few layers of living cells arising from the apical bud in the primary growth. These cells actively divide and produce more wood, which results in a diametral increase in of the tree. Wood can be subdivided into sapwood and heartwood.

Sapwood conducts water and inorganic nutrients, while heartwood is an old sapwood that has been chemically altered by the accumulation of resins and extractives, leading to a loss of water-conducting properties [4, 27].

Within a single tree, morphologies often vary with the climatic conditions of the growing season, as illustrated by the growth rings in the wood. At the same time, fibers growing at different ages and positions in a plant will have different morphologies [454, 460].

5.3 Anatomical Characteristics of Wood

Anatomically, wood can be defined as a heterogeneous set of different types of cells, oriented in the longitudinal direction of the stem. These cells have specific properties and perform different vital functions, such as conducting liquids, transformation, storage and transport of substances, and support of the plant [457].

Both in hardwoods, the group to which eucalyptus belongs, and in conifers, the trunk and branches are the commercially used parts of the tree. These are made up of specialized tissues that perform the functions of protection (bark tissues), conduction of sap and nutrients, and reserve and filling (parenchyma), in addition to support, basically performed by the tissues of the central region of the trunks and branches.

These cells, although tiny, are consolidated into a very resistant material, which can be used in civil structures, furniture, packaging, poles, and paper. The cellular structure of wood significantly affects its processing as the characteristics of the obtained product [57, 409, 413, 425].

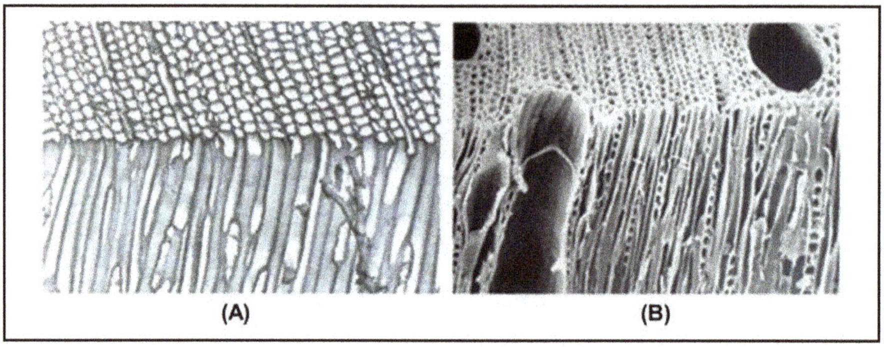

Fig. 5.2 Typical cross-sections of coniferous (**a**) and hardwood (**b**) trunks at the microscopic level. (Adapted from [406])

Figure 5.2 presents the basic diagrams of coniferous and hardwood woods representing their typical cross-sections at the microscopy level.

Despite being heterogeneous, woods have few structural elements: fibers, tracheid, vessel elements, parenchyma cells, cells producing resins and gums (extractives), among a few others. Cell shape varies according to its function and tree species. Conducting and support cells are dead cells due to lignification of their cell walls and have lumens filled with sap, air, or reserve materials [427].

The conifers woods have in about 90–95% of tracheid, which are fibrillar cells with the functions of support and sap transport, and 5–10% of parenchyma cells. In hardwoods, the wood structure is more complex with three basic types of cells that make up most of these woods: fibers, which are long and thin and have a thick cell wall; the parenchyma cells, which are small, generally isodiametric having thin walls; and the vessel elements, which are superimposed forming the vessels that conduct sap [242, 320, 425, 427, 436].

5.3.1 The Supporting Tissues: The "Fibers"

The supporting tissues occupy the central region of the trunks and branches, corresponding to their largest volumetric fraction. They are formed by elongated cells generically called fibers in hardwoods and tracheid in conifers. The fibers are smaller than the tracheid, hence the name short-fiber pulp for hardwoods and long-fiber pulp for conifers.

In addition to length, tracheid differ from fibers in that they have a dual function (sap conduction and support) while fibers are solely responsible for supporting the plant. These anatomical elements are present, although in varying proportions, in all wood species. In many of them, the fibers correspond to about 50% of the total volume and are the main constituents present in short-fiber pulps [103, 331, 385, 415, 457]. These anatomical elements are the points of interest for the manufacture of pulp and paper.

5.3.2 Sap Conduction Tissues: The Elements of Vessels

In hardwoods, the conducting tissues are formed by superimposed hollow cells looking like a barrel. Such cells, called vessel elements or simply vessels, have lengths ranging from about 200 to 600 μm, are shorter than fibers, and are arranged on top of each other forming long capillaries (vessels) through which the saps flow.

Figure 5.3 shows the cross-sectional microscopy of the eucalyptus wood, where these vessels stand out among the fibers destined to support, and the ray cells (parenchyma) destined to the reserve of materials. With thin cell walls (2.5 and 5 μm), the vessels are also much wider than the fibers, with diameters ranging from 60 to 250 μm. In most commercial eucalyptus species and clones, the proportion of vessels in the wood volume varies from 10% to 20%, which corresponds to about 3–25 vessels per mm^2 [79, 316, 318, 325, 330, 331, 385, 414–417, 555].

As a conduit transporting water and nutrients within the tree, vessel elements are essential parts of the cellular matrix in the hardwood wood. The vessels number and dimensions in eucalyptus vary between species, trees of the same species, age, and position on the trunk, both in the pith-bark and base-to-top directions [79, 84, 317–320, 329–332, 344, 358, 385, 397, 417–424, 555].

The vessel elements do not cause major problems in the tissue paper quality. However, their presence in the pulp is critical when producing these papers, as they can produce large amounts of dust in the production and converting machines. However, in the production of printing papers, vessels can cause serious problems related to print quality.

Thus, the vessel elements frequency and dimensions are of great importance when producing printing paper [41, 42, 56, 57, 68, 81, 82, 84, 100, 217, 316–336, 397, 406, 425, 555]. Table 5.1 presents values usually reported for the vessel element dimensions in kraft eucalyptus pulps.

In eucalyptus, vessel elements have a relatively sparse distribution, but most of them are surrounded by parenchyma cells. These primary wood components perform different functions during tree growth and vary with wood species.

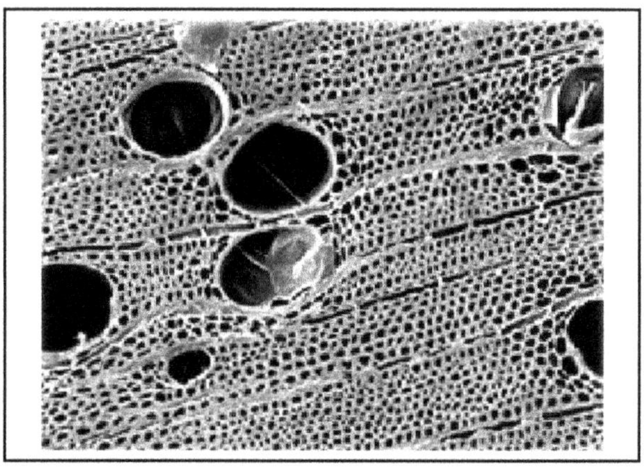

Fig. 5.3 Cross-sectional microscopy of eucalyptus wood showing the different types of cells. (Adapted from [1408])

Table 5.1 Characteristics of the vessel elements of some eucalyptus species [84]

Eucalyptus species	Vessel diameter, μ		Vessels area, μ^2	Number of vessels, by μ^2	Wood density, g/cc
	Radial	Tangential			
Eucalyptus grandis	140	93	11378	8.8	0.401
	137	93	10891	9.9	0.357
	132	94	10524	10.8	0.332
	145	109	13320	12.0	0.404
E. grandis × *E. camaldulensis*	157	111	14818	12.8	0.507
	156	117	15414	12.6	0.528
E. grandis × *E. tereticornis*	164	115	15789	10,0	0.614
Eucalyptus maculata	139	113	13643	10.1	0.653
	154	119	15746	8.1	0.674
	148	119	15036	9.8	0.671

In vessel elements, the primary cell wall is partially reinforced or almost completely covered by a lignified secondary wall. The chemical composition of the vessels is similar to fibers presenting slightly higher cellulose and hemicellulose contents than the fibers [416].

5.3.3 *Reserve Tissues (Parenchyma Cells)*

Storage cells are living, isodiametric, and thin-walled cells which can be observed as relatively continuous bands, found in the radial direction, from the wood. On the trunk, they appear in greater quantity along the outermost layers, being smaller and more numerous in short-fiber woods than in long-fiber woods. They are responsible for storing reserve material and radial transport in trunks and branches [425, 427].

In the woods used for pulping, the proportion of parenchyma cells is between 20% and 50%. Such cells consume chemicals in pulping and do not contribute to paper resistance but make it difficult to drain the fiber suspension [40, 426].

Due to the low strength of their cell walls, they tend to be destroyed during pulping or pulp preparation processes for papermaking. If not destroyed, they will be present in the fines fraction of unrefined pulp [40, 385, 415].

5.4 Chemical Characteristics of Wood

The main chemical constituents of wood are cellulose, hemicelluloses, and lignin, which are distributed in a multilamellar way in the cell walls.

The wood has an average composition, which may vary depending on the species, planting site, position on the trunk, and age of the trees, of 40–50% cellulose,

25–30% hemicelluloses, and 25–35% lignin, in addition to small amounts of low-molecular-weight compounds ranging up to 6–8% [40, 47, 49, 435, 428–436, 588].

The chemical composition is an aspect in the wood characterization because, for the same pulping process, this chemical characterization impacts both the process yields and the behavior of the fibers in papermaking, as well as the quality of the finished product [31, 199, 352, 437–439, 528].

5.4.1 Cellulose

It was during the 1830s, Anselm Payen, an eminent French chemist of his time, identified cellulose as a separate entity in wood. Before that, wood had been considered a homogeneous chemical compound. Within a few years of Payen's discovery, wood fibers were being used in papermaking [7].

Despite the existing differences between the contents and composition of lignin, hemicelluloses, and extractives between softwoods and hardwoods, when only cellulose is evaluated, the differences are minimal. The cellulose molecule is the same despite some differences in crystallinity proportions but nothing that is significant [57, 842].

Cellulose is a long linear polymer with the chemical formula $(C_6H_{10}O_5)_n$, where (n) is the degree of polymerization (DP), that is, the number of repeating units of anhydrous glucose (β-D-glucose) joined together by (1→4) β-glycosidic bonds to form disaccharide units (cellobiose), the smallest repeating unit in cellulose. The native cellulose consists of about 10,000 glucose units [38, 57, 79, 433, 434, 442, 444, 452, 453, 588, 736] (Fig. 5.4).

Cellulose is the most abundant biopolymer on the planet, being the smallest crystalline building block in a wood fiber [38, 57, 79, 434, 450–455, 588].

Cellulose molecules are aligned in a helix due to the intramolecular and intermolecular hydrogen bonds holding the individual cellulose chains together [47, 445, 452, 484, 485]. This structure makes cellulose a very stable hydrophilic polymer, difficult to dissolve in common aqueous solutions and with no melting point.

The supramolecular structure of cellulose is formed by two distinct regions: crystalline and amorphous. The presence of long chains and large amounts of

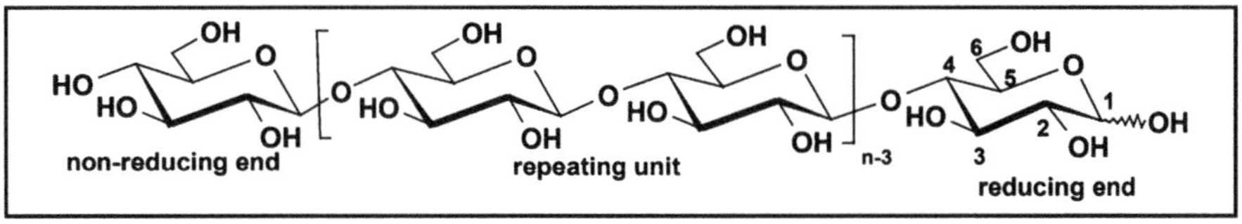

Fig. 5.4 Diagram of cellulose molecule showing (1→4) β-glycosidic linkage to form disaccharide units (cellobiose) [842]

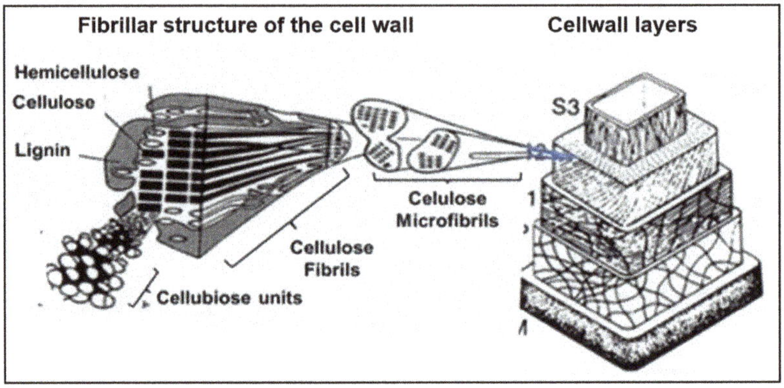

Fig. 5.5 Scheme of the cellulose aggregate in the cell wall matrix. (Adapted from Per Hoffmann's illustration in [572, 842])

hydrogen bonds formed between the chains increase the tendency of cellulose to have a crystalline structure [440, 456, 842].

In general, the amorphous region corresponds to 25–35% of the weight while the crystalline regions represent 65–75%. The proportion between the crystalline and amorphous regions is called "degree of crystallinity." Generally, an increase in crystallinity leads to an increase in cell wall strength accompanied by a decrease in fiber hydration capacity [412, 446, 842].

The linear format of cellulose allows the formation of molecular aggregates, called fibril (elementary fibril). They measure 1–5 μm in length and have a diameter ranging from 2 to 30 nm [57, 447, 448, 450–455, 842].

The fibrils are grouped into microfibrils that form sorts of strings visible in electronic microscopy, with approximately 10–60 nm, which corresponds to 4–5 fibrils joined together and arranged in a square cross-section.

The microfibrils are also arranged in a lamellar structure, surrounded by hemicellulose and lignin, called macro-fibril [37, 57, 427, 440, 447–450] (Fig. 5.5).

Fibrils and their aggregates, due to their ordered structure, have excellent mechanical properties and efficient charge sharing at the atomic level.

This resistance is mainly due to the cellulose; however, the lignin and hemicellulose matrix that fills the spacing between the fibrils also contributes to the compressive strength during cell wall tensioning [47, 427, 433, 441–446, 588].

5.4.2 *Hemicelluloses*

The hemicelluloses are the second main component in wood fibers. They differ from cellulose in that they have a branched, non-crystalline structure, lower weight molecular, and a lower degree of polarization (about 100 in long-fiber woods and 200 in short-fiber woods). Hemicelluloses have branches in the form of acid radicals (acetic acid, glucuronic acid, or galacturonic acid) or sugars. They are less resistant than cellulose and are more easily hydrolyzed. Some acidic sugars, such as galacturonic acid and glucuronic acid, are also constituents of hemicelluloses [88, 329, 427, 433, 434, 442, 456, 1213].

One, two, or several types of monomers usually form the backbone of hemicellulose polymers. Most hemicelluloses also have short branches containing different types of sugars than those in the backbone [434].

Hemicelluloses are formed by monomers of simple sugars with different constitutions that aggregate either in the central chain or inside branches that may also contain acidic radicals. In the case of monomers with five carbons, such as xylose and arabinose, the hemicellulose is classified as pentosan and, in the case of monomers with six carbons, such as glucose, mannose, and galactose, they are called hexosans. In pentosans, where the rings have fewer carbon atoms, the monomers are less stable [13, 57, 329, 427, 433, 442, 444, 456].

Hemicelluloses vary widely in structure depending on the tissue and type of plant. In eucalyptus wood, hemicelluloses are mainly composed of a xylan rich in uronic acids, which is reasonably stable in kraft pulping. In coniferous woods, the main constituents of hemicelluloses are mannose, galactose, and xylose [49, 427].

Hemicelluloses are present in the cell wall matrix, oriented parallel to the cellulose molecules surrounding the microfibrils and forming, together with lignin, the cell wall matrix. They act as a coupling agent preferentially coating the cell wall pores or as an intermediate between cellulose and lignin [38, 57, 427, 433, 444, 456, 487]. They influence the orientation of cellulose fibrils and regulate tensions between primary and secondary cell walls. Hemicelluloses can even influence lignification processes, being important for the mechanical properties and load bearing of the cell wall [27, 121, 1406].

In hardwood and coniferous woods, the hemicellulose content is around the same. Hardwoods have between 20% and 30% of total hemicelluloses and conifers have 15% and 35%. However, there are significant differences between the types of hemicelluloses present in each of them [57, 79, 133, 433, 434, 452, 456]. In coniferous wood, the main hemicelluloses are O-acetyl-galactoglucomannan, or simply glucomannan or mannan, with a galactone-mannose ratio of 3:1, and arabino-4-O-methylglucuron-xylan, or simply xylan. The mannan content in wood is 20%. This hemicellulose is considered more associated with cellulose molecules while xylan, which corresponds to about 5–10% of wood, is free or associated with lignin molecules [38, 433, 488, 575] (Fig. 5.6).

Eucalyptus wood consists of approximately 13% of 4-O-methyl-glucuronoxylan, which represents approximately 18% of the carbohydrate fraction and 71% of the total hemicellulose content. Consequently, pulping conditions that favor the dissolution of this hemicellulose have a significant impact on the yield of the process [57, 214, 433, 489, 490].

Xylans correspond to the highest percentage of hemicelluloses in eucalyptus kraft pulp [57, 115]. The hemicellulose content in the pulp is an important parameter to be monitored, as it depends on the initial content in the wood and the ability of the pulping process to extract or reprecipitate them [57, 214, 489, 491].

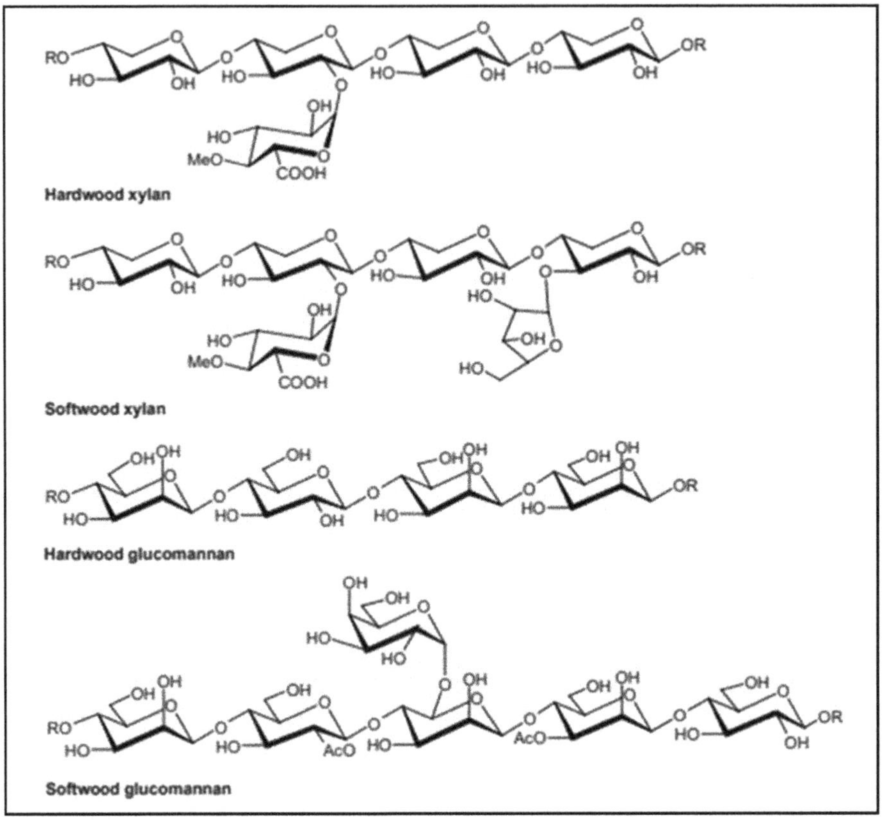

Fig. 5.6 Molecular structures of xylans and glucomannans in hardwoods and softwoods [842]

Wood hemicelluloses make distinct biomechanical contributions, with glucomannans increasing the compression modulus of elasticity and xylans contributing to a dramatic increase in elongation under tension [121, 1406].

No other fiber chemical component in bleached eucalyptus kraft pulps has as much influence on paper properties as hemicelluloses because, due to its hydrophobic characteristic, it has a strong relationship with the physical properties of dry paper. The refining behavior of a given chemical pulp is defined by its cellulose and hemicellulose contents [57, 113, 115, 121, 126, 214, 347, 406, 427, 437, 474, 489, 493–501].

Pentosans, with their amorphous and hydrophilic structure, contribute to many intrinsic properties of the fiber, such as the ability to absorb water, making the cell walls more plastic and flexible, contributing to increase the fiber's bonding capacity. Pulps with higher hemicellulose contents are easily refined producing more resistant and smoother papers, but with lower specific volume and porosity [47–49, 57, 62, 79, 113, 115, 126, 214, 308, 347, 433, 437, 467, 474, 478, 488–512].

Refining energy consumption decreases with hemicellulose content increasing in the pulp because its abundant terminal groups are more accessible to water molecules compared to cellulose [478, 504, 508, 513, 514].

For some types of paper, such as absorbent, decorative, filter, and impregnation base paper, a low hemicellulose content is desirable [57]. The use of wood with lower hemicellulose content, cooking for lower kappa numbers with higher final alkalinity, or the removal of hemicelluloses using more drastic alkaline bleaching are alternatives to reduce the hemicellulose content in the bleached pulp, which also contribute to the reduction of the yield of the pulping process [57].

In the case of fibers with a moisture content of around 10%, which corresponds to the pulp's commercialization standard, hemicelluloses act as an effective tension transfer matrix. As the fibers become wetter, the matrix begins to lose considerably its stress transfer capability [515].

5.4.3 Lignin

The third component of the cell wall is lignin. This amount varies significantly between plant types, species, plant tissues, regions where plants grow (site), as well as within the same plant.

Lignin is more abundant in softwoods where it can occur in proportions that reach 25–35% of the wood dry weight. In hardwoods, this content is a little lower varying of 20–35% in the eucalyptus [38, 57, 79, 406, 433–435, 452, 736].

Lignin is a complex three-dimensional molecule that acts as a bonding matrix for cellulose microfibrils, being the only aromatic polymer present in wood. It is an amorphous and hydrophobic phenolic material whose function is to group neighboring cells together, which stabilizes the size of the cells and waterproofs the cell walls. Lignin also confers resistance to microorganism attack in addition to giving rigidity to cell walls. Lignin gives mechanical resistance to the wood of trunks, branches, and roots [38, 57, 412, 433–435, 441, 575–579, 736].

The lignin macromolecule is highly flexible, probably due to the large amount of ester linkages, and can easily assume several conformations contributing to the viscoelastic properties of the fiber. Under cell wall tensioning, lignin presents a more viscoelastic response than polysaccharides, moving independently of cellulose during fiber tensioning. Thus, lignin provides compressive and hydrostatic resistance in the spaces between the microfibrils when the fiber is longitudinally tensioned [38, 441].

At the time when lignin is deposited, the ordered structure of cellulose and hemicelluloses is already established and lignin polymerizes in the spaces left in the cell wall matrix. Then lignin structuring is the result of spatial restrictions due to the cellulose-hemicellulose organization previously deposited in the cell wall [38, 571].

Lignin is composed of three different monolignols: p-coumaryl alcohol, coniferyl alcohol, and sinapyl alcohol illustrated in Fig. 5.7. The molecular structure of lignin differs between long-fiber and short-fiber woods. In conifers, lignin is formed

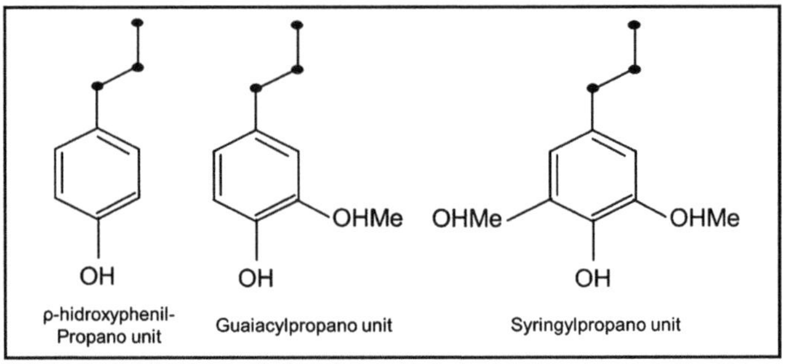

Fig. 5.7 The monolignols structures. (Adapted from [842])

mainly by coniferyl alcohol units, while in hardwoods by a copolymer of coniferyl and sinapyl alcohol units [38, 57, 406, 433, 523–527, 736].

While the lignin content is a notable advantage for the manufacture of charcoal and the use of wood as fuel, the pulp and paper producer want lower lignin content woods with higher amounts of cellulose and hemicelluloses [346, 358, 359, 363, 406]. The lignin content directly affects the wood conversion into kraft pulp as it impacts the reagents consumption, the pulping yield, and the dry solids generation for the liquor recovery system. Thus, the lignin content of the wood is a vital characteristic for the selection process of individuals for cloning or hybridization [406].

As more delignified fibers are more flexible and have a greater specific surface, lignin must be removed from the middle lamella and cell walls producing more resistant papers. In addition, the presence of lignin reduces the pulp brightness and causes paper yellowing due to its reaction with ultraviolet light. However, the delignification conditions must be such as to preserve the structural integrity of the fibers while removing lignin in the maximum amount as possible [199, 474, 573, 574].

5.4.4 Extractives and Mineral Salts

The low-molecular-weight substances in wood correspond to the organic matter commonly referred as extractives, and the inorganic matter summarily obtained as ashes.

Extractives, which account for less than 10% of the dry weight of the wood [40, 452], impart color, odor, and flavor to the wood. While some function as a source of energy (fats and waxes) for biological functions in native wood cells, resin components also protect the wood against microbiological damage or insect attacks.

The composition of extractives is quite variable (aliphatic compounds, fats and waxes, terpinoid oil-resins, and phenolic compounds). The extractives in the eucalyptus kraft pulp are a complex mixture in which the aliphatic components, saturated or not, aromatic and oxygenated aliphatic stand out [40, 79, 486].

Extractives undoubtedly play an important role in the behavior of wood during cooking and pulp bleaching. The increase in wood extractive content leads to reduced cooking yield increasing chemical demand [85, 361, 367, 379, 516, 588].

The ash content consists mainly of salts such as silicates, carbonates, oxalates, and phosphate of potassium, magnesium, calcium, iron, and manganese, as well as silicon [40], which correspond to 0.5–1.0% of the wood.

5.5 The Wood Basic Density

Density is the ratio of mass to volume. In the case of wood, the basic density corresponds to the smallest density, that is, the ratio between the smallest weight (which is the wood dry weight) and its largest volume (which is the green volume or volume saturated in water). The basic density, therefore, uses the ratio between the smallest mass and the maximum volume, corresponding to a direct measure of the fiber collapse potential [99, 318, 344, 359, 467, 493, 529–534, 555].

The basic density results from the set of anatomical and chemical characteristics of the wood, being one of the most important parameters to evaluate the wood quality for pulp production [31, 39, 49, 52, 84, 94, 99, 103, 181, 323, 347, 359, 381, 383, 467, 478, 482, 493, 517, 535–542, 555].

It is the simplest determination of the potential of wood for the production of pulp for papermaking being a low-cost assessment and easy to be determined in the laboratory [25, 31, 39, 52, 95, 103, 181, 207, 382, 383, 467, 478, 493, 535–544, 555]

The basic wood density varies between genera, species of the same genus, between trees of the same species, and between different positions in the same tree. It can therefore vary significantly between plantings due to the influence of seed origin, local characteristics, growth rate, age, spacing, and silvicultural techniques employed [20, 40, 94, 103, 181, 281, 288, 304, 344, 348, 358, 359, 363, 375, 383, 468, 483, 536, 546–556, 561].

The basic density relates to the thickness of the cell wall and thus to the fiber's coarseness [31, 49, 52, 84, 87, 103, 181, 184, 281, 323, 344, 358, 467, 478, 493, 518, 531–533, 539–545, 555–563].

For the interrelationship between fiber length and wood basic density, there are contradictory results. For the same eucalyptus species [52, 103, 181, 564] found higher densities associated with longer fiber lengths, while [39, 555] reported that fiber length had no effect on density from the wood.

A high volume of vessels tends to reduce the wood basic density [333, 344, 467, 531, 547, 555]. Wood with a higher density had a higher frequency of small diameter vessels and, in its anatomical structure, a smaller area of empty spaces [323].

The wood density directly affects the economics of the pulping process because it influences the specific wood consumption which is the most significant price component of the pulping process. Associated with the chemical characteristics of the wood, which condition the gravimetric pulping yield, wood density is the determining factor of production capacity, which is conditioned by the available pulping equipment volumes [25, 31, 49, 87, 94, 307, 349, 352, 367, 369, 438, 555].

It is highly desirable to have wood with the most homogeneous density as possible, as this gives better process yields and better quality to the final product. Therefore, it is important to study the factors affecting the density and, also the pulping processes, looking for strategies to minimize the overall impact from this density variations [383].

Due to bottlenecks and restrictions, some pulp mills prefer to use low wood density (400–450 kg/m^3), medium wood density (480–520 kg/m^3), or high wood density (520–600 kg/m^3). Such restrictions may originate from the pulp mills themselves, from customers, or even from forestry areas (wood availability). These bottlenecks explain why some market pulp suppliers provide pulps that are not similar in terms of performance or quality [406].

Denser woods produce pulps that require more refining energy. These papers drain better in the paper machine forming area producing more voluminous, porous, and opaque papers with less surface smoothness and strength properties [31, 84, 99, 181, 194, 478, 529, 532, 555, 565].

Unfortunately, however, density alone is not a good indicator of pulp quality since two wood samples with the same density are not necessarily technologically or anatomically similar. They may have different processing characteristics or fiber properties [31, 99, 181, 344, 348, 381, 483, 529, 532, 555, 567, 568].

The cell wall density, that is, the density of the wood not including the existing empty spaces, is practically constant for all species (1.451–1.56o kg/m^3). The variation is due to different proportions of crystalline cellulose, the relative amounts of hemicellulose and lignin, and the cell wall micropore size and distribution [43, 103, 478, 531, 569, 570].

Table 5.2 presents values usually reported for the wood basic density of some eucalyptus species.

Table 5.2 Wood basic densities for the most used eucalyptus species

Species	Variation range	References
E. dunnii	0.486 to 0.536 g/cm^3	[518]
E. globulus	0.426 to 0.650 g/cm^3	[20, 31]
E. grandis	0.306 to 0.682 g/cm^3	[20, 27, 84, 348, 358, 483, 493, 518, 521, 522]
E. nitens	0.393 to 0.498 g/cm^3	[29, 519]
E. saligna	0.400 to 0.610 g/cm^3	[358, 518, 521]
E. urophylla	0.415 to 0.572 g/cm^3	[518, 521]

Chapter 6
The Fiber Cell Wall

The various tissues that make up wood have their origin in the differentiation of secondary cambium cells. The new cellular elements originated from this secondary cambium develop the necessary characteristics to perform their functions in the tree. In this process occurs the formation of the cell wall, which is a typical structure of plant cells.

In general, the functions of the cell wall are to keep the plant upright, transport saps, and defend itself from the stress of the external environments and microorganisms.

Woods are basically constituted by the cell walls that correspond to the largest proportion of their dry weight. Its main constituents are cellulose, hemicelluloses, and lignin, distributed in a multilamellar and concentric manner. In much smaller proportions, proteins, pectin, extractives (terpenes, fatty acids, organic acids, resins, etc.), and some inorganic components (mineral ions absorbed by the roots to participate in metabolic processes can be found in cell walls [7, 38, 40, 47, 49, 57, 79, 329, 406, 427–436, 442–445, 447, 448, 452–456, 486, 521, 575, 576, 577, 578, 588, 736].

The lignification process is the final phase of cell wall formation, ending its thickening and consolidation. The protoplast of the differentiating cells forms the lignin precursor compounds, which move through the wall and are converted into lignin along the way to the middle lamella (ML) where lignin is deposited in greater proportions, although a part considerable of its content in the cell is in the secondary wall [370, 442, 580, 581, 583].

Due to the lignification process, which waterproofs the cell wall, cells with secondary cell walls are usually dead at maturity [461].

The cell wall provides mechanical strength to the cells, which serve as supporting tissues to allow vascular plants to grow to great heights. The deposition of lignified secondary walls on tracheal elements not only strengthens these water conduits to resist the negative pressure generated during transpiration, but also makes them impermeable for efficient water transport [550].

© The Author(s), under exclusive license to Springer Nature Switzerland AG 2024
V. Manfredi, *Eucalyptus Kraft Pulp Refining*,
https://doi.org/10.1007/978-3-031-47285-5_6

The cell wall is characterized by a structure of polysaccharides, specifically semicrystalline cellulose microfibrils, incorporated into an amorphous matrix containing hemicellulose, pectin, lignin, and proteins, which are organized differently in the various layers that form the structure of the cell wall. The arrangement is such that properties that exist in different directions are maximized so that they meet the tree's requirements while also having the ability to adapt to a changing environment without affecting the structure's long-term performance [27, 429, 430, 432, 441].

Cell walls are important in the characterization of the wood to be used in papermaking. Several quality indices for both pulp and paper quality consider values related to cell walls, such as wood basic density, coarseness, wall fraction, wall thickness, fibrillar angle, cellulose crystallinity, and strength of individual fibers [57].

6.1 Lamellar Structure of the Cell Wall

A plant fiber can be thought of as a reinforced material composed of a hollow, amorphous matrix, the cell wall. This wall is composed of natural short-chain and branched polymers, mainly of hemicellulose and lignin, involving a spiral bundle of fibrils corresponding to agglomerates of cellulose crystallites approximately 10–30 μm in length each [442, 453, 555, 584–588].

Cellulose microfibrils are responsible for the structural reinforcement of the fibers. They are long, and from segment to segment they are tightly adhered to each other, forming more rigid regions known as "crystalline regions" [555, 587, 588].

This microfibril bundle forms a constant angle in relation to the fiber axis being directly connected to fiber's longitudinal and transverse elastic modules. Thus, fiber is a material composed of a rigid phase, with high strength and high elastic modulus, embedded in a ductile matrix with poor mechanical properties.

The reinforced structure of the cell wall can be compared to concrete where steel rebars are equivalent to microfibrils and concrete to hemicelluloses and lignin (Fig. 6.1).

This structure gives wood cells a robust and flexible construction, making the stem and branches of wood able to withstand a wide range of forces acting on them [8, 27, 427, 587, 599, 601].

The cell wall of the fibers is formed by a primary wall (P) and three layers of secondary wall (S1, S2 and S3) that differ in thickness, chemical composition, and fibril alignment, as well as in function of the species and position in the stem [38, 57, 102, 260, 429, 430, 444, 450, 454, 555, 586–590, 593–596].

During the last stage of differentiation, when lignin is deposited, a layer is formed with spherical particles deposited internally to layer S3 and of complex composition. In it, the microfibrils are arranged at a moderate deposition angle [427].

This layer is known as the tertiary wall (W), or simply "warty layer."

Between the fibers there is the middle Lamella (ML) that surrounds the entire primary wall (P) and acts as a cementing substance uniting the cells.

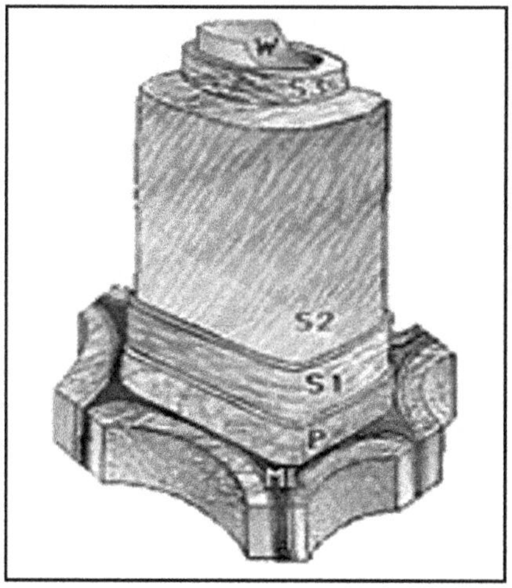

Fig. 6.1 Schematic of the lamellar structure of the fiber cell wall and fibrillar angle. (Adapted from [572])

6.1.1 The Middle Lamella

During the formation of plant tissues, a cambium cell is divided into two (cytokinesis) by the formation of the cell plate that later becomes the middle lamella, which is a thin layer, 0.2–1.0 μm thick, composed mainly of lignin and pectin. About 23–28% of the total lignin in wood is found in the middle lamella [27, 38, 57, 427, 444, 447, 575].

In the middle lamella, lignin forms granular aggregates of lignified matrix that do not show orientation, which suggests that aggregate formation is a property of the cell wall matrix [449, 571].

Toward the lumen, the lignin is successively replaced by increasing amounts of cellulose and hemicelluloses [561].

6.1.2 The Primary Cell Wall

The primary wall (P) is the first layer that develops during cell differentiation. It has two conflicting functions as it must be rigid enough to withstand internal and external stresses and, at the same time, must be flexible to allow for the expansion of the cell wall during its differentiation process [38, 427].

The primary cell wall is a strongly hydrated gel-like compound, composed mainly of cellulose, hemicelluloses, pectins, structural glycoproteins, lignin, and small amounts of phenolic acids. It contains 3–4 layers of cellulose microfibrils with a diffuse orientation, although some degree of alignment can be observed in tissues where cell elongation is taking place [38, 427, 454, 575, 612, 613].

The primary layer is thickened, in order to increase the resistance of the wall, through the action of the living cellular content of the cell which starts to secrete the compounds used to form the wall structure. For this, the reserves present in the vacuoles or accumulated in the cytoplasm of the cells originated from the cambium are consumed [57]

In the primary walls of growing plant cells, cellulose is grouped into long microfibrils, a few nanometers in diameter, and arranged in an irregular crossing pattern, which gives the cell wall its ability to expand during cell growth [57, 300, 427, 446].

The primary cell wall (P) is associated with the fiber's ability to swell when exposed to moisture and can function essentially as a sleeve, restricting the increase in fiber volume [462]. Moisture-accessible regions are also aligned in a parallel arrangement in the cellulose fibrils, explaining their high moisture resistance [441].

6.1.3 The Composite Middle Lamella

Middle lamellae and primary cell walls are two hydrophilic structures present in all terrestrial plants. Later, after the lignification process, they will become indiscernible and can be grouped under the term "composite middle lamella" (CML), which is, as a rule, lost during chemical pulping [27, 47, 260, 589, 593].

The average composite lamella has a thickness of the order of 0.1 μm [47, 593] which represents 7–15% of the fiber solid volume, varying more or less depending on the total cell wall thickness [57, 224]. It consists of 50–70% lignin (about 38–42% of the total wood lignin), 25–40% carbohydrates, most of which are hemicelluloses, plus 2–5% ash. Depending on the stage of cell development, proteins, pectins, xylans, glucans, mannans, galactans, etc. can also be found [57].

6.1.4 The Secondary Cell Wall

During differentiation, once cell enlargement is complete, secondary wall layers are formed, thickening the cell wall. The addition of new materials, which establishes the topochemical organization and interactions between cellulose, hemicelluloses, and lignin, occurs inside the primary cell wall, reducing the diameter of the cell lumen [27, 432].

The secondary wall (S) is a multiple wall, formed by three different layers. The layer (S1), more external and adjacent to the primary wall, the layer (S2), much thicker than the other two and occupying an intermediate position, and the layer (S3), more internal, which is the last one to be formed [27, 57, 575, 599].

The cell wall provides compressive, tensile, and shear strengths to the wood. The secondary wall, which contains about 40–50% cellulose, corresponds to most of the wood biomass. In addition to cellulose, secondary walls are composed of hemicelluloses and lignin, organized around the structure of cellulose microfibrils forming the cell wall matrix [27, 57, 432, 575, 550].

Hemicelluloses are necessary for the normal assembly and mechanical strength of secondary walls. The reduction in hemicellulose content causes a severe decrease in cellulose deposition, secondary wall thickening, and mechanical strengths [214, 550]

All wood polymers, within the secondary wall, exhibit a preferential orientation along the cell wall, being higher for cellulose, whose microfibrils are deposited in concentric layers, and lower for lignin, which presents a small degree of orientation, probably induced by structural constraints at the time of its deposition [441].

Layers S1 and S3 are relatively thin compared to layer S2, but nevertheless play a critical role in increasing the cell's transverse modulus of elasticity by acting to reduce the deformation of layer S2 under compression and tension [27].

6.1.4.1 The S1 Layer of the Secondary Wall

Adjacent to the composite middle lamella, it has a higher concentration of lignin than in layers S2 and S3. Its cellulose fibrils are arranged at an angle of 50–70° in relation to the fiber axis, forming a set that resembles a bundle of ropes tied to hold the S2 layer from the outside [37, 47, 57, 191, 411, 427, 456, 462, 555, 585, 588, 603].

6.1.4.2 The S2 Layer of the Secondary Wall

The S2 layer of the secondary wall is the thickest one, corresponding to 40–90% of its volume. Thus, it is the layer with the highest amounts of cellulose and hemicellulose in relation to layers S1 and S3. As it is the thickest layer, it has the greatest influence on fiber properties [27, 37, 38, 47, 69, 143, 191, 441, 462, 575, 585–589, 601–604].

In this layer, the microfibrils are deposited in a more oriented way than in the others, following a helical pattern aligned with the fiber axis. This alignment characterizes the microfibrillar angle (MFA) providing great resistance to traction or longitudinal compression to the fiber, but little resistance to axial or transversal compression, which facilitates its collapse and significantly influences the properties of the paper. In other words, the smaller the fibrillar angle, the stronger and stiffer the fiber is [27, 47, 121, 190, 427, 448, 450, 458, 462, 488, 566, 594, 605, 610, 888].

The angle of deposition of microfibrils, the fibrillar angle, varies depending on the species, age of cambium, position in the cell wall, and growth rate; it dictates some of the characteristics of the plant, such as rigidity [27, 437]. At younger ages, the wall has more open angles in this layer (25–35°), while at older ages the cambium forms more closed angles (15–20°). Therefore, the woods closer to the pith have more open angles in relation to the fiber axis [57].

The S2 layer, being the thickest of the cell wall, is the one with the highest content of cellulose microfibrils. They are aligned with a typical angle of 10–40° to the axial axis of the fibers [37, 57, 47, 191, 260, 441, 462, 555, 585, 588, 603]. In eucalyptus fibers, in special, the microfibrils have small fibrillar angles, relatively close to the fiber axis, ranging from 8 to 15° [57].

6.1.4.3 The S3 Layer of the Secondary Wall

The inner layer is the S3 layer, which is similar to the S1 layer, with a small thickness (<0.1 μm) and an arrangement of the microfibers almost perpendicular to the fiber axis (60–90°), being more a mooring to contain the layer S2 attached to the cell wall [27, 37, 47, 57, 191, 260, 427, 441, 456, 462, 555, 585, 588, 603].

About 2–6% of the cell's total lignin is found in this layer [57].

6.1.5 The Warty Layer or W Layer

With lignification, the cell wall becomes more hydrophobic and rigid, with the entire protoplast being consumed and deposited internally to the S3 layer to form a warty layer consisting of spherical particles with a diameter of 0.1–0.3 μm [43, 427]. This layer is known as the tertiary wall, or simply "warty layer."

6.1.6 The Lumen of the Cell

The space inside the cell wall structure is the lumen (L) which, after cell differentiation is completed, becomes a hollow nucleus. In tracheid, the differentiation of conifers and in the vessel elements of hardwoods, the lumen corresponds to the space through which the saps circulate or, in the tissues that form the core of the wood, they are used as reserve spaces, mainly of extractives from the wood, which gives the typical color of the central regions of the trunk, the heartwood.

Fibers with large lumens tend to flatten like ribbons collapsing during pulping which results in good strength properties for the produced paper [57, 456].

Chapter 7
The Pulps (Fibers) Characteristics

The term "fiber" describes a single, elongated, thick-walled plant cell. Fibers are the main constituent of cellulosic pulp and, therefore, their characteristics have a great influence on the performance of papermaking methods as well as on the properties (quality) of the finished product.

Each type of paper has its specific demands regarding the fiber's characteristics [26, 406] which depend on several factors such as wood species, age of the tree, growth conditions, and pulping process [36].

Although the pulp and paper manufacturing processes have a significant impact, the basis for obtaining a high-quality product will always be the use of high-quality and uniform fibrous material. The basic factors involved are the ability of the fibers to collapse and to establish connections between them. The physicochemical properties of carbohydrates, especially the hydroxyl groups content. Fibers with a higher content of carboxylic groups, for example, are more likely to establish bonds [32, 56, 57, 82–84, 89–91, 99–102, 105, 113, 126, 199, 217, 218, 298, 292, 343, 366, 437, 443, 466–470, 496, 502, 511, 555, 588, 653, 659–666, 798].

7.1 Anatomical Characteristics of Fibers

The pulp morphological characteristics include the fiber length, fiber width, cell wall thickness, and lumen diameter, as well as the content and types of fines, and, in the case of hardwood pulps, the content and dimensions of the vessel elements.

The fiber coarseness, which is a relation between the fiber length, the cell wall thickness and the fiber width, the fiber average length and the cell wall intrinsic strength, which depends on the wall thickness being strongly influenced by the used pulping process, are traditionally indicated as relevant characteristics of fibers for papermaking [20, 78, 80, 83, 91, 165, 283, 292, 335, 393, 397, 555, 654, 667–669].

© The Author(s), under exclusive license to Springer Nature Switzerland AG 2024
V. Manfredi, *Eucalyptus Kraft Pulp Refining*,
https://doi.org/10.1007/978-3-031-47285-5_7

These characteristics vary greatly between the species and age of the wood as well as the pulping process type and operating conditions. There are large variations both between and within trees. The anatomical element dimensions vary both with the period of growth and with the age of the tree and growing place [26, 57, 377, 406].

There is a great variation in sizes and formats among the main woods used as raw materials in obtaining cellulosic pulp, as shown in Fig. 7.1.

Such anatomical differences significantly interfere in the rheological behavior of the pulps, which directly affects their refining and influences on the characteristics and properties of the obtained papers. Table 7.1 presents average characteristic values for some eucalyptus species.

7.1.1 Fiber Length

The size and shape of the fibers are considered, for the most varied types of papers production, as important parameters in pulp quality. Thus, fiber length measurements have been a major focus in many raw material researchers about wood and pulp quality [42, 56, 80, 84, 85, 87, 96, 168, 169, 202, 208, 267, 281–283, 301, 393, 397, 635, 638, 640, 645, 658].

In the case of length, as well as for other fiber characteristics, both the average values and their distributions must be considered, being the weighted average length [126], which is compatible with the coarseness measure [633], the best representation of fiber length.

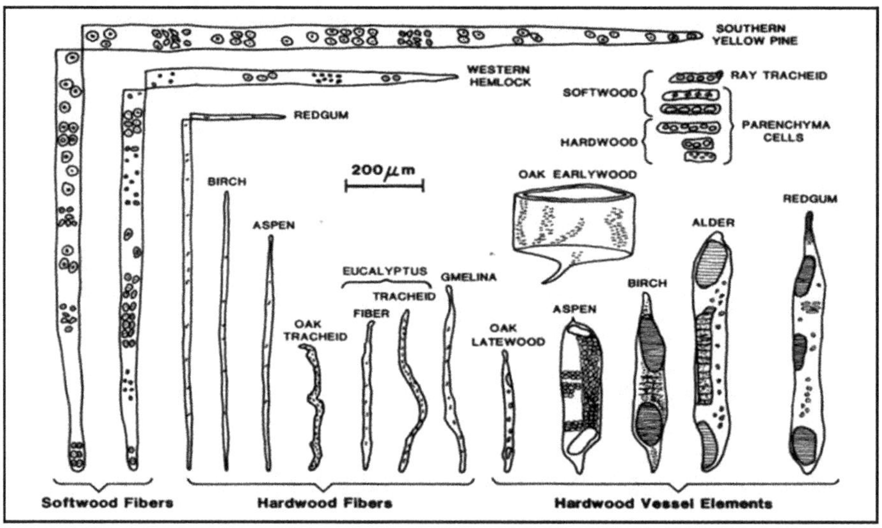

Fig. 7.1 Main types of anatomical elements in woods [331]

Table 7.1 Anatomical fiber characteristics of some eucalyptus species [27, 58, 84, 133, 184, 194, 217, 321, 348, 483, 493, 518–521, 532, 565, 635, 670, 1124]

		Diameters		
Eucalyptus species	Fiber length, mm	Fibers, μm	Lumen, μm	Cell wall width, μm
E. camaldulensis	0.780–0.931	15.20–18.00	7.00	4.00
E. citriodora	1.080	18.90	8.10	5.40
E. cloeziana	0.676–0.931	16.00–18.00	7.40	5.30
E. deanei	0.860–0.880	12.00–18.20	9.00	4.50
E. deglupta	0.864–1.296	21.54–15.89	7.41–11.91	3.39–5.53
E. dunnii	1.090–1.120	13.20–20.00	7.10	3.00
E. fastigata		12.30–14.40	750	2.01–2.59
E. fraxinoides		12.40	7.60	2.40
E. globulus	0.580–0.103	11.00–21.00	7.90–10.60	2.21–4.10
E. grandis	0.566–1.360	10.20–23.90	7.70–17.30	2.10–6.20
E. macarthurii		11.40	5.80	2.80
E. maculata	1.133–1.174	15.60–15.90		3.10–5.50
E. nitens	0.610–0.970	11.20–20.0	6.40	2.40
E. oreades		11.80	7.50	2.20
E. regnans	0.632–1.060	15.60		1.98
E. rostrata	0.55–0.56			
E. saligna	0.945–1.300	13.20–24.00	6.55–12.90	2.70–5.22
E. smithii		12.30	7.10	2.60
E. tereticornis	0.828–0.830	14.80	6.30–6.34	4.20–4.23
E. urophylla	0.780–1.220	8.60–24.53	2.90–11.70	2.80–5.10
E. viminalis	0.520—0.830	9.27—20.00	4.36–8.82	2.87–6.05

The main characteristic of eucalyptus wood is its low fiber length, which typically ranges from 0.5 to 1.5 mm, while the fiber length of conifers ranges from 2.5 to 4.5 mm [27, 36, 58, 84, 194, 217, 321, 348, 428, 433, 483, 493, 518–521, 532, 565, 584, 624, 864, 865].

The increase in length is normally associated with an increase in the pulp strength level and improvements in printability characteristics [42, 56, 84 166, 202, 208, 267, 283, 295, 393, 593, 638–640]. The explanation for this behavior is that longer the fibers greater is the probability of establishing contacts with other fibers and, thus, of increasing the bonded area between the fibers.

However, and it is important to highlight, the effect of fiber length is, apparently, less important than traditionally attributed. This is evidenced when other characteristics of the fibers that also influence the bonding capacity between the fibers are considered. Among them are the fibrous population, coarseness, cell wall thickness, fiber diameter, hemicellulose content, and intrinsic cell wall resistance [42, 80, 207, 376, 478, 493, 634, 642], in addition to the wall fraction and fiber flocculation capacity.

7.1.2 Fiber Diameter

The fiber diameter varies from 10 to 50 μm depending on the raw material used, being up to 25 μm for hardwoods and 25–50 μm for softwoods [428, 433, 584]. These values vary according to the species, tree age, and position on the trunk, tending to increase in the pith-bark direction.

7.1.3 Cell Wall Width

Short fiber woods generally have cell wall thicknesses on the order of 0.01–0.40 mm [614]. Most eucalyptus species have irregular cell wall thickness [122]. The smallest variation in cell wall thickness is found in clones [534].

Cell wall thickness is considered a fundamental morphological fiber property with significant effects on their papermaking properties.

Thinner fiber cell walls are more flexible favoring the establishment of a greater number of fiber to fiber bonds in the paper structure making it denser, smoother, less porous, and opaque and with greater mechanical resistance, in addition to greater drainage resistance [42, 56, 78, 80, 84, 96, 105, 108, 132, 153, 190, 194, 199, 205–208, 216, 262, 266, 281, 294, 295, 307, 310, 318, 335, 343, 378, 397, 437, 467, 468, 478, 531, 534, 555, 626, 634, 640–650].

For cell wall thickness, both the mean value and its distribution must be considered. However, cell wall thickness alone, as well as fiber length, is not an adequate indicator to predict pulp behavior because fibers with the same wall thickness can have different diameters [42, 132, 194, 615]. In this case, the fiber coarseness, or the fiber cell wall fraction, generates more accurate estimates.

7.1.4 Fiber Coarseness

The fiber coarseness corresponds to a measure of linear fiber density, expressed in mass per length (mg/100 m is the most traditional unit). It represents the relationship between cell wall thickness, fiber length, and cell wall width. In this way, the morphological properties of the fibers determine their coarseness value, which will depend on the species and age of the wood, the conditions of the planting growth ("site"), the position in the tree, and the type of operating conditions used in pulping process [49, 84, 105, 122, 164, 205, 207, 255, 266, 298, 377, 389, 406, 467, 529, 624, 653, 669, 672–675, 1113].

A derived morphological characteristic is cell wall density or packing density, which is calculated based on fiber coarseness and cell wall cross-sectional area. The formula for this is [115]:

$$\text{Cell wall density}\left(\frac{\text{g}}{\text{cm}^3}\right)=\frac{10(\text{coarseness})}{(\text{wall cross sectional area})}$$

Where:

Fiber coarseness is expressed in mg/100 m
Cell wall cross-sectional area is expressed in micrometers

Coarseness is a relatively easy parameter to determine and has often been used as an indicator of the fiber cell wall collapsing ability (fiber flexibility) and, therefore, has a significant effect on fiber surface area (fiber bonding). A coarseness variation significantly influences all pulp properties as well as the refining energy demand [58, 83, 84, 115, 128, 164, 166, 172, 1414, 203, 206, 208, 266, 267, 283, 298, 315, 377, 389, 397, 467, 534, 633, 637, 650, 653, 676, 677, 1338]. For pulp fibers, coarseness values ranging from 5 to 11 mg/100 m have been reported [20, 27, 115, 217, 406, 493, 514].

Traditionally, the greater the fiber coarseness, the lower the fiber collapsing capacity and lower the fiber flocculation tendency. Tear resistance and refining energy requirements also increase, while wet and dry strengths and solid retention in paper machines are reduced. This retention reduction is due to the higher volume of voids in the paper structure [23, 49, 83, 91, 115, 117, 153, 164, 166, 267, 298, 389, 534, 618, 656].

However, as for fiber length and cell wall thickness, coarseness should not be used alone for predicting paper quality [26, 219, 377, 548],

7.1.5 Fibrous Population (Number of Fiber per Gram)

The number of fibers per unit of mass, that is, the "fibrous population," is derived from the fiber length and coarseness being traditionally expressed by the number of fibers per gram. The fiber population is related to the weight of each individual fiber and, by extension, to the coarseness and wall fraction of the fibers, which are fiber characteristics easy to be determined [115, 406].

For the same wood, the number of fibers per gram of pulp varies with the cooking yield, which is higher and the yield being lower [105, 153, 298, 469, 496, 660].

In the case of eucalyptus pulps, whose fiber length is relatively uniform but fiber coarseness widely varies, the fibrous population is strongly influenced by fiber coarseness, ranging from 12 to 30 million of fibers per gram of pulp. This allows good margins for the selection of market pulps to be used in paper production according to the desired paper properties in the final product [57, 115, 217, 406, 618].

The number of fibers per gram is directly and positively related to formation, surface smoothness, opacity, surface strength, and tensile, popping, and bending strengths, as well as porosity and water retention index, wet fiber flexibility, wet sheet strength, wet elongation, and Schopper Riegler degree of unrefined pulp [56,

57, 202, 209, 217, 267, 352, 468, 493, 529, 567, 618, 660, 662, 669, 672, 680–685, 868].

A larger fibrous population means smaller fibers and, therefore, a higher specific pulp surface. In this way, for same basis weight papers, more surface to establish fiber bonding, as well to light scattering, will be available.

7.1.6 Pulp Fines

A pulp particle is defined as fine when it is able to pass through a 75 μm diameter round hole or a 200-mesh screen of a fiber classifier [686–689]. The best methodology for measuring fines in a pulp is the dynamic jar test, developed by Dr. Britt. It is based on the percentage of the dry weight of a pulp passed through a 200-mesh wire (70 micrometer openings) under constant conditions of time, temperature, and consistency [115].

Fines are recognized as differentiated components of the fibrous raw material as their properties differ considerably from that of the fibers. They are formed throughout the entire production process, from the pulping of the raw material to the paper formation stage. Its management is fundamental for the pulp and paper industries, because in high or extremely low proportions, the fines influence the paper's drainage, structural, and strength properties [690, 1339].

According to their shape, fines are classified as lamellar fines or fibrillar fines. Lamellar fines, which have an irregular shape, are formed by parenchyma cells and by fragments or pieces of the cell wall from the parenchyma cells, vessel elements, or even from the fibers themselves. Fibrillar fines are filamentous, flexible, and rich in cellulose as they basically result from the release of cellulose aggregates from cell walls.

Due to their shape, the lamellar fines are directly related to the pulp drainage resistance in paper machines. On the other hand, fibrillar fines, with a high specific surface, contribute to fibers bonding. However, fines are most often seen as a problem rather than a solution [115, 688, 691, 692].

7.2 Fiber Characteristics Related to Their Uses

Wood pulp fibers mainly consist of cellulose fibrils embedded in a matrix of lignin and hemicellulose. They can be defined as tubular structures with multilayer wall, and which have a set of properties, such as flexibility, collapsibility, intrinsic strength, and strength of the connections between fibers, which define the consolidation capacity of the paper's internal structure.

It is evident that, as fibers are the most abundant element in the pulp, their characteristics, especially their bonding and collapsing capacity, are decisive factors, directly interfering with production costs and paper quality [32, 34, 41, 57, 80–102, 104–108, 154, 255, 343, 384, 919].

The wood species, the type and the operating variables of the pulping process, and the preparation of stock in papermaking determine the chemical composition, structure, and general conditions of the fibers, which directly influence the final paper quality as well as its production costs [42, 57, 84, 90, 96, 122, 126, 208, 217, 437, 666, 693–696].

Although the fiber wall properties can be modified through different pulping processes, they also strongly depend on the intrinsic fiber structure, that is, the chemical and physical cello wall structure [620, 698, 699].

Pulp strength consists of at least three components: individual fiber strength, fiber-to-fiber bonding strength per unit area, and total fiber-to-fiber bonding area [622].

All fibers have advantages and disadvantages. The papermaker must find the optimal point to meet his needs, producing paper with the characteristics demanded by the market [511, 623].

7.2.1 *Fibrillar Angle*

The fibrillar angle (MFA) is defined as the helical angle formed by cellulose microfibrils in layer S2 of the secondary wall in relation to the cell longitudinal axis (Fig. 7.2). It is an important controlling factor of the mechanical properties of the fiber. If the medium fibrillar angle is known, we can predict whether the fiber is

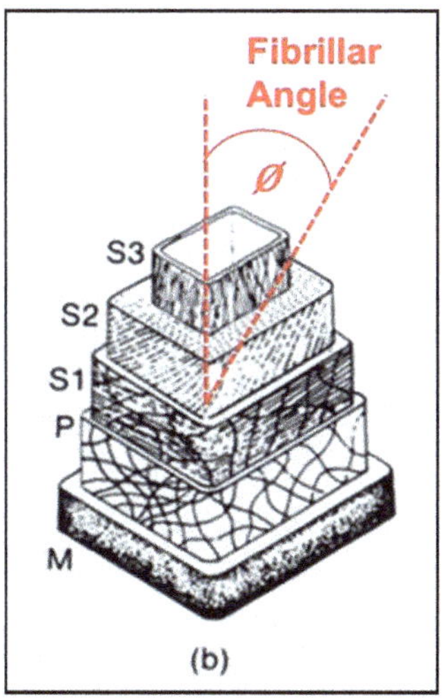

Fig. 7.2 Schematic model of the cell wall of the wood fibers [572] presenting the definition of fibrillar angle

more easily elongated or not, which is important in refining as one can estimate if the fiber will be more easily broken or elongated [191, 589, 620].

Fiber collapse is fundamental to increase both fiber flexibility and fiber bonding capacity. It depends on the fibers cross-sectional elasticity modulus that is directly influenced by the fibrillar angle. With the reduction of the fibrillar angle, the elasticity modulus increases up to a maximum value obtained when all fibrils are aligned in the longitudinal direction. In this situation, fibers have maximum resistance.

When the fibrillar angel increases, fibrils are loaded more and more in their transverse direction, which have lower mechanical resistance compared to the axial chain direction [38, 47, 191, 589, 620].

The axial loading of a sample with a small fibrillar angle implies the instant loading of the cellulose chains. The inter fibrillar matrix is deformed by shear, resulting in a more elastic and brittle mechanical behavior followed by an abrupt fracture. Thus, the paper tensile index increases with the reduction of the fibrillar angle [178, 191, 460; 603, 604].

The fibrils arrangement with lower angles in the fibers secondary cell wall provides tensile strength, while the larger angle in the primary cell wall provides compression resistance along the cell axes [427, 571].

7.2.2 *Interaction Between Fiber Anatomical Measures*

7.2.2.1 Aspect Ratio and Runkel Index

These are two parameters traditionally used to predict the quality of wood for the manufacture of woody products in general. They are classic and, therefore, we chose to present them in this book.

The aspect ratio factor, also defined as a slender index, is the relationship between the length and the diameter of the fiber (Eq. 7.1), while the Runkel index measures the cell wall thickness in relation to the thickness of the cavity, or lumen of the fiber (Eq. 7.2). These two indexes are traditionally related to the fiber flexibility.

$$\text{Slendermess Index} = \frac{\text{fiber length}}{\text{fiber diameter}} \tag{7.1}$$

$$\text{Runkel Index} = \frac{2(\text{cell wall thickness})}{\text{lumen width}} \tag{7.2}$$

Wood fibers have relatively high "aspect ratio." Typical lengths are in the order of 1–3 mm and the typical widths are 10–50 μm. It is generally stated that if the ratio is lower than 50–70, the fiber has no value for quality paper production [426, 624].

Fibers with higher Runkel indexes are more rigid and less flexible and form bulky papers, with low grip areas, than fiber with lower indexes. In the manufacture of paper with short fibers, the desired Runkel index values are 0.5–1.0 [219, 426, 563, 624, 700].

For eucalyptus woods, the Runkel index increases in the radial direction of the marrow to the shell.

7.2.2.2 Flexibility Index

The flexibility ratio measures the relationship between the lumen and the fiber diameters. It expresses the actual proportion of lumen in the cross section of the fibers, in a percentage, as shown below [69].

$$\text{Flexibility Ratio} = \frac{\text{Lumen Width}}{\text{Fiber Diameter}}$$

Based on the flexibility index, pulp can be classified into different distinct groups [75]:

- High elastic fibers: represent woods with flexible index of over 75%. They are low density woods, usually below 430 kg/m^3, having thin walls and large lumen. These fibers have good bonding capacity as they collapse easily producing good superficial contact.
- Elastic fibers: have flexibility ratio between 50% and 75%. They are medium basic density woods presenting similar dimensions for wall thickness and cellular lumen. The fibers collapse partially given a relative contact and moderate bonding between.
- Rigid fibers: fiber flexibility indexes between 30% and 50%. Cell walls are thicker with medium to high fiber density. They have little superficial contact and therefore poor fiber-bonding capacity.
- High rigid fibers are those with flexibility rate lower than 30%. This is usually applicable to excessive mature trees. The fibers have very thick walls with narrow cellular lumen. The superficial contact is small giving low bonding capacity between fibers.

7.2.2.3 Cell Wall Fraction

The "cell wall fraction" is the percentage relation between the cell wall thickness and the fiber radius. In other words, the wall fraction corresponds to the percentage of the cross-sectional area of the fiber occupied by the cell wall.

The higher the cell wall fraction, the smaller the fiber population and more rigid the fiber. Greater normal forces will be demanded to meet same internal shear level during refining [115, 196, 406, 650, 1338].

Figure 7.3 illustrates the concept of wall fraction and interaction with the diameter, cell wall thick, and coarseness of the fibers. The fibers (A) and (B) have the same diameter but the fiber (A) is thicker and, therefore, has greater coarseness.

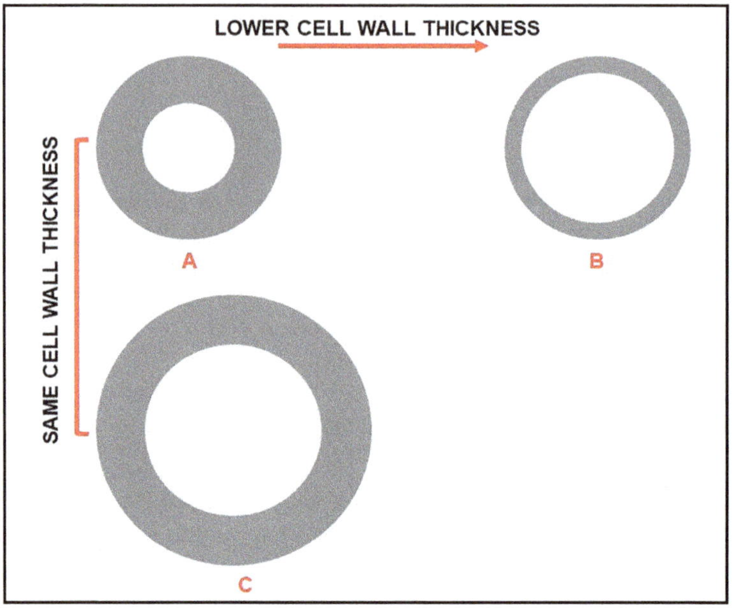

Fig. 7.3 Interactions between fiber cell wall thickness and fiber coarseness

It is easy to identify, even visible, that fiber (B) is easier to collapse (more flexible) than fiber (A). It is that of the lowest coarseness which highlights the coarseness importance in defining the fiber collapse capacity and flexibility.

Similarly, but comparing the fibers (A) and (C), having fraction the same wall thickness, we identify that fiber (C) with the greatest coarseness has the highest capacity to collapse. This is contrary to the expected when only coarseness value is used, in this evaluation. The fiber (C) has greater collapsing and flexibility capacity because it has lower cell wall fraction.

Fiber wall fraction is best suited to predict fiber collapsibility and flexibility. The cell wall fraction is directly correlated with fiber collapsibility (intrinsic fiber resistance – Young's modulus) [406, 478]. The papers manufactured with smaller diameter fibers and larger wall thickness, that is, with larger wall fraction, usually lead to the highest opacity [294, 295, 478].

7.2.3 *Fiber Swelling*

Also reported as cell wall swelling, this fiber characteristic is directly responsible for the fiber's collapsing capacity, that is, their flexibility. It is probably the most important fiber characteristic for both pulp processing and the assignment of the desired properties in the finished product.

Swelling is generally considered as a process in which the dimensions of a material are expanded as the material cohesion is reduced given softening and greater flexibility to the material [433].

This property is influenced by the fiber surface water affinity associated with the chemical composition of this surface, fiber morphology, and fiber blanket structure [627].

In living wood, two types of water are present: free water, which predominates in the cell's lumens, and the water directly linked to sites of water sorption, which is found mainly in the hemicelluloses. Only one-third of the hydroxyl groups on the surfaces of cellulose microfibrils is available as sorption sites. The other two-thirds of these groups are involved in the interaction between and within the cellulose chains [27, 628].

The cell wall can be considered as an ionic gel with fixed anions. The gel contains linked acid groups that cannot move, and counter-ions to these groups must remain within the gel to obtain electrical neutrality. This leads to a higher ion's concentration within the gel than in the surrounding liquid causing an osmotic pressure difference that is balanced by the water absorption diluting the loads inside the cell wall. In this process, there is an increase in cell wall volume [124, 229, 433, 450, 603, 632, 657, 701–704, 706, 708, 709, 715, 1338].

Swelling is a direct result of water penetration in the fiber cell wall capillary system with simultaneous interaction between the polar liquid, that is, water and hydrogen bonds formed by hydroxyl groups (–OH) and carboxyl (–COOH) of cellulose and hemicelluloses, which are the constituents of the cell wall [1339]. The swelling phenomenon begins at the moment the pulp is immersed in water being called as free fiber swelling

The fibers initially soak on their surfaces, following the penetration of water through the cell wall by filling the free spaces of the amorphous regions and reaching the fibers lumens. The thermodynamic driving force for water sorption is the enthalpy gain when water molecules are adsorbed in cellulose, and the generated internal stress breaks the hydrogen bonds between the cellulose surfaces. This breakdown of hydrogen bonds in the radial direction (from surface to lumen) causes cell wall delaminating [124, 450, 703].

The regions between microfibers are permeable to water, ions, and small macromolecules, while cellulose microfibrils having a high crystalline structure are not permeable to water [433].

The cell wall acquires a plasticity level similar to the caused by refining and fiber flexibility increases. The process occurs until the equilibrium point in which the forces created by the osmotic pressure equal the cell wall mechanic forces, that is, the fiber swelling is limited by the cell wall structure which prevents the unlimited fiber expansion [72, 187, 337, 429, 450, 511, 601, 603, 684, 701–704, 706, 708, 709, 715, 721, 736, 733, 1338].

In summary, the fiber swelling is the result of three factors: a non-ionic contribution, the osmotic pressure created by the loaded groups on the fiber wall, and the restrictive forces caused by the network structure of cellulose fibrils aggregates in the different layers of the cell wall.

This fiber hydration is determined by the fiber chemical composition which is strongly influenced by cooking and bleaching operations as they affect the pulp hemicelluloses content as well the cell wall degradation. Different pulp techniques modify the concentrations of cell wall loads and significantly affect cell walls swelling [57, 115, 126, 159, 218, 300, 433, 440, 443, 456, 466, 468, 705, 711, 707, 716–720, 735, 1355].

The fiber hydration is controlled by:

- The number of acidic groups present on the fiber cell wall and their dissociation constants [429, 433, 440, 450, 702–707, 721–734]
- Fiber cell wall elasticity modulus [38, 443, 450, 603, 771, 711, 725, 735, 736]
- Type and quantity of counter-ions on the fiber cell wall. Counter-ions are those that maintain electrical neutrality with the pulp suspension [450, 701, 733, 736]
- Ionic concentration of surrounding suspension liquid [433, 450, 705, 712, 707, 736],
- pH of the surrounding suspension liquid [450, 705, 715, 736]
- Refining degree [736].

7.2.3.1 Acidic Groups in Fiber Cell Wall

The maximum swelling level depends on the type and quantities of ionizable carboxylic groups present on the cell wall. The greater this amount, the greater the swelling potential of the cell wall [429, 433, 440, 450, 702–707, 721–735].

Water molecules are strongly attracted to the ionizable groups in the fibers such as carboxylic, sulfonic, phenolic, or hydroxylic acids from the non-cellulosic wood components. These ionizable groups can also be created during the impregnation of chips, cooking, or pulp bleaching. These groups are the main contributors in defining the ionic exchange cover of fibers [387, 444, 702, 703, 723, 728, 730, 734, 735].

Hemicelluloses are the main source of charged groups in the fibers. Joint with cellulose absorb up to four to five times more water than lignin in water saturation conditions at temperatures around 20 °C [47, 703].

The number of charged groups is higher in unbleached pulp than in bleached ones. They are located inside the cell wall and on the surface of the fibers. In this way, pulps with higher hemicelluloses content and larger specific surface, such as the eucalyptus pulp, are more easily hydrated by alkaline treats than long fibers [733, 737].

Carboxylic groups play an important role in fiber modifications to obtain high-quality products. During paper manufacturing process, the introduction of charge alters the pore size and pore distribution inside the cell wall of the fibers [124]. Carboxyl groups are responsible for the retention of several wet additives in the fiber suspension [214, 588, 702, 703, 716, 728, 730, 738–742]

7.2.3.2 Ionic Force (Counterions)

Depending on the chemical environment, the acidic groups of fibers may be dissociated or not. If the ionic form is altered, the fiber swelling, as well as the distribution of the pores on the cell wall, will also change [433, 443, 450, 701–703, 708, 733, 745, 749].

In its non-dissociated form, carboxylic groups have H^+ connected as a counter-ion, that is, ion that accompanies an ionic species to maintain electric neutrality. This H^+ may be exchanged for some metal ions present on the cell wall or surrounding medium by increasing the absorption force and significantly contributing to fiber hydration capacity [702, 413, 745].

Chemical pulp usually contains a mixture of several metal ions such as calcium, magnesium, sodium, and small quantities of aluminum, potassium, and iron. The ion content of the pulp metal is in the same order as the amount of acid groups, as they are present mainly as counterions to these acidic groups [702, 707].

The type of counterion also affects fiber swelling. Di- and trivalent cations inhibit hydration of fibers rather than univalent cations. Kraft pulp fibers increase in volume (fiber swelling) more easily in the following order: $Na^+ > Li^+ > Ca^{++} > Mg^{++} > H^+ > Al^{+++}$ [73, 300, 433, 702, 706–708, 721, 736, 733, 745–749].

When the pulp is in sodium form, carboxylic acids are almost completely in dissociated form, having sodium ions as counterions. The pulp features cell wall pores with larger diameters and more fiber swelling [433, 706, 736].

The change of counter-ion, from Na^+ to Ca^{++} or H^+, reduces fiber hydration as it reduces the degree of dissociation of carboxyl groups in fibers. Minimum swelling is obtained when the pulp is in its hydrogen form and maximum swelling when Na^+ ions are present as counterions [73, 733, 747, 749].

Due to the greater cell wall hydration, and consequent fiber flexibilization, refined Na^+ pulp showed greater tensile resistance than refined pulp in Ca^{++} form [733].

Fiber functional groups and ion interactions can be divided into non-specific and specific ones. Non-specific interactions are the electrostatic attraction or repulsion between ions and the charged matrix of the cell wall. Specific interactions include, for example, metal cation complexation with functional groups, as well as adsorption of polarized surfaces [433].

Normally, pulps are considered in the Ca^{++}form [733]. Industrially, the electrolytes concentration during refining is influenced by the closure degree of the paper machine water circuit [704, 721, 733].

7.2.3.3 Ionic Concentration of Surrounding Liquid

The chemical environment in the fiber surrounding liquid also has a strong influence on fiber hydration. Acidic group dissociation progressively increases with alkaline-acidic pH variation and, as a large number of counterions are created in the cell wall, osmotic pressure is generated causing additional migration of water to the cell wall as a way to reduce this pressure [124, 433, 450, 456, 616, 701, 702, 705, 707, 712, 715, 721, 750, 1338].

7.2.3.4 pH of the Surrounding Medium

The suspension pH is the most important factor determining the ion exchange rate between the fiber and the external environment. As the pH value increases, carboxylic acids dissociate and repulsive forces in the loaded groups lead to increased swelling. Thus, it may be understood that the ion transfer kinetics between the fiber wall and the surrounding medium is widely governed by the rate of fiber charge variation [450, 705, 707, 752].

Varying the fiber suspension pH, the dissociation degree of carboxyl groups can be controlled and, therefore, the charge-induced osmotic pressure can be managed [701–703, 707, 753]. At pH below 3, carboxylic and phenolics groups in the cell wall are not dissociated. This justifies why there is virtually no adsorption of metal cations which does not favor fiber hydration [632]. When pH increases above 3, the dissociation of carboxylic groups begins and adsorption capacity increases to a maximum level in the pH 7–9 range when carboxylic groups are completely dissociated.

The maximum swelling occurs at a pH interval of 8–9. If pH increases above this value do not allow new increases in Sodium adsorption level and swelling decreases. A pH value of 9 is the best condition for the swelling of the fiber [433, 632, 701, 702, 704]. Thus, to increase the volume of the fiber, the suspension pH is often adjusted above the neutral conditions as a way to convert almost all carboxylic groups into a discounted form [701, 737].

7.2.3.5 Cell Wall Elasticity Modulus

The dissolution of cell wall materials and their consequent hydration reduce the elasticity modulus of fiber that becomes more flexible due to higher collapsing capacity. The degree of swelling (fiber hydration) also depends on the mechanical properties of the cell wall. The swelling is restricted by the network distribution of cellular microfibrils, the presence of hydrophobic lignin, and areas with hydrogen bonds preventing unlimited fiber expansion [38, 450, 603, 711, 736, 771].

7.2.3.6 How Fiber Swelling Is Measured

The amount of water retained on the cell walls is significant and may be much larger than the dry weight of the fibers, due to its large affinity with carbohydrates, especially with hemicellulose [115, 502, 503]. Water may also be linked to hydroxyl groups or other charged groups of wood polymers, with counterions influencing water quantity. Water may also be associated with cell wall pores formed when removing lignin and hemicelluloses during wood pulp production or created during pulp refining or adhered to fiber surfaces [433, 650, 754]

The amount of water in fibers can be determined through the water retention value (WRV) or the fiber saturation point (FSP) [62, 71, 124, 406, 433, 594, 708, 721, 756, 771, 1388].

The fiber saturation point refers to the point where all free water was removed from the pores and from the cell wall cavities except the linked water in these walls [62, 63, 220, 221, 430, 456, 711]. The higher the FSP values, the more difficult it is to remove water from the pulp [456].

WRV correlates with fiber saturation point (FSP) and, by operating simplicity, has been widely used to estimate the water content on fiber cell walls. The WRV technique is fast, easy to run, and precise providing reproducible results [62, 63, 115, 124, 186, 300, 406, 434, 650, 706, 733, 756, 771, 759–766, 768].

The WRV consists of the relationship between the amount of retained water and the weight of the dry pulp, in a percentage, after submitting the pulp sample to the action of a centrifugal force over a certain time period. In general, the centrifuge works between 3000 and 5000 rpm according to the purpose of the tests. Imagine that even after submitting the wet pulp to this enormous G-force, there is still a significant amount of water adhered to the fiber cell wall. The more the retention of water, the greater the WRV [406, 771]. The WRV determination formula is as follows:

$$WRV = \frac{\left(\text{peso}\, \textit{após}\, \text{centrifugação} - \text{peso seco}\, \textit{em}\, \text{estufa}\right) * 100}{\left(\text{peso seco}\, \textit{em}\, \text{estufa}\right)}$$

There is a linear correlation between WRV and the volumetric variation of the cell wall after hydration (fiber swelling). Thus, the water retention value has been used to determine the volume of the cell wall pores [71, 124, 127, 406, 708, 769, 1338].

When the WRV is high, the pulp retains a lot of water, which can impair drainage and pressing capabilities in the wet part of the forming paper [406].

It is easy to understand that pulps with the highest volume of lumen and the largest number of fibers (larger specific surface) have larger WRV results. Small pores retain water more strongly than large pores due to the larger superficial tension forces associated with small pores [115].

On the other hand, most pulp extractives are hydrophobic. High extractive pulps have less capacity to retain water on the fiber wall [115] having lower water retention values.

7.2.3.7 Fiber Hydration Effects

Fiber swelling has a significant influence on aspects related to paper production and the quality of the finished product. The cell wall delaminates forming new surfaces (pores) reducing the fiber's intrinsic resistance [57, 214, 430, 466, 770, 709]. This is quite important because this intrinsic resistance contributes to all mechanical paper properties.

In addition to assisting in the refining energy reduction, fiber swelling promotes fiber straightening and increased mechanical resistance, but impairs water removal in pressing and reduces the specific volume, light scattering, and the paper Elmendorf tear [6, 57, 159, 161, 399, 433, 450, 588, 702, 716, 717, 721, 761, 762, 771–798, 913].

Fiber swelling can be changed by mechanical or chemical treatments used in mass preparation, such as enzymatic treatments and refining or chemical modification [703, 721, 751, 783, 892]. The effects on the fibers caused by refining or the ion exchange are identical and therefore probably governed by the same mechanisms. Chemical treatment is more uniform in action and does not lead to cutting fiber as can occur in refining. As a result, a chemically treated pulp may have lower fiber "cuttings," cell wall fibrillation and fine generation than a refined pulp [707].

7.2.4 Fiber Bonding Capacity

The mechanisms controlling the fiber bonding capacity can be classified into three different groups:

(a) Molecular contact area, which also includes the inter diffusion that are nanometric scale measures, increasing the molecular contact area, that is, a multiplier of the intermolecular bonding mechanisms (van der Waals forces, hydrogen bonds and Coulomb interaction). Inter diffusion will take place when polymer chains partially spread into the connecting region from one surface to another surface. This is facilitated because the swollen surface is considered a swollen hydrogel [707, 732, 795, 1013] and the presence of microfibrils partially released on the fiber surfaces. Those microfibrils come from the external fibrillation promoted by refining and increase the contact area between the fibers.

    The molecular contact area between the fibers is the most important factor in defining the resistance of the connections between the fibers as it is a multiplication of all molecular mechanisms. An increase in contact area leads to a directly proportional increase in the connection force [593, 732, 1013, 1014].

(b) Intermolecular bonding mechanisms: hydrogen binding, van der Waals forces, and Coulomb interaction. Although some authors point out that van der Waals forces are more important, hydrogen bonding is the most common reference to a fiber connection mechanism, which is not surprising because carbohydrates have many OH groups. These connections are estimated to be responsible for at least 60% of paper tensile resistance [732, 1006, 1013, 1014].

    These connection mechanisms have their origin in the hemicellulos's carboxylic groups. Under alkaline conditions, during kraft pulp production, these acidic functions dissociate and fibers retain negative charges [732].

(c) Mechanical bonding is a bonding mechanism that occurs on a micrometer scale such as mechanical interlocking and capillary bridges. Interlocking is the entanglement of surface fibrils released during refining causing the mechanical entanglement of fiber surfaces [732, 1013].

The capillary bridges correspond to the links between the fibers promoted by the capillary forces. They define the resistance of the wet pulp mat [1012] (Fig. 7.4).

The driving force for contact development is the Laplace pressure (Eq. 7.3) due to capillary bridges. When the two surfaces touch each other, the peaks of both topographical surfaces become more and more deformed. This deformation leads to a force opposite to the force pressing the surfaces together, and the degree of contact between the surfaces is determined by the balance between these forces [732].

The meniscus in the capillary bridge causes an adhesion pressure, the so-called Laplace pressure (LP),

$$LP = \gamma\left(\frac{1}{r_m} + \frac{1}{r_b}\right) \approx \frac{\gamma}{r_m} \tag{7.3}$$

The Laplace pressure (LP) is determined by the air-water surface tension ($\gamma \approx 70$ mN/m) and the two radii of curvature (m) in the menisci (r_m and r_b). As generally $r_b >> r_m$, the equation can be approximated by (γ/r_m). So, the key parameter determining the Laplace pressure on fiber surfaces together is the distance between the surfaces ($2r_m$) and large capillary forces occur when the distance between the surfaces is small and the meniscus (r_m) resulting from the liquid is small [732].

When water is retracting, the menisci move inward decreasing their radii and increasing capillary pressure. Thus, as the water is removed, the capillary bridges become stronger reducing the distance between the surfaces and increasing the contact area between the fibers [732].

The specific surface of short fibers is greater than that of long fibers, which makes them more easily hydrated by alkalis while giving them a greater fiber bonding [314, 737]. The bonding capacity between fibers is a function of the average contact area between them, which determines the number of connections and the resistance of these connections [42, 43, 78, 126, 154, 159, 208, 215, 217, 218, 282, 393, 468, 470, 474, 555, 588, 593, 672, 693, 732, 758, 784, 792, 888].

The contact area between the fibers, also known as the relative bonding area (RBA), is defined as the bonded fraction percentage of the total fiber surface being

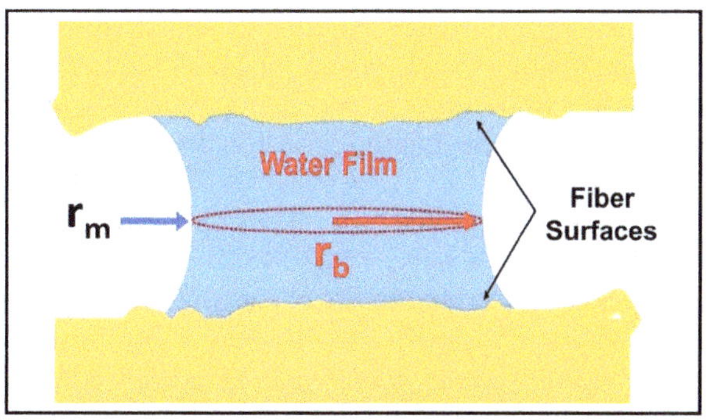

Fig. 7.4 Capillary bridge of water between two fiber surfaces. (Adapted from [732, 1012])

defined according to Eq. 7.4, where A_t is the total area available for bonding and A is the non-adherent area on the formed paper [470, 693, 1007, 1009].

$$\text{RBA} = \frac{(A_t - A)}{A_t} \tag{7.4}$$

The most common measurement method is the light scattering [302, 785], which assumes that $S = c.A_t$, where S is the light scattering coefficient of the paper, c is a constant depending on both wavelength and fiber properties, and A_t is the surface area available for scattering. If S_0 is the scattering coefficient for a completely unbonded sheet, then [1009].

$$\text{RBA} = \frac{(S_2 - S)}{S_0} \tag{7.5}$$

As S is inversely and linearly proportional to the paper apparent density ρ_a providing a method to correct S and ρ_a for the fiber shape, which results in Eq. 7.6, showing that all data can be fitted by a straight line when corrected values are plotted against each to other [1010, 1011].

$$\text{RBA} = \frac{\rho_a}{f.\rho_f}(1 - r) \tag{7.6}$$

In Eq. 7.6, f is defined as the finer filling factor, that is, the ratio between the fiber wall area and the area of a rectangular delimiting box around the fiber, and ρ_f is the cell wall density considered as 1500 kg/m^3. The $(1 - r)$ value, based on measurements on several pulp samples, equals ¾ [1009].

The relative binding area (RBA) varies according to the amount and type of pulp fines as well as the fiber's flexibility and collapse which, in turn, are controlled by their dimensions, where coarseness, as indictive of the wood basic density, the average length and the transversal section of the fibers the main conditioning factors [78, 393, 406, 467, 470, 693, 701, 732, 784–786, 822, 1007, 1008].

The hydrophilic nature of the fibers causes them to absorb a greater volume of water producing a gelatinous hydrate and, therefore, a greater capacity for cell wall swelling and fibrillation which increases the fibers' ability to form bonds between themselves impacting the paper properties. The structural paper properties, paper strength, and paper hygro-expansion are impacted [115, 199, 406, 678, 701, 732, 793–796].

In papermaking, free water flows through the forming fabric while the fibers are trapped forming a bulky sheet [462]. The fibers are closer to each other and the capillary forces bring the fibers and fibrils closer together generating a more intimate contact. If the molecules involved in the formation of hydrogen bonds are from different fibers, these fibers unite with each other through forces very similar to those of crystallization, with a resistance that depends, among other factors, on the number of bonds formed between them.

Numerous bonds, including hydrogen bonds, are formed between fibers during the paper consolidation process [113, 213, 343, 633, 1006]. The greater the amount of hydrogen bonds, the greater the paper resistance.

In a fiber web, adhesion between wet fibers is usually explained by the presence of capillary forces which hold the swollen fibers together provided there is enough free water wetting the fiber wall. Such capillary forces are in fact the main interactions between swollen fibers in the range of 30–45% solids content [477].

As the solid content increases, the capillary forces are gradually replaced by hydrogen bonds. Hydrogen bonding and van der Waals forces occur only in the area of molecular contact between the fibers, that is, when the bonded surfaces are at a small distance (below a few Ångstroms) [477, 732, 733].

In pressing, the pulp is compacted by compression, eliminating part of the water retained in the fibers. As the water content reaches approximately 50% of the mass, nearby fibers bond together forming a network not only due to mechanical entanglement but also due to chemical interactions. By means of additional heating in drying, the fiber content is brought to about 90–95% of the mass by the end of the paper production line. The molecular segments, which were mixed in the wet stage, become entangled and interpenetrated [462, 678].

7.2.5 Fiber Collapsibility and Fiber Flexibility

Fiber collapsibility is associated with wet sheet compaction reflecting the ease with which fibers fit and conform to each other in the paper structure. Thus, the collapsibility of fibers is related directly to the connection area between fibers.

Fiber collapse depends on the moment of inertia of its cross section, that is, on its diameter and cell wall thickness, and on its transverse elasticity properties and the cell wall fibrillation degree. Fibers with thicker walls, that is, with a higher wall fraction, are more rigid and resistant to collapse [42, 56, 53, 57, 84, 87–90, 99–102, 105, 218, 294, 298, 319, 467, 478, 534, 555, 603, 653, 662, 669, 673, 675, 683, 797, 798–801].

Amorphous wood polymers significantly contribute to fiber properties in the transverse direction than in the longitudinal direction [413]. Collapsed fibers are more flexible and have a larger area available for establishing inter fiber bonds.

More collapsed fibers will produce a denser paper having a greater difficulty in draining and lower porosity, opacity, softness, and ability to absorb water in addition to higher elasticity and resistance levels (tensile, rigidity, rupture, bending, compression, and surface strength) [26, 43, 87, 99, 115, 178, 184, 190, 195, 202, 203, 207, 208, 218, 266, 299, 310, 319, 377, 468, 529, 534, 588, 610, 651, 774, 788, 800, 958].

The fibers collapse and flexibility varies with the type of wood, with its age, and position in the trunk and decreases with an increase in the basic wood density [529, 958].

Collapse and flexibility tend to increase with refining treatment due to fiber swelling caused by cell wall hydration [57, 78, 126, 195, 202, 203, 208, 218, 226, 262, 588, 693, 710, 803–811].

Flexibility is the opposite of stiffness and is highly and positively related to fiber bonding, fiber lumen collapse, and water retention index [43, 115, 413, 433, 456, 642, 710, 800–802, 810–816].

Flexibility is determined by a combined effect of fiber length, fiber thickness, and fibrillar angle. Fibers with smaller cross-sectional moment of inertia collapse more easily. In other words, they are more flexible [242, 797]. For homogeneous materials, flexibility (F) is related to bending stiffness (S), area moment of inertia (I), and modulus of elasticity (E) as given in Eq. 7.7 [797].

$$\frac{1}{F} = S = E.I \tag{7.7}$$

Fiber flexibility correlates very well with paper wet strength and wet elongation that are two important properties for machine runnability [115, 1809].

7.2.6 Fiber Bonding Strength

The quality of fibers bonds can be described by the specific bond strength (SBS) referring to the bonding force required to break a unit area of fiber bonding (Eq. 7.8).

$$\text{SBS} = \frac{\text{bond strength}}{\text{bonded area}} \tag{7.8}$$

Thus, the measurement of the specific bonding force consists of measuring the force, or energy, and the fiber bonding area which can be defined as the geometric area of overlap of two fibers where the fibers are in molecular contact [313, 693, 817]. Bond strength relates to the hydrogen bonds that hold particles and anatomical elements together. Factors affecting these bonds will influence the specific strength of the bonds [215, 218, 474, 593, 787–790, 817–822].

To meet the desired bond strength, it is necessary to have a wide fiber molecular contact and a high interaction in this contact area. The mechanical treatment of the fibers, the use of chemical additives, or the introduction of adequate fines can alter the bonding capacity between the fibers [78, 90, 159, 218, 343, 693].

The extension of the fiber connections is related to the contact area between them and to the specific resistance of these connections, that is, the resistance per unit of area of fiber contact. The influence of the fibers intrinsic strength is greater as higher is the connections strength [126, 218, 313].

The increase in specific resistance of the fibers bonding by reducing, for example, the coarseness or fiber cell wall fraction is a potential way to reduce the paper basis weight without impairing the paper optical or mechanical properties [208]

7.2.7 Intrinsic Fiber Strength

Individual fiber strength is traditionally considered as an important factor in paper mechanical strength definition. This resistance can be reduced chemically with the cell wall constituent degradation, especially cellulose, during the pulping process [47, 83, 311, 474, 824, 901, 1003].

Fiber degradation can also occur by mechanical action causing damage to the cell wall structure. Although connected with chemical degradation, the mechanical one is the main cause of fiber intrinsic strength reduction [47].

Degradation, which can be homogeneous or heterogeneous, has different effects on fiber and paper properties depending on the extent and location of degradation. Homogeneous, or random degradation, is caused by the pulping process and promotes a small reduction in strength. The heterogeneous, or localized, degradation caused by complementary mechanical action significantly weakens the fiber cell walls [47, 575, 825, 1002].

Zero gap tensile strength tests, performed with both dry and wet samples, provide good fiber intrinsic strength indications. The values obtained, in which theoretically only the energy required to break the fiber is measured, provide reliable results when estimating the intrinsic resistance of the fibers [47, 83, 91, 130, 232, 249, 311, 470, 474, 633, 823–828].

The resistance with zero spacing depends on the fibrillar angle, the presence of cell wall defects, the pulp viscosity, and the lignin content in pulp. This resistance increases linearly with the fiber's coarseness (wall fraction) and does not depend on the presence of fibers deformations [42, 115, 233, 298, 376, 826, 904, 928, 1000].

For those who do not have this measuring equipment, there is always the possibility of estimating this value through the tearing resistance test adopting a low level of refining and expressing the result in a fixed value of paper compaction (specific volume) [115]. In the case of properties dependent on fiber strength, good interaction between properties and paper density is expected, especially when intrinsic strength is altered by fiber hydration mechanisms [26, 78, 83, 172, 191, 203, 218, 262, 468, 588, 651, 693, 801–811].

7.2.8 Flocculation Tendency

Fibers in suspension tend to agglomerate, through chemical or mechanical mechanisms, forming floc. The fibers are aggregated as a result of inter-fiber friction caused by fiber bending without the need for attractive forces. When subjected to high turbulence and high shear, the flocs are dispersed and broken up [234, 830–834].

The presence of these flocs influences aspects related to pulp processing and the quality of the finished product. A good example of the effect of flocs on paper quality is its formation which improves with a reduction in flocs size. The flocculation mechanism determines the floc properties and therefore the pulp suspension behavior [999].

The flocculation tendency, as well as the anatomical characteristics of these flocs, depends on the pulping process used, the pulp consistency and viscosity, the turbulence level, and the flow velocity of the suspension. The chemical and anatomical characteristics of fibers, in particularly those related to their flexibility and bonding capacity, are also important [34, 141, 156, 166, 209, 217, 615, 644, 656, 742, 836–839, 868–882].

Longer fibers present a greater flocculation degree, forming larger and more stable flocs. Shorter fibers have smaller flocs and produce smoother papers with better formation and greater opacity [26, 46, 141, 156, 166, 185, 202, 266, 615, 656, 868, 873, 874, 879, 882, 889, 1371].

By fixing length, increasing the stiffness of fibers leads to increase the floc size and stability [156, 166, 234, 873, 874, 879, 882, 890].

Hardwood fibers show a good tendency to flocculate in aqueous suspension, forming flakes with typical diameters in the range of 2–3 mm [343, 838, 1119]. Evaluating anatomical measurements of commercial pulp flocs [868] determined values smaller than 2.5 mm for eucalyptus pulps and between 4.4 and 5.2 mm for softwood pulps. In mixtures of these pulps, flocs diameters ranging from 2.0 to 5.6 mm were determined.

When stripping sheets of commercial pulp bales, the flocculation process corresponds to a size reduction process in which the dry pulp sheets separate into aggregates that interact to reorganize the internal fibers and group the loose fibers of the pulp surface to form its final stable size and shape [837].

In the result pulp suspension, rotating movements of adjacent fibers will easily result in mutual collisions. This mechanical interaction between fibers is the main factor responsible for flocculation [34, 145, 234, 881, 994, 995, 1371]. There are two mechanisms, mechanical and chemical, which can cause particle agglomeration in dilute suspension forming flocs. Such mechanisms can be related as follows:

- Fibers are grouped by the relative movements of rotation and translation, with the mechanical entanglement of the fibers giving rise to cohesive forces forming the flocs.
- If the flocs are weak, hydraulic forces disperse the flocs.
- Due to dynamic equilibrium, transient flocs are formed and dispersed, with the floc size decreasing and shear level increasing.

- If the flocs are resistant, large coherent flocs are formed leaving few free fibers in the suspension.

This flocculation tendency can be described by the agglomeration factor, which is proportional to the product of the consistency and the square of the fiber length [166, 885, 889, 992, 993, 1371]. Depending on their consistency, fiber suspensions can be classified as [876] follows:

• Low consistency (C_m = 0–8%): when the suspension is a fluid medium formed by fibers and water

• Medium consistency (C_m = 8–20%): suspension created by filtering a low consistency suspension

• High consistency (C_m = 20–40%): formed by mechanical pressing from a suspension of medium consistency

• Ultra high consistency (C_m > 40%): formed by evaporative drying process

At low consistency, a suspension is a two-phase fluid, and at medium and high consistencies, it changes to a three-phase heterogeneous mixture of water, fibers, and air. With high air content, volumetric concentration (C_v) is used instead of suspension consistency, which are related as presented by Eq. 7.9.

$$C_v = C_f \left(\frac{1}{\rho_f} + \frac{X_w}{\rho_w} + V_L \right) \rho_b \tag{7.9}$$

Where:

C_v volumetric concentration of the suspension
C_f mass fraction (consistency) of the suspension
ρ_f fiber density (kg/m^3)
ρ_w water density (kg/m^3)
ρ_b apparent density (mass per volume of suspension) (kg/m^3)
X_w amount of water absorbed by the cell walls of the fibers (kg of water/kg of fibers)
V_L volume per unit mass of fiber lumen (m^3/kg of fiber)

The volumetric concentration defines the pulp suspension volume occupied by the fibers giving an idea of the space available for fiber rotation and translation movements. The higher the fiber concentration, the greater the number of contacts between the fibers that allow higher flocs formation.

So, there is a critical fiber concentration (CFC) in which the fibers are on the limit between totally free movement and the beginning of the contact between the fibers (Fig. 7.5).

The volumetric concentration defines the fiber suspension volume indicating the space available for fibers rotation and translation movements.

The higher the concentration, the greater the possibility for fiber collisions. This concentration is called volume concentration (CV), and in cases where CV < CFC, inter fiber collisions are relatively rare, while in the case of CV > CFC, such collisions are increasingly frequent [34, 878, 985, 991].

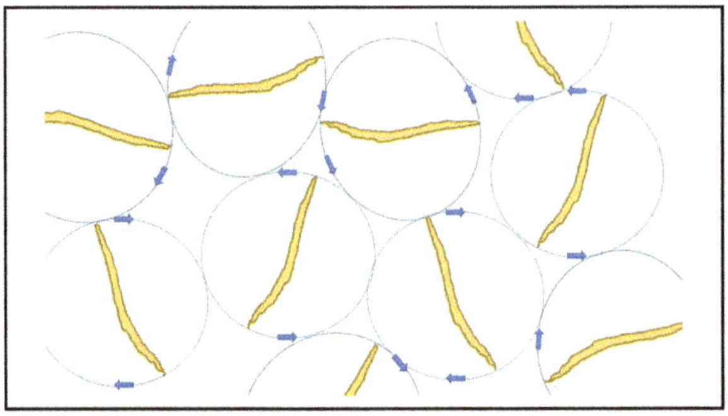

Fig. 7.5 Schematic illustration of the volume required for individual fibers to rotate freely without colliding with neighboring fibers. (Adapted from [34, 885])

These contact regimes were described by an agglomeration factor (AF) defined as the number of fibers in a volume covered by a single fiber length. The agglomeration factor, also named as crowding factor, can be calculated from the pulp consistency (C_f), the fiber length (l), and fiber coarseness (w). If two fiber suspensions have the same crowding factor, they are expected to have the same tendency to flocculate. The agglomeration factor is defined as follows (Eq. 7.10) [234, 650, 885, 1125, 1371]:

$$\mathrm{AF} = \frac{2}{3} C_f \left(\frac{l}{d} \right)^1 \tag{7.10}$$

Or, for hydrated pulps, more conveniently as

$$\mathrm{AF} = \frac{5 C_f l}{w} \tag{7.11}$$

Where:

AF agglomeration (crowding) factor
C_f volume fraction of fibers, %
l fiber length, m
d fiber diameter, kg/m
w the fiber coarseness

For suspensions with three or more contacts per fiber, AF can be related to the number of contacts per fiber (n_c), by the theoretical expression below [885, 990].

$$\mathrm{AF} = \frac{4.\pi (n_c)^3}{3(n_c - 1)} \tag{7.12}$$

Initially, the value of AF = 1 was considered as the agglomeration factor from which collisions between fibers begin. Subsequently, it was observed that at AF $\approx$ 60, the fiber suspensions have about three contacts, each restricting the fiber mobility, ceasing the shearing action, and fixing the fibers. In this way, the fibers are fixed and bent generating elastic flexion that creates normal forces at the contacts. The resulting friction force confers mechanical resistance to the flocs network [882, 885, 989].

When AF < 60, the flakes are loosely aggregated, but when AF > 60 (stiffness limit), the flakes adopt mechanical strength. Their sizes and resistances are conditioned by the characteristics of the flow [340, 584]. Within the range of values of 1 < AF < 60, the gel agglomeration number was identified, corresponding to AF = 16 (connectivity limit). Below this value, the suspensions exhibit essentially diluted behavior, while above it, the fibers interact but without being completely immobilized [234, 584, 987, 1125].

So, the agglomeration factor is highly dependent on fiber length and suspension consistency, being a useful indicator of the rheological state of a fiber suspension, especially in processes where fibers can have their length altered [166, 628, 650, 985] such as in the refining treatment.

7.3 Effect of Pulping, Pressing, and Drying Processes

The type of cooking (conventional or modified kraft), delignification rate (kappa number), bleaching method (with or without elemental chlorine, with or without oxygen delignification), draining, pressing, and drying influence the bleached pulp quality as they impact the chemical and anatomical properties of fibers, thereby reducing the fiber strength [31, 57, 82, 84, 194, 214, 348, 429, 437, 474, 504, 567, 666, 692, 903, 954, 966, 968, 977, 981–984, 1124] due to the removal of cell wall constituents and microfibrillar elements.

These modifications affect the fiber flexibility and collapse directly influencing the fiber bonding capacity. The cell wall becomes more fragile with the degradation of carbohydrate chains and less resistant with more anionic groups and fines content, among other alterations [26, 32, 35, 36, 76, 82–84, 87–106, 115, 126, 178, 218, 298, 343, 406, 434, 437, 467, 469, 474, 496, 502, 555, 588, 622, 660, 662–665, 690, 692, 804, 964, 968, 977].

Most of the primary wall and the secondary wall layer S1 of the fiber is removed giving access to the secondary wall layer S2 containing fibrils that are exposed by refining.

Once material is removed, fiber properties such as fiber population, coarseness, cell wall thickness and, as a direct consequence, individual fiber strength are reduced according to the pulp kappa reduction [351, 83, 91, 105, 298, 437, 454, 502, 514, 660, 905, 979, 980].

7.3.1 Pulping Processes (Cooking and Bleaching)

From the same wood, the quality of the pulp is predetermined during cooking. In addition to a high delignification (low kappa number), pulp uniformity and whiteness are important to be obtained [26, 31, 83, 84, 91, 105, 218, 437, 492, 496, 501, 977].

The process yield, which depends on the wood chemical and anatomical composition, is a critical factor in defining its economy. The pulp cellulose content has a strong positive correlation with the yield, while the extractives and lignin contents result in a negative correlation. The total cost of pulp production decreases with increasing wood cellulose and hemicellulose contents [25, 31, 344, 367, 555, 971–976].

The alkaline condition during pulping favors the carbohydrate degradation, reducing their quantities in cell walls. Due to the lignin and carbohydrate removal, the fibrils become more open and loose and the cell wall more porous. This degradation leads to reductions in cooking yield, pulp viscosity, and individual fiber strength, which make the fiber more flexible due to cell wall collapse impacting drainage, wet strength and optical properties, and structural and dry paper strength [38, 48, 57, 83, 91, 105, 113, 115, 126, 153, 196, 214, 298, 429, 430, 433, 437, 442, 443, 450, 469, 489, 496, 502, 653, 660, 666, 711, 955, 964–970].

At high temperatures, above 140 °C in a kraft cooking, native lignin exhibits substantially increased mobility increasing the likelihood of adjacent cellulose microfibrils or small surfaces of cellulose fibril aggregates coming into close contact and reorganizes to form a larger structural unit. Thus, cellulose aggregates increase [37, 434]. Carbohydrate degradation, which is greater in conventional kraft cooking than in modified processes and more intense in bleaching stages with oxygen and its derivatives, is the main effect of the process on the bleached pulp strength [42, 43, 105, 308, 437, 474, 489, 495, 502, 588, 606, 622, 660, 666, 824, 920, 954, 955, 959–964].

For the same wood, and up to a certain delignification level, the pulp resistance increases with the kappa number reduction, which is natural since the removal of cell wall constituent forms pores uniformly distributed throughout the cell wall thereby increasing fiber hydration. The cell wall loose it's intrinsic resistance being more flexible that increases the fiber bonding ability. The paper obtained will be stronger [466, 469, 496, 502, 588, 650, 660, 663, 664, 666, 711, 753, 930, 943].

There is a significant effect of kraft cooking yield on fiber flexibility. For laboratory cooks, the stiffness decreases by a factor of four with decreasing yield from 85% to 50% [958].

Hemicelluloses are partially degraded during kraft pulping, during which most of the total yield loss occurs. This removal of hemicelluloses, along with the removal of lignin, plays a significant role in the formation of pore structures in the cell wall [83, 105, 126, 224, 298, 413, 434, 437, 443, 450, 466, 468, 469, 496, 502, 575, 588, 650, 660, 663–666, 711, 753, 930, 943, 957, 1335, 1360].

Hemicelluloses possibly spread after lignin removal, behaving as fillers on new surfaces (pores) formed by the lignin removal [48, 313, 930].

The digesters must be operated in such a way as to guarantee minimum degradation and removal of the hemicelluloses present in the wood. Pulps with lower hemicelluloses contents require more refining energy to reach the target resistance values [73, 437, 488, 493, 494, 510, 660, 696, 954–956].

For eucalyptus pulps, the only remaining hemicelluloses, after the kraft pulping, are xylan structures containing small amounts of uronic acids [444, 433, 513]. The kraft pulping process also creates carboxylic groups such as hexenuronic acids that are not present in native wood.

Hemicelluloses are also related to the consumption or chemical reagents in the bleaching process. Therefore, they affect not only yields and pulp uses but also impact costs generating operational bottlenecks and interferes with the final product quality [406, 513]. The content of hexenuronic acids in the unbleached pulp shows a positive correlation with the content of pentosans in the wood [439].

Part of the dissolved hemicellulose is adsorbed on the fiber surfaces at the end of cooking which leads to an increase in the final product strength properties [38, 279, 370, 952, 953].

The collapsing of cell wall pores depends on the type of fiber and its treatment history before the invention of paper machine [754].

The modulus of elasticity is the most significant property affected by pulp pressing. The increase in the modulus of elasticity leads to greater stiffness due to the increase in paper density, giving a greater number of contact points between the fibers that distribute the applied load on paper [129, 195].

7.3.2 *Pulp Drying*

In a paper machine the water removal from the fibers cell wall begins in pressing when a solids content of 20–25% reaches, after pressing, about 45–50% [754].

The water to be removed in drying is the most difficult. The process consumes a large amount of energy and causes irreversible collapse in the cell wall structure. In addition to structural cell wall defects it is possible, in some cases, cell wall rupture occurs [633]. When water is removed during papermaking and the sheet gradually increases its dry content, hydrogen bonds are formed between neighboring fibers, thereby increasing the consolidation of the paper structure (Fig. 7.6).

There are two "dryings" to consider.

The first is drying in the paper machine, where the final removal of water occurs as the last stage of papermaking. In this, the formation of bonds between the fibers consolidates the paper structure that acquires the desired final quality. The second drying to consider is done by the commercial pulp suppliers producing the pulp bales that will be shipped to customers.

In this case, the pulp is dried to reduce transport costs and reach the commercial specification of air-dried weight, that is, the standard commercial unit. This drying has a significant effect on the quality of the fibers influencing both their processing

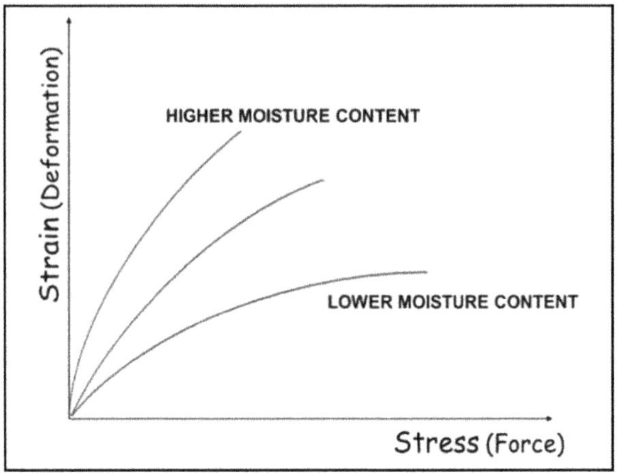

Fig. 7.6 Effect of the moisture content on paper strength properties

and the quality of the finished product. As an example, commercial pulps on the market have water retention rates (WRV) ranging from 90% to 130%, while never-dry pulps have WRV ranging from 150% to 220%, resulting in a huge difference in the pulp (fiber) behavior in pulp processing [406].

When pulps are dried, there is an irreversible loss of their hydration capacity. This phenomenon is known as hysteresis and causes pulp hornification restricting fibers rehydration by closing large pores and a large quantity of small pores in the cell wall [71, 97, 124, 186, 249, 310, 355, 430, 450–452, 466, 505, 594, 650, 699, 702, 708, 761, 775, 892, 929, 930, 936, 939–951].

Hysteresis corresponds to the formation of stable bonds between adjacent fibers generating crystalline regions along the cellulose fibrils. The fibers become more rigid giving more difficulties to generate connections between them [58, 105, 348, 406, 443, 506, 650, 708, 775, 931, 1124].

When this dried pulp is rehydrated, both water absorption and retention are significantly reduced affecting fiber swelling, fiber bonding, refining energy consumption, drainage capacity, collapsibility, strength, and optical properties of the paper.

Prolonged rehydration and increase in the temperature of the pulp disaggregation process contribute to breaking these bonds between the inner layers of the cell wall [511, 694, 931–934].

The refining response of both dried and rehydrated pulps derives from the properties of the undried pulp [185].

Comparing the commercial pulp with a never-dried pulp, we have that the latter does not have restrictive internal connections (hornification) and, therefore, presents a greater number of open pores in the cell walls and absorbs water more easily than the commercial pulps. Its fibers are more flexible and consumes less refining energy to reach the target values so as to achieve the desired paper properties [71, 450, 505, 929, 930].

7.3.3 *Other Effects of the Pulping Processes*

7.3.3.1 Fiber Damage and Deformations

In wood, the fibers are relatively straight except for some deformations due to growth stresses. In pulps, the fibers are rarely straight presenting some level of rupture or damage due to dislocations or micro compressions located in the cell walls [193].

This fiber damage is not always visible but can impair the transfer of internal tension between the cellulose fibrils weakening the fiber cell walls [47].

Industrial pulps, produced from the same raw material, are generally weaker than laboratory pulps. This is due to the mechanical damages they suffer along the fiber line that generate fines and fibers deformations. Such deformations can also be generated in the mass preparation steps for paper production. In chemical pulping, the fibers undergo extensive mechanical treatment, for example, they rub against pipes and valves, they are dewatered, they receive chemical reagents in dynamic mixers, and they are stripped or are mistreated by the fluidization rotor of medium consistency pumps. Therefore, fiber deformations increase along the production line as more energy is added in pumping at medium and high consistency, in presses and in mixers, damaging the structure of cell walls and decreasing their intrinsic resistance [26, 43, 47, 178, 263, 313, 330, 406, 510, 606, 622, 634, 824, 898, 903–905, 916–928].

Reaction, discharge, and wash temperatures; chemical dosage and concentration; washing; and other factors in the pulping process damage the fibers. This behavior is evident in pulps dried by the "flash" process, which present greater curling when compared with pulps dried in a hot air mattress [72].

The morphological properties of the fiber cell wall affect the development of fiber deformations that occur preferentially in regions of the cell wall with lower density of elastic components, that is, where hemicellulose and lignin are present in smaller amounts [27, 47].

Among the deformations of the fibers, we have undulation (curl) and folds (kinks), in addition to fiber cutting, which is the most severe fiber damage during refining. The waves (curls) are smooth curves of the fibers, while the kinks change direction abruptly forming an angle [47, 634].

Fiber shape factor (S) is typically reported as a measurement of curl or kinks [115, 143, 406, 913–915, 917, 1355]. It is defined as the maximum fiber extension length (I) and the actual fiber length (L). It is often presented as a percentage [330] (Fig. 7.7).

Fibers without deformations affect the paper structure improving strengths, stiffness, elongation, and load transfer efficiency by spreading breakage damage over a wider area and a greater number of fiber bonds that need to be "broken" [26, 47, 143, 406, 546, 815, 912, 928, 1121].

Deformations influence the properties of the finished product, which vary according to the extent of fiber deformations. Pulps with fibers showing curl tend to form sheets with greater elongation, specific volume, absorption capacity, and porosity,

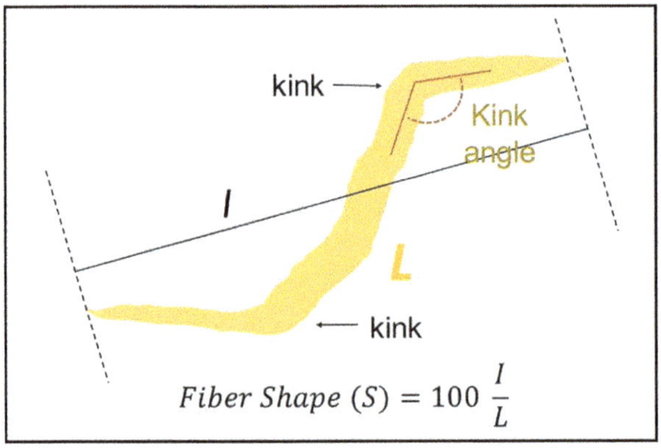

Fig. 7.7 Illustration of fiber curl and kink

but have lower resistance because they lead to a poor distribution of forces in the paper structure [26, 43, 73, 96, 115, 178, 266, 233, 248, 249, 263, 264, 266, 469, 489, 496, 501, 606, 607, 622, 737, 824, 897–910, 912–916, 925, 928].

Therefore, one way to add value to chemical pulp for paper production is to minimize deformations in the fiber cell wall. This means that the pulps not only present higher initial values of tensile strength and elasticity modulus but also lower refining energy demand to reach a given objective value of strength [328, 337, 896, 897, 1120].

7.3.3.2 Fines Formation During Refining

The fines present in unrefined commercial pulp are referred to as primary fines. Basically, they are parenchyma cells, vessel elements, and fragments of cell walls generated by friction during the pulping and pulp bales disaggregation. Secondary fines are generated by refining.

The physicochemical properties of fines differ from the fiber properties. They have a much larger surface area being more swollen and influence paper properties in different ways [62, 406, 456, 688, 771, 893, 894, 1339].

Primary and secondary fines originate from different morphological entities and, therefore, have different characteristics.

The cell wall structure of primary fines is much less resistant than the fiber cell wall being more affected by hydration. In this way, primary fines can swell two to three times more than fibers [771].

Secondary fines originate mainly from the surface of the fibers that undergo defibering during the refining action. They are fibrillar, flexible, and have a higher water retention rate. Regarding carbohydrate composition, they are similar to fibers, but the relative lignin content is higher in secondary fines, but it is lower than primary fines [699].

In general, fines fill empty spaces in the paper structure reducing its porosity, delaying water drainage, and increasing its specific surface [62, 406, 456, 688, 771, 893, 894, 1339].

For the same fines content, secondary fines increase the strength of the fibers bonding more than primary fines, mainly in pulps with higher yield where the fibers are less flexible [62, 688–699, 931].

Currently, the pulp fines fraction is considered as raw material for the production of new materials, such as micro-fibrillated cellulose and nanocellulose [1339].

Chapter 8
Refining the Chemical Pulps

The fibers coming directly from pulping process do not have sufficient bonding characteristics to guarantee an adequate level of paper resistance to the final product. Thus, they need to be physically modified to improve the fiber bonding ability. In the papermaking refining does this treatment being the first and most important step [17, 102, 299, 331, 401, 640, 661, 720, 744, 946, 1065–1069, 1136]. The selection of the fibrous material to be used and the refining stage correspond to the most important management ability to produce the required paper quality under a competitive production cost [169, 298, 299, 511, 555, 583, 766, 875, 1066–1068, 1104, 1113]. In many cases, it can be said that "the paper is made in the refiner."

The refining promotes a greater contact area between the fibers and treats the fibers surface to further improve the bonding of fibers [353, 769, 1121].

It is evident that a manufacturing process that has existed for nearly 2000 years has already undergone a series of modifications and optimizations, mainly justified by the importance of the manufactured product and by the need to use alternative raw materials in order to meet the growing demand for the product. Even so, the basic process remains the same, despite the great evolutions in the equipment used, the refining of the fibers continues to be carried out by successive cycles of impacts on the fibers in the presence of water. These impacts cause irreversible structural modifications in fibers. It can be said that the development of refining was the decisive factor in the development of the paper manufacturing process, in such a way that the history of paper is intertwined with the history of refining.

As refining is a mechanical treatment, it is an energy intensive process. In a paper machine, refining is the second largest consumer of energy, surpassed only by the drying stage. Refining is also the first in importance for defining paper machine runnability, the production costs and the product quality [20, 27, 299, 402, 687, 720, 721, 946, 983, 1036, 1067, 1068, 1076, 1077, 1111, 1144–1148, 1409, 1413].

© The Author(s), under exclusive license to Springer Nature Switzerland AG 2024
V. Manfredi, *Eucalyptus Kraft Pulp Refining*,
https://doi.org/10.1007/978-3-031-47285-5_8

8.1 The Refiners Evolution

When the first papermakers pounded their raw materials to make it suitable for paper production, they did the same type of mechanical work on the fibers today done by the most modern refiners. This operation is indispensable for the paper manufacture.

A continuous improvement in fiber treatment techniques was observed during the Islamic expansion. Slowly, new techniques were introduced to improve the results of the disaggregation of the fibrous material, improving the production capacity and maintaining the paper characteristics. Materials used by the Chinese were not available to Middle Eastern manufacturers, who instead used flax and other fiber substitutes, as well as the introduction of the hammer mill (triphammer) and circular grinding wheels [11, 12, 28].

An important aspect was the introduction of mechanical improvements to the mills used in the treatment of raw materials, especially those related to the use of hydraulic energy through waterwheels and wind energy substituting the human or animal traction.

In Europe, the mechanical treatment of fibers with hammer mills (stampers) was introduced in Italy during the thirteenth century. In these mills, a battery of pylons was lifted by a camshaft driven by a water wheel. The pestles were coated with metallic accessories of different formats (nails, rivets, nodules, etc.) to reduce the raw material (usually rags) to fibers in suspension [12, 28] (Fig. 8.1).

Circular grinding wheels, typical in grain mills, began to be used to treat vegetable fibers destined to make paper. These millstones rotated around a vertical axis moved by a long radial arm that was pulled circularly by animal traction or water wheels [28] (Fig. 8.2). Fiber treatment is performed in high consistency level [262].

In Holland, at the end of the seventeenth century, a machine appeared that disintegrated rags at a much higher speed than hammer mills or circular grinding wheels, making in four or five hours the same amount of paste that a hammer mill produced

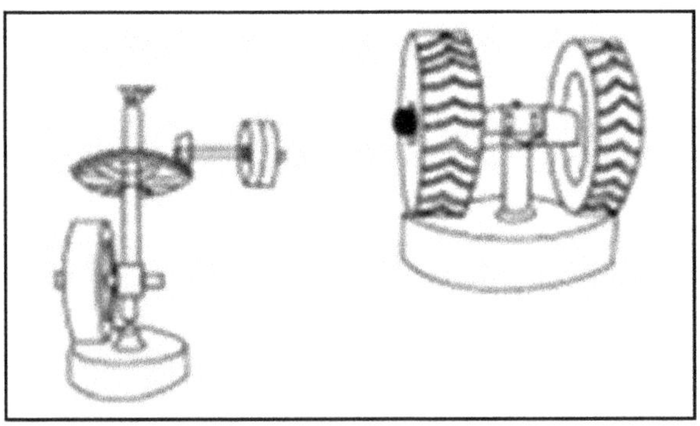

Fig. 8.1 Circular grinding wheels used for the treatment of fibrous material. (Adapted from [28])

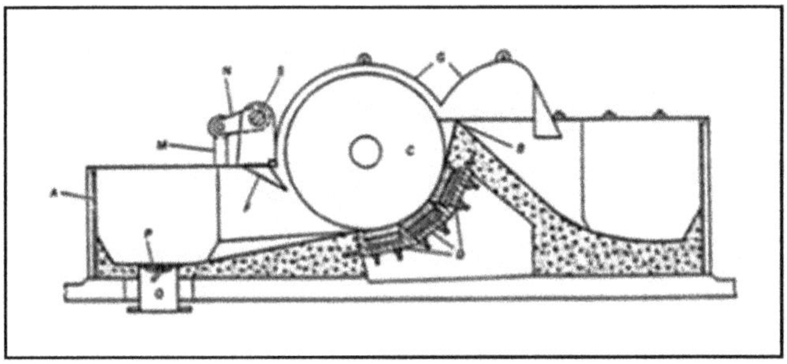

Fig. 8.2 The Hollander beater. (Adapted from [219])

in twenty-four hours. The so-called "Hollander mill" spread quickly and has reached our days without significant changes in its functioning way [2, 10, 28, 1094].

The fiber treatment principle consists of a solid rotating roller equipped with metal bars rotating on a stationary base plate. The pulp in suspension passes between the mobile and fixed bars, around a chute and again between the bars (Fig. 8.2). Under continuous circulation, the fibrous material is subjected to mechanical action in different production campaigns until the pulp is ready to be transformed into the desired paper. The equipment was operated in atches cicles with relatively low yield compared to the size of the equipment.

To overcome such deficiencies, refiners were developed which, even operating at low consistencies, occupy lower space and allow continuous treatment. The refiners use opposing grooved surfaces and are of fast rotation subjecting the suspended fibers to mechanical impacts at a much higher frequency than in the Hollander beater [877, 886].

In 1856, T. Kingsland introduced the disc refiner equipped with a horizontal rotating disc. The entire surface of the discs was provided with grooves. Although this equipment did not achieve commercial success, due to restrictions on engineering materials and manufacturing techniques [1100], its great merit was the introduction of the concept of continuous production, paving the way for the development of current refiners.

Conical refiners appeared in 1858, idealized by Joseph Jordan and Thomas Eustice, maintaining the concept of continuous production. They have a conical rotor rotating inside a conical stator. In these refiners, the pulp in suspension, also at low consistency, is pumped to the smaller end of the cone and is squeezed between the rotating and stationary cones. The spacing between the refining surfaces can be controlled very precisely by adjusting the axial position of the conical rotoer to obtain the desired modifications on fibers.

Due to the layout of the bars and channels, they have better pumping characteristics than disc refiners, offering better hydraulic capacities [710, 872]. Such equipment gained increasing acceptance and dominated paper production until the introduction of the modern disc refiners in the 1950s.

The operating principle is the same for both refiners. The energy transference occurs between the intersecting bars, with the spacing between the bars being greater in the conical refiner [23, 63, 178, 778, 1150]. The treatment is abrasive because it results from the action between the fibers and the bars leading to the generation of fines and high external fibrillation due to shear forces, greater in conical refiners [709, 1040].

Most conical refiners used refining cones (trimmings) with deep grooves, about 20 mm deep. These cones are very expensive and their replacement was a complex operation. However, with deep channels, the suspended slurry flows more freely through the refiner without receiving treatment, which is one of the reasons why conical refiners have higher refining energy consumption [1100].

Over time, disc refiners, with lower energy consumption than conical refiners, gained market acceptance. The conical refiners also went through adjustments and optimizations. Disc and conical refiners have become involved as a choice for papermakers who, depending on the type of paper to be produced, opt for one or another type depending on each specific need.

The pressurized double disc refiner was introduced by E. D. Jones in the 1950s [1094]. The refiner consists of a free moving rotor, with discs on both sides, which rotates between stationary plates, also equipped with refining discs, thus providing two refining surfaces [877].

For a given refiner size, speed, and throughput capacity, the twin disc refiner offers lower refining intensity and higher energy efficiency [983, 1096, 1097]. This means that several conical refiners, for example, can be replaced by a single double disc refiner. For example, to treat 1.0 t of slurry, under an intensity of 1.0 Ws/m and refining energy of 100 kWh/t, four conical refiners operating at 220 kWh/t are required. The same work can be carried out by a 24-inch double disc refiner, with 175 kWh/t, corresponding to consumption of 45 kWh/t [1096].

With an elaborate system for adjusting the movements of both the stator and rotor discs, double disc refiners have since replaced conical and single disc refiners, gaining industry preference [1100, 1151].

A modification carried out in the refiners, with the introduction of multiple rotating and stationary elements to increase the total number of bars, makes it possible to significantly reduce the refining intensity, allowing for more uniform treatment and with lower refining energy consumption. These refiners are known as "multiple discs" [21, 877, 983, 1096, 1152, 1153].

Refiners have always been the target of continuous development (Fig. 8.3) aimed at reducing energy consumption and fiber treatment heterogeneity, as well as production capacity improvement. Production capacity is limited by the available net power as well as the pulp flow rate through the refiner. Indeed, increasing capacity along with reducing energy consumption and heterogeneity is an important vision to consider in refiner design [456].

Currently, no other equipment, or stage of the mass preparation process, has been studied as much as refiners. This is justified because they are closely related to the final papermaking result whether they are focused on the product quality, on the paper machine runnability, or on the production costs (Fig. 8.4).

Disc refiners have been widely used by several researchers to understand the low-consistency refining forces and mechanisms [368, 1227]. In addition, disc refiners have also been used to obtain micro-fibrillated cellulose [1232].

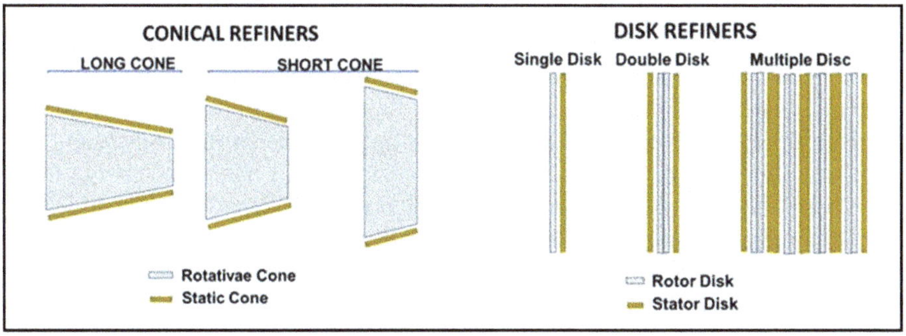

Fig. 8.3 The main types of refiners available for pulp processing. (Adapted from [1164])

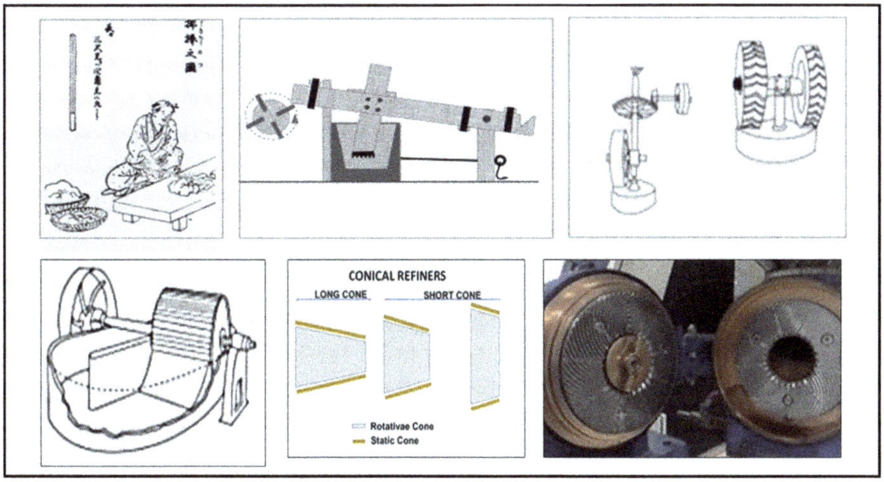

Fig. 8.4 Evolution of the raw material treatment along time. (Adapted from [28, 614])

More recently, a new refiner model has been commercialized, the cylindrical refiner consisting of cylindrical plates, which combines the advantages of the cylindrical refining zone of the Hollander beater with the continuous operation of the refiners. With compact design, it reports to the cylindrical refiner advantages such as a gentle and stable fiber treatment, with low energy consumption and simple equipment maintenance because the refiner segments are easily accessible [5, 880, 1413].

8.2 Inside the Refiner

The disc refiner, which can be compared to a radial centrifugal pump, has two main regions. A working one, where the fiber suspension circulates between the refining discs, and a much narrower one, comprising the suspension inlet and outlet pipes [1129, 1130, 1155].

The pulp in suspension enters the refiner axially, that is, in the center of the discs. The rotating disc imparts energy to the fluid causing it to move out of the machine in the radial direction at a relatively higher pressure. As the pulp in suspension moves outward, the rotational movement of the rotor transmits energy to the suspension, which is compressed in the narrow space between the discs (gap). The abrasion suffered by the fibers during this process modifies the characteristics of the fibers up to the desired level [401, 1092, 1151, 1156].

The spacing between the discs can be increased or reduced by changing the energy, or rather the forces applied to the fibers, depending on the extent of the desired refining.

8.2.1 Pulp Suspension Flows Through the Channels

Even at low consistency (2–6%) the fibers do not remain individualized but in the form of flocs [343, 664, 772, 1017, 1018, 1051, 1057, 1075, 1087, 1112, 1128, 1155, 1157–1159] that are transported along the disc's channels. These channels must be wide enough to allow the flocs to rotate in the channel. The flow pattern in the channels has an important role for the refining homogeneity, as well as, for the flocs transfer and capture by the refining positions along the bar edges.

The driving force to move the suspension through the refiner, controlling the capture and orientation of the flocs in the refining zone, is provided by the refiner itself. Higher pressure difference reduces the flow rate in the rotor while increases the flow rate in the stator disc [1033, 1036, 1151, 1163].

Rotational motions are observed both in the rotor and in the stator groove. The rotational movement becomes stronger as the flow moves towards the periphery, being strongest in the upper part of the channel.

The refiner works pressurized, with the outlet pressure being greater than the supply pressure. Due to this pressure difference, centrifugal flows occur in the rotor channels, while flow reversals occur in the stator channels, generating spiral reflux vortices, enhancing the internal circulation. This is important for refining, as the pulp suspension recirculates ensuring that the flocs are treated in the working region between the discs. Intermediate streams arise between the discs [686, 1087, 1151, 1164–1167].

The rheological behavior of the fiber suspension is a consequence of several parameters such as the uniformity of the bar edges and surfaces coverage, the distance between discs (gap), the rounding, or wear of the bar edges, and the characteristics of fibers, especially in terms of their flocculation tendency [112, 765, 1168].

The forces created by the rotor movement depend on the speed of rotation and the radial position in the disc (or cone, in the case of conical refiners). Due to the difference in linear velocity, the forces at the refiner's periphery are greater than those near the inside radius, increasing the suspension acceleration. Thus, acceleration, pressure differential, suspension pressure, and radial suspension flow are greater in the rotor channels than in the stator channels, which tends to create backflow in the stator channels where the backflow velocity increases with refiner velocity and decreases, to a lesser extent, with refiner flow rate [112, 1051, 1084, 1087, 1129, 1165, 1170, 1233].

The pulp flow through the channels is much greater than that through the space between the discs. This usually causes uneven refining and leads to significant energy losses with suspension pumping. Vortex formation, to facilitate feeding, is also impaired. The mean pressure essentially depends on the configuration of the discs, the length and width of the bars, the width and depth of the channels, the bars crossing angle, the spacing between the discs, the apparent viscosity of pulp suspension, and the rotating speed of refiner [1079]. Therefore, the selection of discs as well as the maintenance of constant conditions during the refining is of great importance for the refining results.

As the pressure difference in the refiner increases, the pulp suspension rotational movement increases, the velocity at the bottom of the channels decreases, while the velocity increases at the top of the channels. Speeds decrease in the rotor and increase in the stator, the differences being greater for the rotor compared to the stator [1151, 1165]. The refiner's rotational speed, the flow rate, and the consistency of the fiber suspension also control the flow through the refiners [1151, 1156, 1167, 1334].

The rotational movement changes the circumferential velocity of the fluid inside the refiner and, as a result, a suspension rotating movement is created within the channels. The rotational movement, important for placing the fibers on the edges of the bar and inducing shear forces so that the fibers are treated as the pulp suspension moves toward the outlet, is greater (stronger) in the upper part of the channel, while at the bottom, the fluid starts to rotate toward the end of the channel (Fig. 8.5) [1151, 1165].

The suspension rotational speed in the upper part of the channels is different in the rotor and stator discs. In the rotor disc, it increases toward the output, whereas in the stator, it decreases as the suspension moves toward the origin [1151]. Small recirculation, identified as secondary and tertiary flows, is also observed [1087, 1130].

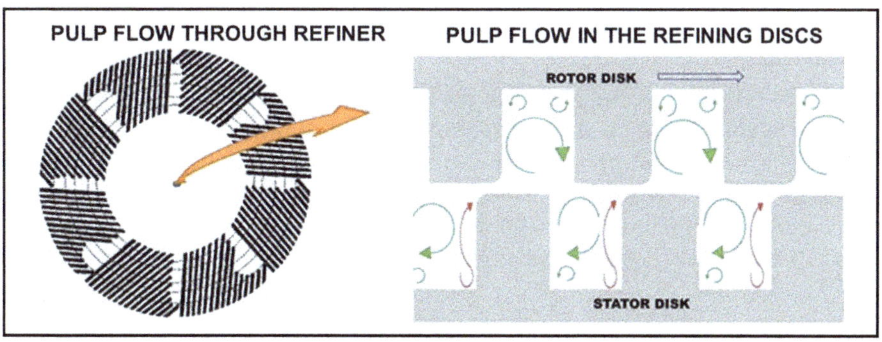

Fig. 8.5 Simplified diagram of the pulp suspension flows inside the refiner. (Adapted from [1020, 1063, 1087, 1130, 1151, 1166])

The reduction in the apparent viscosity of pulp suspension, which occurs through refining, strongly affects the intensity of secondary flows and the internal pulp circulation. This reduction on apparent viscosity favors the rotational movement that extends to the bottom of the channel [190, 1087, 1165, 1234].

8.2.2 Fiber Trapping

In a low-consistency refiner, the suspended fiber flocs are captured between the rotor and stator bar edges and worked into the narrow gap between the rotor and stator discs. No refining takes place if the fibers are not captured by the edges of the disc bar edges.

In the space between the discs, there are thousands of fibers simultaneously subject to forces, whose intensity and direction vary continuously with the operational conditions used. Rotation movements inside the channels allow the flakes to be exposed and captured by the bar edges. The flow in the space between the discs is important as most of the refining takes place in this narrow region where the fibers are captured and treated due to the movement of the rotating disc [63, 511, 1091, 1121, 1151, 1163, 1235–1237].

In the refining area, the flocs are subject to high static pressure, floc rotation, and high shear force. When the disc bars intersect, the pressure increases generating a pressure peak that is relieved as they move apart. This rapid increase in pressure increases the refining action on the fibers that are captured by the bars and deformed under the action of applied forces [178, 261, 744, 1017, 1099, 1151, 1160, 1163].

The number of captured flocs by the bar edges (fiber trapping) is determined by factors such as the number of bars crossing each other, the speed of these crossings, the channels geometry, consistency, and turbulence of the pulp suspension. The reduction in fiber trapping increases the refining severity [178, 636, 710, 765, 839, 866, 867, 1018, 1117, 1127, 1238, 1240].

Most of the flocs are trapped within the space between the discs and a small fraction remains wrapped around the leading edge of the bars. The number of fibers accumulated on the edges increases with the pulp suspension speed and decreases with the refining degree. The fiber flocs that are captured in the region of the rotor bar can be released becoming part of the internal flow in the stator or they can be captured again or even become part of the external flow in the rotor [650, 1121, 1166, 1241, 1242].

Larger flocs are more easily captured and are more subject to the fiber cutting action during refining, while smaller flocs, such as those of eucalyptus pulp, being less resistant, tend to break during refining and flow more easily through the refining zone [222, 766].

8.2.3 The Forces in Refining

In the refiner, the fibers are mechanically treated by high frequency cyclic deformation generated by the bars intercrossing. The cyclic load must overcome the compressive strength and friction stress of the fiber cell wall to cause it to delaminate. The fibers elasticity modulus and coarseness are the fibers parameters resisting to compressive loads needed to fibrillate the cell wall [63, 178, 650, 1032, 1127, 1243].

Under deformation, the flocs tend to stretch in the flow direction and regain their original shape when the stress is released. This mechanism occurs by the presence of water being mainly due to the viscoelastic nature of the cell walls of fibers that dissipate the compression and deformation forces. As this occurs repeatedly in a refiner, the stresses imposed on the fibers eventually lead to plastic deformations, or defects, in the cell wall structure where the structural bonds (hydrogen bonds) are broken and replaced by water molecules, hydrating the fibers [63, 178, 222, 445, 650, 772, 816, 839, 933, 1017, 1028, 1032, 1082, 1085, 1127, 1123, 1243–1245].

Refining begins with the removal of water from the fibers when the flocs are captured by the leading bar edges. The fibers, compressed by the bar edges, are sheared heavily between the bar's surfaces, in a combination of compressive and abrasive forces. As the applied pressure exceeds the structural elastic limit of the fibers, plastic alteration of the cell wall of at least part of the fibers occurs which is permanently modified. Finally, after crossing the disc bars, the release of mechanical pressure allows the water reabsorption by the broken cell walls after which they are dispersed back into the suspension flowing through the refiner [436, 452, 454, 456, 546, 614, 650, 687, 688, 772, 816, 839, 1016, 1017, 1032, 1034, 1036, 1083–1095].

Three distinct forces are imposed by the bars on the fibers – a corner force (F_c) exerted on the edges of the bars, a normal force (F_n) due to the flocs compression between the bars, and a shear force (F_s) produced by the movement of the bar's surfaces over the flocs. These forces generate tensions that, depending on the position and duration of the applied forces, generate irreversible physical deformations in the fiber structure (Fig. 8.6) [299, 315, 436, 456, 511, 577, 589, 650, 668, 710, 866, 884, 1016–1036].

The normal force holds the fibers between the bar surfaces being related to the pulp flocs compression between the stator and rotor bars [42, 589, 595, 1077].

In addition to the force holding and compressing the pulp, two other primary forces act on the fiber flocs: a corner force from the bar's edges and a sliding friction force on the surfaces of the bars.

These forces are the main parameters causing important changes on fiber characteristics [348, 511, 668, 866, 1026, 1033, 1087, 1227, 1240].

The corner force is produced by the bar edges. The flocs compression initially results in fiber rearrangement causing floc densification as they expand in the x-z plane when compressed in the y direction. The rearrangement of fibers orients then in the compression direction [1021].

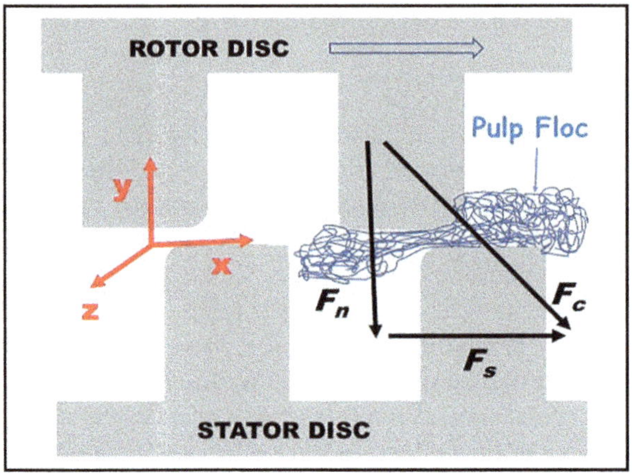

Fig. 8.6 A simplified illustration of forces in refining

The tension generated by the corner force is as much short duration and is highly concentrated, being greater as shaper the bar edge is. Depending on the impact intensity it may undergo internal cell wall delamination and the damage caused will increase as the cyclic sequence of impacts continues with refining and may even cause the fiber rupture [456, 589, 595, 671, 1035, 1036, 1074, 1077, 1087, 1246]. The tension forces in the space between the discs applied in the direction of the fiber axis are responsible for the fiber shortening [1036, 1087, 1246].

The shear force, which corresponds to the friction force between the bar's surfaces and the flocs, originates from the tangential component of the corner force and from the friction of the fiber by the bar's surfaces during the passage of the fibers. These processes result in shear fatigue causing the outermost layers of the cell wall to delaminate. This effect is one of the primary objectives of refining as the fiber with a large surface area will allow more bonds with other fibers forming a more consolidated and resistant paper [456, 589, 595, 1032, 1036, 1074, 1077].

Most of the fibers in the space between the discs are not in contact with the refiner bars but with each other resulting in friction. The shear action is intense as the disc bars are moving. After crossing the bars, the mechanical pressure relaxes when water can be absorbed by the fibers and fibrils before the turbulence redisperses them in the suspension [1087, 1121].

The bar forces multiplied by the distance the bar is moved during this application represents the energy consumed during the bar crossing, that is, the refining energy consumption [650, 1026, 1032, 1036, 1083–1085].

The spacing between the discs (gap) controls the applied forces [876, 1033, 1087, 1159, 1248, 1249]. The magnitude of these forces is a function of the disc configuration and rotating speed, the distance between discs, the distance to the center of the discs (due to the difference in linear velocity, the forces on the periphery of the refiner are greater than those close to the inner radius), and the intensity of the impacts. The amplitude of fiber's modifications depends on the fiber characteristics [650, 866, 1033, 1084, 1085].

The distribution of forces on fibers is similar to that in a mill, that is, a small number of fibers carry the load in a stress current and therefore undergo deformation. So, only a small number of fibers are treated in each cycle and, consequently, new cycles must be applied to treat the remaining fibers [836, 1190].

8.2.4 Cell Wall Modifications of Fibers

The presence of water is fundamental in the process, since the structural modifications of fibers occur due to cell wall hydration increasing the probability of bonds between them. Therefore, the fiber's ability to absorb water makes for easier fiber refining. A low amount of lignin, which does not absorb water, and a high amount of hemicellulose are therefore beneficial for refining [44, 57, 217, 219, 299, 445, 454, 588, 702, 704, 721, 761, 762, 774, 776, 1184].

Under the mechanical action, the fibers are revolved and compressed from different directions. This action promotes the fiber cell wall delamination due to the breaking of the cross-links between the fibrils. This allows the osmotic forces to produce hydration dividing the cell wall into several layers that remain concentric but mechanically independent. The primary wall, the *S*1 layer, and part of the *S*2 layer of the secondary wall are ruptured, enlarging the cell wall pores which start to retain a greater amount of water and the removed material is released into the pulp suspension.

As these S1 layers are the main cell wall structures restricting against fiber swelling due to their high fibrillar angle, the fiber transversal section can expand from 15 to 20 times [62, 63, 73, 105, 124, 125, 190, 387, 433, 445, 450, 452, 466, 505, 514, 594, 657, 678, 688, 770, 774, 788, 815, 866, 930, 1024, 1026, 1042, 1043, 1045, 1049, 1056–1064, 1159].

Refining also increases the surface charge of fibers. The carboxylic groups are surrounded by water molecules increasing its flexibility and conformability, influencing the wet end chemistry, retention of cationic papermaking additives, fiber flocculation and paper mechanical [220, 456, 701, 726, 1250].

All this fiber modification, which occurs through the cell wall delamination, is the refining objective as it makes the fibers more flexible and increases the fiber's bonding capacity making it possible to obtain, after drying, the structure and desired properties of the produced paper [63, 202, 218, 266, 280, 355, 568, 588, 650, 657, 676, 761, 778, 803, 815, 866, 878, 929, 983, 1024, 1026, 1033, 1040, 1190–1192].

In extreme cases, such modifications lead to the rupture of the cell wall causing a reduction in the average fiber length. This is traditionally referred to as "fiber cutting." However, the cutting itself does not occur during refining because the average fiber length reduction is caused by the lowering on cell wall intrinsic resistance causing its rupture under tension (Fig. 8.7).

The cellulose chains in the cell wall are rich in hydroxyls groups, which are potential functional groups in the formation of hydrogen bonds. During refining, the amount of these hydrophilic groups increases, especially the hydroxyl ones among

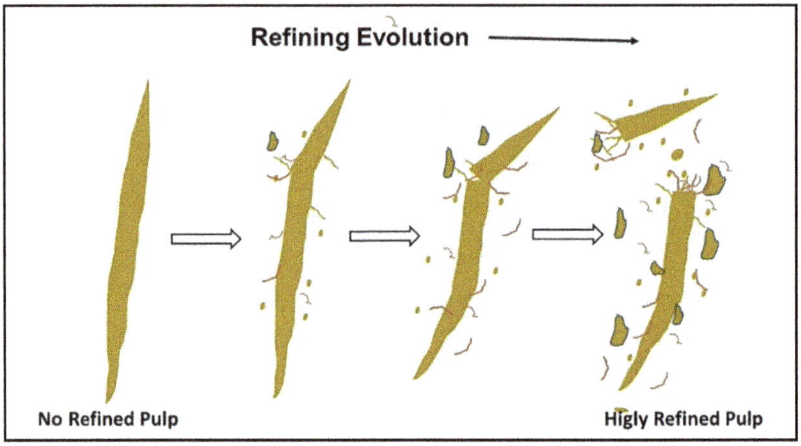

Fig. 8.7 Refining effects on the fibers as a function of the amount of applied energy. The same effect occurs when the refining intensity is increased

the cell wall polysaccharides, which results in higher fiber flexibility due to greater hydration of the cell walls. The greater fiber flexibility reduces the flocs average size, lowering the coefficient of friction between the fibers [46, 446, 1039, 1251].

During refining, the turbulence of the pulp suspension and the shearing forces on the cell wall, the flocs are deformed and disaggregates. With the reduction of the fiber's intrinsic resistance that improves the collapse and flexibility of the fibers, the flocs are formed again but smaller than the initial ones. The apparent viscosity of the pulp suspension is also reduced [178, 289, 454, 866, 1099, 1165, 1166, 1252–1254]. With the reduction of the apparent viscosity, the coefficient of friction between the fibers decreases and the suspension flows easily through the channels of the discs reducing the flocs capture by the bar edges.

The reduction on the apparent viscosity of suspension makes the pulp layer between the discs thinner. The smaller flocs are no longer able to maintain the spacing between the discs. This reduces the amount of pulp protecting the disc bar surfaces being one of the reasons why the external areas of the discs are the ones that most frequently present problems (damages) with discs wear.

During refining, chemical actions also occur. For example, removal of hydration material from the cell wall of fibers (molecular fibrillation) does not occur by classic way of dissolving compounds or changing the molecules' structure but rather by absorbing water with material removal (carbohydrates and lignocellulose) due to bond breaks (H-bond) mainly in the amorphous regions of the cell wall [52, 105, 231, 306, 468, 511, 546, 663, 702, 917, 937, 1051, 1166, 1215].

Refining alters the composition of the interfaces between the cell wall and the surrounding medium rearranging the molecules in monopolar formations in the places with mechanical action in the presence of water. The result is the separation, by the entry of water, between the microfibrils and fibrils and even their release into the surrounding medium [445]. The degree of removal is stronger at the beginning of treatment and gradually decreases being more pronounced in unbleached pulps [63, 105, 153, 221, 231, 306, 468, 511, 546, 663, 702, 917, 937, 1166, 1214, 1215].

Although refining has a great impact on the ultrastructure of cell walls, it is not able to break down cellulose fibril aggregates [434].

8.3 Refining Effects on Fibers

In papermaking, refining is the only unit operation that modifies the morphology of fiber. The characteristics of the fibers are adjusted to allow obtaining the desired structural, optical, and resistance properties in the produced paper [178, 454, 589, 747, 1051, 1087, 1091–1093, 1101, 1142, 1340].

It is necessary to weaken the connections that unite the structural elements in the fiber's cell wall reducing their intrinsic resistance. In practice, this is done in the refining process, and the extent of treatment depends on the pulp characteristics and the operating variables used in refining [178, 378, 710, 747, 801, 1125, 1087, 1172, 1372].

The fibers collapse and become more flexible and, associated with the release of superficial fibrils, which increase the surface contact between fibers, improves the fiber bonding capacity. This improves the paper's mechanical resistance and reduces their specific volume, porosity, and opacity values, in addition to the drainage of the fiber suspension [191, 203, 216, 242, 262, 588, 676, 678, 797, 803, 809, 919, 1015, 1255–1258, 1340].

During refining, the fibers are modified in a controlled manner in order to obtain the best relationship between operating costs, behavior of the paper machine, and properties of the produced paper. The extent of modifications in the fibers is determined by the applied power and the geometric parameters of the discs, which characterize the refining intensity [34, 35, 63, 108, 124, 129, 167, 178, 191, 218, 219, 221, 233, 261, 262, 289, 299, 301, 314, 337, 389, 443, 469, 470, 479, 511, 546, 566, 588, 594, 606, 650, 661, 708–710, 715, 720, 744, 747, 761, 784, 803, 817, 836, 866, 909, 928, 1015, 1027, 1034, 1035, 1040, 1049, 1063, 1096, 1097, 1100, 1107, 1109, 1119, 1120, 1122, 1128, 1131, 1132, 1146, 1161, 1168–1190, 1338–1340, 1410].

The effect is accentuated by the release of microfibrils from the surface of cell walls, which increases the width of pores and matrix lamellae on the outer parts of the cell wall, loosening its structure and generating fines and exposing material (fibrils) previously internal to the wall. The increase in the specific surface creates new contact areas for establishing connections between fibers [93, 105, 163, 183, 201–203, 215–219, 233, 262, 298, 299, 301, 313, 314, 343, 348, 353–356, 412, 433, 434, 456, 466, 473, 511, 546, 555, 588, 611, 617, 637, 644, 650, 663, 676, 702, 704, 709, 761, 774, 782, 783, 803–809, 817, 822, 828, 917, 919, 937, 942, 944, 966, 1001, 1020, 1022, 1051, 1061, 1065, 1075, 1096, 1120–1123, 1145, 1147, 1170, 1175–1177, 1124, 1182, 1184, 1188, 1192–1208].

The water retention value (WRV) measures the hydration of fibers, which is the basic effect of refining. WRV generally increases linearly with the specific refining energy being linearly related to the flexibility and bonding capacity between the fibers [43, 159, 546, 1075, 1117, 1217–1220].

As its variation ranges is wide, the water retention value can be used as an indicator for the operating control of the process, where the slope of the WRV curve versus refining specific energy can be used to characterize the refining efficiency [406, 1117, 1220].

The dissolution of cell wall materials, which for long fibers reaches values of the order of 0.5–4.0%, is more pronounced for unbleached pulps [663, 702].

The effects of refining are somehow related to complete fatigue of the cell wall which causes fiber breakage or delamination of cell wall layers [63, 93, 105, 153, 156, 161, 223, 470, 919, 937, 941, 1022, 1040, 1051, 1057, 1075, 1096, 1100, 1111, 1120–1122, 1194, 1207]. All these effects are due to fiber cell wall hydration.

It is important to remember that hydration and swelling are processes with specific kinetics requiring a certain time period for water to penetrate all available spaces. Therefore, fiber swelling occurs not only during the refining operation but also during the simple contact of the fibers with water (free swelling) and quickly [455, 1172].

In practical terms, the effect of refining is the fibers' flexibility and the consequent delamination of their cell walls [63, 93, 105, 153, 156, 161, 203, 217–223, 261–264, 298–301, 306, 348, 433, 450, 454, 470, 505, 511, 589, 588, 664, 702, 803, 804, 866, 919, 937, 941, 1022, 1040, 1051, 1057, 1075, 1096, 1100, 1111, 1120–1122, 1124, 1194, 1207].

The similarity of fiber hydration effects allows us to assume that the degradation of the material constituting the cell wall during pulping processes is equivalent to a chemical refining. Chemical refining results from fission reactions of carbohydrate polymers which can occur in the amorphous regions of the cell wall located between the hemicellulose micelles but predominantly with the cellulose molecule itself.

When strong oxidants, such as ozone and oxygen, are used in pulp bleaching, chemical shearing of the wood carbohydrate chains occurs [443, 666, 862] with consequent cell wall hydration reducing its intrinsic resistance and making the fibers more flexible.

Similar effects are obtained by the action of enzymes [109, 182, 863, 956, 1386, 1391, 1400–1402] and or even ultrasound treatments [324, 1401, 1404, 1407] and even by freezing the fibers.

In this way, we can classify traditional refining as mechanical refining, refining by chemical action (pulping, bleaching and enzymes), and refining by ultrasound and freezing, which are not topics of this book but which are worth recording.

8.4 Classification of the Refining Effects

The effects of refining on the fibers are classified, depending on the place of occurrence, into primary and secondary. The primary effects correspond to structural changes in cell walls that differentiate refined from unrefined fibers. Secondary effects are those effects caused by refining, but which can also originate in other stages of the pulping or pulp preparation process.

Primary Refining Effects

- Hydration of the fibers cell walls causing the delamination of their structure and consequent fibers swelling (internal fibrillation)
- Total or partial removal of the primary wall releasing secondary fines (external fibrillation)
- Fiber flexibility increase
- Average fiber length reduction
- Relative fiber bonding area (RBA) increase

Secondary Refining Effects

- Fiber cell wall delamination and displacement
- Fiber stretching and/or fiber compression: fiber collapse
- Partial solvation of surface hemicelluloses (gel formation)
- Dissolution of colloidal material from the cell wall with the redistribution of hemicelluloses from the inside of the fiber to the outside of the cell wall
- Fiber straightening reversing deformations
- Fiber deformations

Traditionally, these effects are defined based on observations under a microscope, that is, modifications in the fibers that are visually observed which correspond to the internal and external fibrillations of the fibers and the fines formation that are evidenced by the "cutting" and shortening of the fibers and fiber straightening (elimination of fiber deformations) [80, 219, 262, 299, 428, 747, 803, 817, 1041, 1049, 1051].

8.4.1 Internal Fibrillation

Hydration of the cell walls, produced by the action of cyclic compressive forces inside the refiner, occurs by breaking the cross links (hydrogen bonds) between the cell wall structural elements. The fibrils move relative to each other and the cell wall collapses and takes on a ribbon-like shape becoming more flexible in the transverse direction. In paper technology, this phenomenon is called internal fibrillation which results in fiber swelling [124, 125, 221, 347, 355, 433, 454, 505, 588, 594, 688, 761, 778, 788, 816, 866, 1024, 1032, 1039, 1060, 1062, 1190, 1172, 1227, 1242, 1251].

Internal fibrillation increases with refining energy allowing the production of more resistant papers but with lower levels of drainage, opacity, specific volume, and porosity [26, 34, 62, 63, 73, 222, 251, 261, 310, 353, 387, 443, 511, 657, 678, 688, 785, 797, 222, 928, 916, 1013, 1014, 1040–1046, 1049, 1121].

The increase in internal fibrillation of refined fibers is usually determined as swelling growth which is characterized by the water retention value (WRV) [63, 387, 456, 433, 459, 691, 657, 744, 756, 866, 1261].

8.4.2 External Fibrillation

External fibrillation begins when the fibers are compressed between the bars disc and subjected to shear stress leading to the removal of particles from the outermost cell wall layers allowing water to enter the cell wall. The fibrils of the innermost layers are released into suspension (secondary fibrillar fines) or remain partially attached to the fibers surface. When they remain attached, they give a fibrous ("fluffy") appearance to the surface of the fibers which considerably increases their specific surface [63, 73, 125, 202, 208, 347, 443, 452, 454, 588, 678, 688, 778, 815, 866, 918, 1018, 1024, 1026, 1032, 1040, 1042, 1043, 1049–1056, 1121, 1159].

The most significant effect of external fibrillation is the increase in the specific surface of the fibers, which represents the area available for the fiber connection increasing the paper's structure consolidation allowing an increase in its mechanical resistance. The greater the paper consolidation, the lower the pulp drainage rate, the opacity, the specific volume, and the porosity of the produced paper [63, 125, 195, 202, 208, 300, 387, 454, 479, 657, 688, 815, 822, 866, 1024, 1032–1034, 1041, 1042, 1051].

Although long controversial the effect of external fibrillation is now considered to be important when considered in conjunction with internal fibrillation [805, 1042, 1051, 1121, 1200].

8.4.3 Fiber "Cutting"

When wood started to be used as a source of fibers long fibers were the predominant material and, for this reason, most treatments were perfected aiming both the properties development and the average fiber length.

The long fiber "cutting" is used to reduce the agglomeration number of the fibers resulting in smaller flocs, aiming to improve paper formation. However, with the growing in the use of short fiber woods, the fiber shortening is often an undesired side effect of refining [21, 26, 108, 161, 337, 456, 715, 816, 830, 1036–1041, 1246, 1251, 1255, 1257].

Fiber rupture ("cutting") occurs when the bar shearing force on the fiber is large enough to completely disrupt the cell wall causing transverse subdivision of the fibers [1073]. Fiber "cutting" is an effect always associated with other refining results, such as the formation of lamellar fines. Being a refining effect, it can be managed by technological approaches that favor one refining effect over another.

In the case of fiber "cutting" this management must be done by refining intensity. More severe refines tend to "cut" the fibers while softer refines tend to increase the fiber bonding while minimizing the "cutting" action [178, 289, 761, 816, 1077, 1091, 1248].

8.4.4 Curling and Straightening of Fibers

One of the main objectives of refining commercial pulps is to recover the fiber bonding capacity lost during the pulp drying process. During refining, water penetrates the cell wall pores removing creases, wrinkles, and micro compressions straightening the fibers. Although not all deformations are reversed, there is an increase in the fiber bonding capacity with great effect on paper [26, 161, 178, 196, 215, 214, 228, 233, 330, 387, 459, 410, 430, 466, 588, 619, 702, 721, 781, 793, 826, 866, 892, 916, 917, 924, 928, 944, 945, 949, 1116, 1120, 1152, 1229, 1262–1268].

Fiber straightening plays an important role in paper properties as it improves load-carrying capacity as well as stress distribution in the fiber network. There is an increase in the modulus of elasticity and paper tensile strength [72, 73, 196, 337, 505, 619, 904, 928, 945, 1040, 1152, 1265, 1268–1270].

To achieve optimal benefits, chemical pulp refining should be conducted in such a way as to increase internal fibrillation and fiber straightening while keeping external fibrillation and fines generation at an adequate level [778]. Low-consistency refines favor fiber straightening wile high-consistency refines encourage retention of axial micro compressions [248, 263, 410, 470, 879, 1271].

8.4.5 Formation of Fines

The primary fines are those present in the unrefined pulp and have their origin in the parenchyma cells of the wood, in fragments of vessel elements, in and small fragments removed from the fibers cell wall by attrition during the pulping process. During refining, the pulp fines content increases. The fines generated during refining are known as secondary fines [300, 447, 1120, 918, 1042, 1206, 1339].

Secondary fines are fibrils and fragments from the P and S1 cell wall layers released by the external fibrillation in addition to macro molecules released by molecular fibrillation that go into the suspension and can influence various properties of the paper [26, 52, 63, 73, 1191, 1196, 1208–1211].

Compared to fibers, secondary fines have a high specific surface, high swelling capacity, and low drainage capacity. For the same pulp, there is a constant linear correlation between the fiber's swelling degree of the fines fraction and fibers. The swelling degree of the fines is about twice that of the fiber fraction [771, 892].

Comparing to primary fines, the secondary fines have higher content of fibrillar elements (about twice as much), which contribute to creating a closer contact

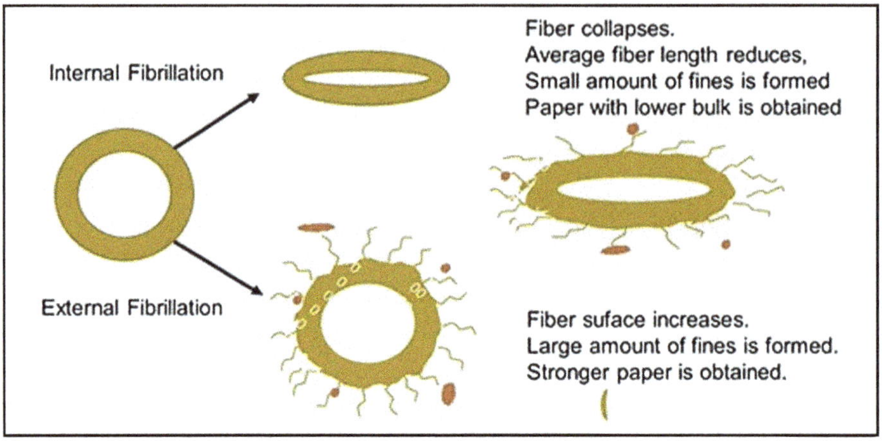

Fig. 8.8 Refining actions on fibers

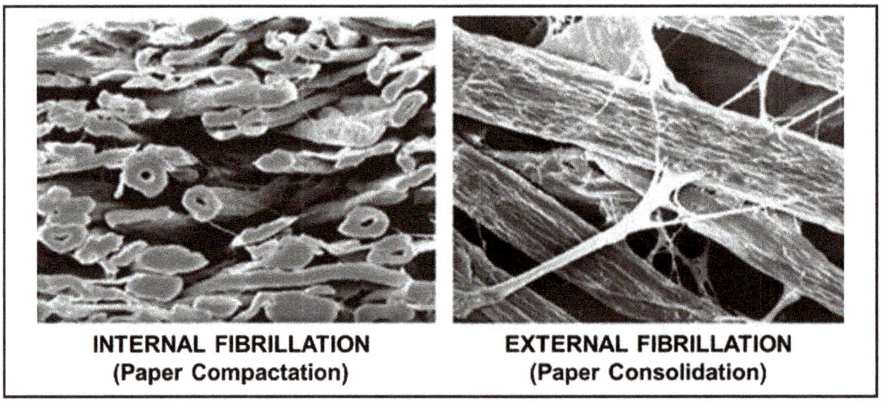

Fig. 8.9 Effects of refining on fibers. (Adapted from [115])

between the fibers during paper formation, generating a denser structure and, thus, increasing the transfer of tension resulting in higher wet paper expansion [26, 73, 90, 115, 180, 406, 456, 601, 686–688, 699, 771, 892, 1339].

Figure 8.8 presents a simplified scheme of the refining actions on the fibers and their direct impacts on the paper characteristics as shown in Fig. 8.9.

8.5 Refining Effects on Paper Properties

In addition to visually based classification, another way of classifying refining effects is by changes in fibers that influence paper properties and paper machine operating characteristics. With this classification criterion, the refining effects are as follows:

Increases fiber surface
Increases fiber flexibility
Decreases cell wall intrinsic resistance
Reduces fiber length
Generates secondary fines

Paper properties related to its use depend on the fiber properties as well as the structural changes introduced during its processing [62, 377, 566, 725, 741, 1008, 1051]. The main fiber-related factors influencing paper properties, which are usually estimated during a typical evaluation of market pulps, are presented in Table 8.1.

During refining, internal fibrillation promotes fiber collapse and flexibility, while external fibrillation generates fibrillar fines and increases the specific surface of the fibers. These actions contribute to joining adjacent fibers that conform to each other during paper forming, pressing, and drying [790, 1199, 1272, 1277].

When in excess, both internal and external fibrillation cause negative results in pulp drainage as well as in the physical and optical paper properties [406, 778, 1040, 1094]. A more intense refining is not the most adequate solution.

Refining increases the fiber flexibility and conformability which means an increase in the binding capacity between them. From a practical point of view, that is, the effects of refining on machine runnability and on the quality of the finished product, this is the reason why we refine pulps. In this way, the effects of refining can be summarized as consolidating and compacting the paper structure.

The fibers, which in the unrefined state can be visualized as tubes, are continuously compacted and sheared by the disc bars during refining. As a result, the internal bonds in the cell wall, those that maintain the fiber structure resistance, are broken allowing the release of microfibrils from the fibers' cell wall exposing new

Table 8.1 Pulp and paper properties and the basic pulp characteristics determining these properties [218]

	Factor influencing the paper properties	
Paper properties	Dominant factor	Secondary impact
Density (specific volume)	Wet fiber conformability	Fines content
Tear index – Elmendorf	Fiber-to-fiber bonding	Fiber intrinsic strength
	Fiber length	
Burst index	Fiber-to-fiber bonding	Elongation (stretch)
Tensile index	Fiber-to-fiber bonding	Fiber length
		Fiber intrinsic strength
Elongation (stretch)	Fiber-to-fiber bonding	Fiber curl[a]
Folding endurance"	Fiber-to-fiber bonding	Fiber length
		Fines content
Gurley air permeance	Wet fiber conformability	
	Fines content	
Freeness	Wet fiber conformability	
	Fines content	

[a]Fiber curl = curling, micro-compressions and kinks

surfaces that absorb more water and consequently fibrillate more easily. In this way, we can summarize the effects of refining as being the hydration of fibers that leads to the compaction and consolidation of the paper structure.

8.5.1 *Refining* Consolidates *Paper Structure*

During refining, in addition to making the fibers more flexible, the cell wall outer layers are totally or partially removed for generating fines that increase the fiber specific surface (external fibrillation), contributing to an increase in the number and strength of bonds.

As the fibers collapse, becoming more flexible, the bonding area between them increases consolidating the paper structure. This is because refining improves the fiber bonds capacity that positively influences the mechanical strength properties of the paper including its wet strength as well as elongation and surface properties [26, 63, 73, 111, 115, 161, 203, 215, 262, 280, 315, 343, 353, 406, 477, 479, 511, 568, 588, 611, 651, 657, 676, 721, 766, 769, 709, 778, 803–809, 895, 1054, 1094, 1097, 1104, 1172, 1184, 1214, 223, 1272–1276].

Refining also forms fines that, when shaped into fibrillar form, contribute significantly to increasing the fiber bonding capacity. The direct results are higher levels of elongation and resistance in the paper.

The mechanical resistances increase with refining until their maximum values, after which they tend to reduce due to the reduction of the cell wall intrinsic strength [315, 1094, 1276].

8.5.2 *Refining* Compacts *Paper Structure*

During refining, the fibers become more flexible, collapsed, and externally fibrillated and have their average length reduced. These fiber modifications contribute to the agglomeration factor (crowding factor) reductant the tendency to flocculate with produces smaller flocs.

These modifications contribute to improving paper formation, especially for long fibers [26, 136, 156, 163, 166, 465, 470, 511, 769, 782, 830, 878, 885, 1097, 1278, 1279].

With the progressive reduction of the intrinsic resistance of the cell wall, caused by the hydration of the cell walls, the fibers collapse, becoming more flexible, which increases the contact area between the fibers, reducing the free surface (fiber-air contact area). The same amount of fiber mass will occupy less volume and smaller channels will be formed in the internal paper structure. The direct effect on paper properties is the reduction of specific volume and light scattering, while the paper becomes smoother [63, 111, 115, 215, 216, 218, 262, 406, 468, 479, 511, 555, 568, 611, 617, 676, 688, 803–809, 888, 895, 1001, 1150, 1186, 1203, 1214, 1255, 1273–1275, 1281].

In addition, refining forms fines that, when in a lamellar format, can block the narrow internal channels of the paper, increasing the resistance to the passage of air (less porosity) and reducing the water absorption capacity of the paper [111, 406].

8.5.3 Paper Fracture (In Plane Strength)

The increase in the contact area between fibers allows greater resistance and frequency of connections between fibers (external fibrillation) leading to greater elongation and tensile strength values. These two paper properties define its "Failure Envelope" which is increased [125, 263, 406, 478] and higher resistance values are obtained. The work done on the fiber, which is equal to the energy stored by elongation, can be calculated [589, 595]:

$$U = \frac{\sigma_x^2}{2E_{\text{long}}} = \frac{1}{2}E_{\text{long}}\varepsilon_x^2 = \frac{I}{\text{V}} = \frac{I}{M}d_f \tag{8.16}$$

Where:

U work done on the fiber
σ_x voltage generated in the fiber
E_{long} fiber longitudinal Young's modulus
ε_x^2 effort applied to the fiber
I energy transferred by the impact of the rods to the fibers
V fiber volume
M fiber mass
d_f fiber density

Although there are many fibers bounds, they are weaker than fibers themselves. For this reason, the main mode of tension failure is connection cell wall breaks. This can be verified by pulling apart a paper strip and looking closely at the fracture line. Fibers will be seen protruding outward illustrating fiber pulling out of the paper structure. For this reason, the mechanical performance of paper is closely related to the mechanical characteristics not only of the fibers but also of the bonds between the fibers, their intrinsic strength, and fiber length [462, 1409].

8.5.4 Refining Effect on Tearing Strength

The paper tear strength corresponds to the energy needed to propagate the tearing line through the paper, that is, the sum of the energy needed to break the fibers and the energy needed to pull the fibers that did not break out of the paper structure when along the break line [86, 215, 256, 257,272, 274, 282, 732, 1282]. This is expressed in Eq. 8.17 [272].

$$\text{Tear} = \left(N_R.E_R\right) + \left(N_P.E_P\right) \tag{8.17}$$

Where:

N_R total number of fibers along the failure line that rupture
E_R average rupture energy
N_P total number of fibers along the failure line that pull out of the network
E_P average stripping energy associated with fiber pullout

Maximum rupture will occur when $N_P \times E_P$ is a maximum, which makes it clear that the greater the fiber length (due to the possibility of establishing more connections) and the coarseness of the fibers (indicating the cell wall intrinsic resistance), the maximum the tear resistance value with lower refining energy. Therefore, the development of tear strength with refining varies with the type of fiber. The work done to pull a short fiber from a sheet is less than that required to break the fiber (the reverse of the situation for the long fiber). Thus, as refining increases the fiber bonding, the proportion of fibers that are fractured instead of being pulled out also increases due to the fiber intrinsic resistance reduction during refining [218, 256, 283, 904, 1409].

Long fibers, which are more flexible than short fibers, react first to the effect of intrinsic strength reduction. Maximum tear strength occurs at low refining energy and tends to reduce continuously with the refining sequence. On the other hand, for short fibers there is an increase in tear resistance with the evolution of refining, due to the increase in the fiber bonding capacity compensating the cell wall intrinsic resistance reduction. At the most accentuated refining levels, when the reduction of the intrinsic resistance is more accentuated being the most important factor, the tear resistance starts to show decreasing values with the refining energy [45, 218, 242, 249, 256, 274, 283, 298, 1313].

In cases where fiber coarseness is small, the cell wall intrinsic resistance will also be small leading to a downward trend in the evolution of the tear index, that is also observed in high refining energy application. The same tendency is observed for burst strength [63, 315] (Table 8.2).

Table 8.2 Effect of refining on the pulp and paper properties

Refining effect on pulp and paper properties	
Decreasing values during refining	Increasing values during refining
Freeness	Schopper-Riegler degree
"Bulk"/specific volume	Folding endurance
Porosity	Apparent density
Surface roughness	Elongation
Water absorption (Cobb)	Vessel picking resistance
Opacity	Internal strength (z direction)
Tear strength (Elmendorf) (After slow increasing in fist steeps of refining)	Burst strength (Müllen) Tensile strength
Thickness	Wet strength

8.5.5 *Refining Effect on Vessel Picking*

The vessel elements' size and frequency, as well as the interaction between vessels and fibers, are traditionally considered to be the main causes of vessel picking problems.

With increasing speed and new generations of printers with five or six colors, the problem has resurfaced [84, 316–320].

To avoid the pulling out of vessels, the objective is to increase the surface resistance of the paper and the number of connections between the fibers so that, due to the presence of a smaller number of vessels, or because of their better anchorage to the paper, it is reduced. the number of particles picked off the paper surface.

The papermakers adopt several ways to select raw material. Pulps with lower vessel size and content are the preferred one.

Alternative processes to avoid vessel elements remova are also used suche as the pressing efficiency increasing and the paper sizing optimization [53, 56, 68, 80, 81, 100, 121, 165, 179, 198, 217, 316–321, 328, 332–334, 338, 343, 393, 417, 424, 664, 807, 865, 875, 929, 989, 1096, 988, 1128, 1221, 1227, 1283–1286, 1288, 1298, 1290].

The conformability of the fibers is increased during refining leading higher fiber bonding (number and strength) better connecting vessel elements to fibers reducing the likelihood of vessel pullouts [320, 338, 1288].

Vessel elements have lower strength properties than fibers and are partially destroyed by refining. Therefore, there is a reduction in the number and average size of vessel elements. They are totally or partially destroyed generating fibrillar fines that are dispersed in the pulp suspension [52, 53, 56, 84, 100, 121, 217, 316–320, 334, 338, 350, 393, 493, 838, 868, 872, 878, 1096, 1104, 1115, 1221, 1289, 1291, 1292] (Figs. 8.10 and 8.11).

8.6 Refining Effect on Paper Machine Runnability

Operational stability is the most important factor in machine runnability. The paper in production must, in addition to meeting the required quality in the finished product, be resistant enough to withstand the tension variations typical of the processes involved [63].

In refining, there is a trade-off between the development of desired paper properties and their adverse effects, such as their negative impact on water removal during the dewatering, wet pressing, and paper drying stages. Removing water from the pulp is more difficult for the fibers that are more refined [42, 93, 102, 111, 136, 163, 179, 389, 399, 456, 479, 511, 588, 661, 702, 709, 747, 782, 1049, 1061, 1087, 1096, 1099, 1100, 1117, 1125, 1161, 1172, 1184, 1191, 1197, 1274, 1275, 1291–1299].

Refining reduces the slurry drainage rate and therefore limits paper machine output and energy consumption because it increases dewatering and drying demand

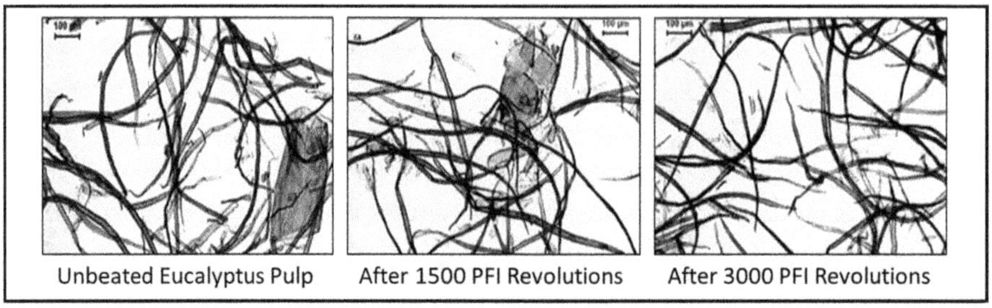

Fig. 8.10 Effect of refining (PFI mill) on fibers and vessel elements of bleached eucalyptus pulp. Optical microscopy

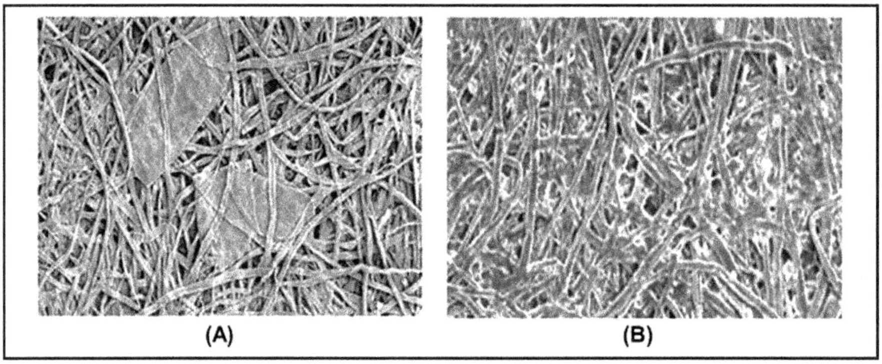

Fig. 8.11 Comparison between the surfaces of eucalyptus paper produced with unrefined pulp (**a**) and refined pulp (**b**). Effect from refining on the vessel structures. (Adapted from [115])

[1048, 1094]. Better drainage reduces the dry content of the paper sheet leading to an increased paper machine runnability and thermal energy savings in the drying section.

The drainage capacity is defined by the type and content of fines, the conformability of fibers, and the pulp mat densification during the paper formation. If the fines are not attached to fibers, they tend to move freely through the fiber mat and to settle into the paper internal channels during the dewatering process [63, 505, 710, 1048].

The fiber suspension drainage resistance increases with the refining. This is due to the fiber's specific surface increase as a direct result from the fines formation, the creation of folds and knots, and increased swelling of the fibers [178, 505, 709, 710, 1048].

Water removal is more difficult in pressing because the water is retained in the cell walls and no longer free water which was removed in the paper formation area. The water retention value (WRV) is an indicator of this amount of water and can be used to estimate machine productivity and energy consumption for drying [42,179, 217, 299, 406, 511, 562, 588, 611, 704, 762, 774, 1184].

However, the most important factor in ensuring paper machine production capacity is wet sheet strength, which is increased by the simple act of refining the pulp. For the same dry content of the fiber web, refining increases the wet strength because it generates cell wall modifications that increase the bonding area between the fibers and the wet paper density [63, 73, 76, 216, 477, 805, 915, 918, 1090, 1179, 1272, 1275, 1306].

8.7 Refining Is a Heterogeneous Treatment

All fiber modifications occur simultaneously. However, refining is a heterogeneous treatment, with part of the fibers passing through the refiner without treatment, while others are intensively impacted [80, 299, 303, 304, 664, 710, 817, 895, 1036, 1037, 1121, 1122, 1149, 1159, 1190].

Due to the characteristics of refining, it is impossible to eliminate an unfavorable effect or concentrate exclusively on a positive effect. However, it is possible to conduct and control refining in such a way that one of its actions on the fibers is favored to the detriment of others as we will see later in this book.

The force distribution on fiber webs in a refiner is similar to that found in beaters, that is, a small number of particles carry the applied load in a tension chain and therefore undergo deformation [836].

8.8 Pulp Refining Ability

The refiner technical and operating characteristics influence its pulp refining ability [49] that is also influenced by the fiber chemical and anatomical characteristics. The pulp ability to be refined can be defined as the amount of energy needed to reach the target pulp properties. The most common parameters used to measure the refined pulp are "drainage rate" and the "water retention value" [49, 657, 773].

The pulp ability to be refined depends on the ionic state of the pulp suspension. The number of charged ions that must be in equilibrium with the cell wall ultrastructure correlates with this ability. If too many charged ions are introduced, the internal ultrastructure will be damaged anticipating refining effects and little, if any, property development would be obtained in refining [433].

An increase in electrolyte concentration decreases the pulp ability to be refined, that is, increases energy consumption to meet the same properties set. For the same applied energy, the increases in electrolyte concentration improve drainage but decrease the strength properties of the paper [701, 702, 704, 707, 712].

Chapter 9
The Refining Models or "Theories"

Although Chinese archaeologists have found evidence that paper was already used in China, Tsai Lung is credited with inventing paper in the year 105 AD. This paper was produced from a suspension of plant residues, initially mulberry bark and later including hemp, old fishing nets, rags, and agricultural residues, which were kept in water for a long period of time and later macerated using a wooden hammer until a mass (pulp) is formed. After heating, the pulp resulting from this maceration was spread in a thin layer on a smooth surface and left to dry in the sun.

In addition to the use of vegetable fibers, another important aspect of this procedure is the impact treatment in the presence of water. This step individualizes and makes the fibers more flexible, making them suitable for paper production, giving the finished product strength and suitability for use.

The application of impacts in the presence of water is the procedure used until the present day. Over time, various resources were used, ranging from manual pestles and others intended for grinding grain, such as hammer mills and circular grinding wheels driven by animal traction and later by water wheels, to the adoption of specific equipment such as the Hollander Beater and the disc refiners.

All these resources, including the most modern ones, consume energy, and the changes introduced in the fibers are irreversible.

Due to its relevance to both production costs and the quality of the finished product, refining has been one of the most studied unit operations in papermaking. The adequate understanding of the process allows, in addition to maximizing results, the elimination of some "mysteries" involving the process.

Over time, models, also known as "refining theories," were developed with the aim of improving the understanding of the process in order to increase its efficiency, optimize its operation, allow the quantification of the process, and allow the comparison between different types of refiners. Such models, or "refining theories," transformed the "art of refining" into science and even allowed the correlation between laboratory results and industrial operation data.

© The Author(s), under exclusive license to Springer Nature Switzerland AG 2024

V. Manfredi, *Eucalyptus Kraft Pulp Refining*,
https://doi.org/10.1007/978-3-031-47285-5_9

The results of specific refining actions can be predicted and obtained, considering the type of pulp and the desired final paper characteristics. This does not mean that any type of paper can be produced from any type of pulp, but that, within the capabilities of each pulp, specific actions can be taken to optimize the results of paper production.

In the sequence, the most cited models in the literature are presented. However, it is worth remembering that the pulp used to manufacture paper is a highly heterogeneous lignocellulosic material. As a result, there are no satisfactory mathematical equations that unambiguously predict pulp properties based on refining parameters [1307]. However, excellent approximations are possible.

9.1 The Jagenberg' Model

The first contribution to process modeling took place at the end of the nineteenth century. Jagenberg (1887) [631], working with the Hollander beater, defined the basic parameters of grinding as the grinding force and the grinding area, that is, the surface area of the bars that apply force to the fibers. This area became also known as the "Jagenberg area."

Jagenberg [631] also introduced the formula for calculating the crossing velocity of the bars (Eq. 9.1), as edge length per unit time (CLs) and related it to the fiber cutting action in the Dutch (hollander mill).

$$CL_s = z_R . z_B . L . \frac{rpm}{60} \tag{9.1}$$

Where:

CLs bar crossing speed, m/s
z number of bars on the roll and on the base, indicated by R and B, respectively
L average length of bar
rpm roll rotational speed

Jagenberg [631] also recognized that the bar contact area, or Jagenberg area, is primarily responsible for the fibrillation action. The quotient between the grinding force (applied energy) and the grinding area is the first reference in the literature on fiber treatment intensity [583, 1330].

Jagenberg worked with Hollander straight bars ($\alpha = 0°$). Later, Kirchner expanded the theory to consider beaters with inclined bars ($\alpha > 0°$) by calculating the crossing velocity using the actual length of the bar [1331]. He also calculated the area of the bar in contact with a counter bar, in addition to the number of crossing points between the beater rotor roll and the base.

Subsequently, Pfarr [1330] evaluated the probability of fiber entrapment at the crossing point of a single bar and calculated the total area of contact between rotor and stator, concluding that the friction and crushing action of the Hollander beater were proportional to the area of beat per second – a product of the contact area of the bar and the peripheral speed of the beater roll [1330, 1332].

9.2 Fibrage Theory

Also working with the Hollander beater, Smith (1922, 1923) [231, 626] observed that it is the corners of the bars that collect and transport the fibers to the treatment zone of the back bars and that the number of fibers collected depends on the consistency of the suspension, the type of fiber (mainly length and flexibility), and the degree of drainage, or freeness, of the cellulose. Smith was the first author to consider fiber characteristics in a refining model [583].

According to this model, the edges of the bars collect fibers forming a fibrous blanket on which they exert pressure. This pressure is expressed as the contact force divided by the total length of the edges of the active bars. In this way, it is possible to estimate a beating pressure that is inversely proportional to the fiber length, resulting in an intensity measure calculated by the quotient of contact force and crossing velocity (918) [231].

$$I = \frac{F}{CL_s} \frac{\cos\alpha_R \cos\alpha_B}{w_{b,R} w_{b.B}} \omega \tag{9.2}$$

Where:

I treatment intensity, N/m^2 or Pa
F contact force, N
CL_S bar crossing velocity, m/s
α bar angle of the rotor (R) and the base (B)
w bar width of the rotor (R) and the base (B)
ω bar velocity, m/s

9.3 Specific Edge Load

In 1958, Wulsch and Flucher [1223] confirmed the concepts presented by Smith and introduced the concept of refining intensity as the quotient between the effective applied load and the linear length of the bars through which this refining load is applied by time unit.

They used the number of rotor and stator bars to present a more universal refining theory, which they originally designated as "Spezifische Kantenbelastung" or specific edge load (SEL) (Eq. 9.3).

$$\mathrm{SEL} = \frac{P_{\mathrm{net}}}{L_s} = \frac{P_{\mathrm{net}}}{\frac{rpm}{60} z_r z_s l_B} \tag{9.3}$$

Where:

SEL specific edge load, Ws/m or J/m or N
P_{net} net refining power
L_{s} bar crossing velocity
rpm refiner rotational velocity
z_{r} number of bars in the rotor disc,
z_{s} number of bars in the stator disc,
l_{B} average bar length

9.4 Number and Severity of Impacts

Another significant advance occurred when Lewis and Danforth [680] proposed, in 1962, that the treatment given to fibers could be expressed in terms of two independent variables:

- The number of impacts between the edges of the bars to which the fibers are exposed
- The severity of each of these impacts, which would determine the type of treatment given to the fibers

Lewis and Danforth were the first to model refining as a process associating the number and intensity of impacts on the fibers. The severity of the impacts determines the type of treatment given to the fibers, and it is well established that low-intensity impacts lead to fibrillation while high-intensity impacts lead to a reduction in the average fiber length [558, 680, 1073, 1109],

At this point, it is worth stressing that, although the literature calls the reduction of the average fiber length as "fiber cutting," the cutting itself does not occur in the refining process. What occurs is excessive hydration at specific points of the cell wall which reduces its intrinsic resistance, causing fiber rupture and generating secondary fines.

The model presents refining as expressed by the number of impacts (Eq. 9.4) and by the severity of impacts (Eq. 9.5) transmitted to the fibers by the edges of the disc bars [558, 583, 880].

$$N = \frac{BEL_R.BEL_S.rpm.C_{f.}k_1}{AA.q} \tag{9.4}$$

$$S = \frac{\left(P_{tot} - P_{nol}\right).RA.k_2}{DD.BEL_R.BEL_S.rpm.C_f} \tag{9.5}$$

Where:

N number of impacts
S relative severity of each impact (intensity)
P_{tot} refiner total power
P_{no}l refiner no-load power
RA total area of the refiner zone
DD disc effective diameter
BEL_R total edge bar length of the rotor disc
BELS total edge bar length of the stator disc
rpm refiner rotational speed
C_f pulp suspension consistency
AA average contact angle
q pulp suspension flow
k_1 and k_2 specific constants

According to Giertz [1088], the effects of an impact depend on its intensity. Some effects, in particular, only happen when this intensity reaches a specific minimum level (minimum impact intensity). The correlation between the refining intensity and its effect on the fibers is presented in Table 9.1.

As the intensity of the refining force increases, the following events should occur in the particles, with decreasing probability [1171]:

- Elastic and/or viscoelastic deformation of the fibers together with a dissipation of friction in water
- Rupture of hydrogen bonds inside the cell wall
- Disruption of the cell wall
- Displacements and delamination and, finally, the "cutting" of the fibers

Table 9.1 Effects of refining intensity increasing on the fibers

Intensity type	Type of pulp treatment	Effect on fibers
O	Internal friction Elastic straining of the fibers Reversible regain	None
I	Breakage of bonds at the molecular level of hemicelluloses – Internal fibrillation.	Fiber became flexible
II	Partial disruption of the fibers cell wall	Removal of the outer layer, exposing inner surfaces of the cell wall – External fibrillation
III	Fibers cell wall breakdown	Crushing Fiber cutting

According to Mayade (1997) [1109], the work of Giertz (1964) [1088] shows that the intensity of the forces acting on the fibers in the refining is not the same, varying according to the position in the refining elements, but can be described by its distribution resulting in an average intensity value that must be considered.

9.5 Probability of Impacts on Fibers

In 1964, van Stiphout [1180] suggested that the probability of fibers being treated during refining is dependent on equipment characteristics and process parameters.

Two basic terms are used in the model: the force on the mass of fibers concentrated in the space between the discs (FP) (Eq. 9.6) and the total mass of fibers passing through a spacing of one centimeter (TF) (Eq. 9.7), as shown below.

$$F_P = \frac{P_{net}}{rpm.D_R.T.N_F} \quad (\mathrm{Ws2/g}) \tag{9.6}$$

$$T_F = \frac{T.N_F.C_f}{q_m.V_B.z} \quad (\mathrm{gm/cm}) \tag{9.7}$$

Where:

P_{net} net power
rpm rotational velocity
D_R rotor disc diameter
T average distance between discs
N_F change of a fiber slipping between discs
C_f pulp consistency
q_m mass of fibers to be treated
V_B volume between bars
z number of bars

Therefore, the force per fiber mass (F_W) is described as in Eq. 9.8.

$$F_W = F_P.T_F = \frac{P_{net}.C_f.V_B.z}{rpm.D_R.q_m} \tag{9.8}$$

To be refined, the fibers need to be captured by the bars. The number of fibers in the intercrossing of the bars is a fraction of the total fibers present in the refining zone of the discs, and the average number of impacts received by the fiber is proportional to the total intercross length to which the fiber was exposed [422].

9.6 Refining Intensity

The model developed by Wultsch and Flucher was described for its current formulation by Brecht and Siewert [1224] who demonstrated the predominance of the impact of edges of the bars on the fibers. The specific edge load (SEL) describes the intensity of the treatment performed by means of equipment with bars, such as a mill or a refiner.

The model assumes that the refining action occurs due to fiber deformation due to bar edge impacts. These impacts are related to the total length of the bar per unit of time, also referred to in the literature as the "cutting total length." We suggest, since there is no cutting of fibers in the refining, that the nomenclature be changed to "operational edges length" (BEL).

The refining intensity, which determines how the fibers are treated, is calculated by dividing the net power applied by the length of the edges of the disc bars, which is a function of the disc configuration and the rotational speed. The net energy of refining, which is applied by the total length of the bars that intersect in a given unit of time, determines the extent of refining, that is, the amount of treatment, which is the quantitative parameter of the model [219, 1117, 1180, 1183, 1220–1231].

This was the origin of the "specific edge load theory" as it is known today.

So, the refining action will always be a balance between the net energy applied, which corresponds to the number of impacts on the fibers, normally expressed in kWh/t, and the intensity of these impacts, normally expressed in Ws/m or J/m [219, 354, 558, 629, 880, 1072, 1117, 1146, 1225].

Thus, refining is characterized by the disc configuration, or in other words, by the length of the disc bars, which, multiplied by the rotational speed of the refiner, provides the rate of edge crossings (operational edge length) (Eq. 9.9). The intensity of the refining action is then calculated by dividing the net specific energy of refining (Eq. 9.10) by the rate of crossing the edges of the bars (Eq. 9.11) [880].

$$\text{OBL} = z_r . z_s .\text{BEL}.\frac{rpm}{60} \tag{9.9}$$

$$\text{SEC} = \frac{P_{\text{net}}}{q_m \cdot \frac{C_f}{100} . \rho} \tag{9.10}$$

$$\text{SEL} = \frac{P_{\text{net}}}{\text{OBL}} \tag{9.11}$$

Where:

SEC specific energy consumption, kWh/*t*
SEL specific edge load, Ws/m
OBL operational bar length, m/s

P_{net} refiner net power, kW
z_r number of bars in the rotor disc
z_s number of bars in the stator disc
BEL bar edges total length, *m*
rpm disc rotational speed, *rpm*
q_m pulp flow, m^3/h
C_f consistency, %
ρ stock density (ton/m^3)

These formulas indicate that the intensity, or severity, of the refining is an inverse function of the bar crossings and a direct function of the net power applied.

Even though they are not exact mathematical expressions, the formulas considered in the model contemplate, simultaneously, three important refining factors: production (consistency and flow through the refiner (t/h), net energy applied (useful power applied and treatment time) (kWh/t), and the refining intensity (configuration and rotational speed of the discs) (Ws/m).

The refining intensity is a machine parameter, that is, the intensity defines the average energy transmitted by the equipment (by length of bar crossings). The model does not consider how the fibers receive energy from impacts. Fibers are considered only indirectly, as their presence between opposite bars is necessary to achieve refiner power.

Traditionally, high refining intensities imply in the reduction of the average fiber length and fines formation, being suggested for the refining of long fibers as a strategy to improve the paper formation. For short fibers, it is recommended to use lower refining intensities [511, 1080, 1096, 1116, 1133, 1224, 1230].

As the model is very simple and easy to be applied, it has been an efficient way to optimize energy consumption and improve fiber quality, in addition to allowing the comparison of the refining action in different equipment. Whether for conical or disc refiners, even with variations in net energy, configuration, and rotational speed of the discs, the same results are predicted if the specific edge load is kept constant [219, 299, 354, 393, 402, 511, 817, 838, 868, 875, 880, 1035, 1036, 1066, 1072, 1076, 1091, 1096, 1104, 1105, 1109, 1112–1117, 1133, 1146, 1213, 1260, 1341].

Due to its simplicity and quality of the results, the specific edge load model is the most widespread and used by papermakers, as well as by equipment suppliers and research institutions [219, 268, 269, 299, 393, 402, 1227, 1235, 1260, 1307, 1342–1344, 1405].

After its initial conceptual disclosure, the model was gradually studied and optimized. Several variants of the model, in some cases considered to be more accurate ones, were presented over the years [234, 268, 511, 595, 650, 668, 710, 1020, 1026, 1027, 1032, 1035, 1036, 1066, 1070–1082, 1227].

Some of these optimizations arose from criticism of the model because its main parameters (specific refining energy and specific edge load) are indirect indications of the treatment to which the fibers are subjected. The model does not directly consider factors linked to pulp characteristics nor to operating variables (consistency, suspension flow, refiner rotational speed) or disc configuration (bar width, channel width, and depth and angle of inclination of the bars) that several authors cite in the literature as being important factors in defining the refining results [668, 710, 765, 839, 1066, 1069, 1079, 1080, 1096, 1097, 1127, 1133, 1164, 1345–1347, 1355].

As a rule, authors who criticized the theory of the specific edge load (SEL) developed theoretical and/or practical studies and presented new models or theories about refining. Unfortunately, some of these models focused on analyzing the "cutting" action on fibers and/or based their analysis on properties that are not related to the paper quality, such as the degree of drainage of the fiber suspension.

A criticism of the specific edge load model (SEL), as discussed in detail in Chap. 10 – Refining Operating Variables, is the fact that the model does not consider the consistency of the suspension whose variation can lead to different refining energy consumptions to meet the same set of properties in refined cellulose.

Regarding the criticisms about disc configuration, we understand that such criticisms are due to different ways used to calculate the bar length of the discs. When this length is calculated as presented in Appendix II (Operating Edge Length – Example of One Determination Method), the effects of disc configuration (bar and channel width, bar inclination angle, disc sectorization angle, and width of the channels) do not cause differences in the results either in energy consumption or in the quality of the refined fibers.

9.7 Frequency of Fibers Treatment

Returning to the concept of probability of application of impacts (van Stiphout, 1964), Leider and Nissan [1072] proposed, in 1977, an analytical model describing refining as a combination of the number of impacts applied and the intensity per impact (energy applied by each impact). The net refining energy is then determined by the number and severity of impacts. The residence time in the refiner and the possibility of a fiber being in the exact position to be impacted are also considered in the model.

The first step is to calculate the number of impacts (N) based on the factors that determine the average number of impacts, as shown in Eq. 9.12. As refining is unpredictable, that is, fibers can run through the entire refiner without being impacted, the term (A_f/A_g) corresponds to the probability of a fiber be into contact with the refiner bars during refining [221, 589, 595, 1040, 1072, 1117].

$$N_m = f\left(z_r, z_s, rpm, \tau^f, \frac{A_f}{A_g} \right) \tag{9.12}$$

Where:

N_m number of impacts
z_r number of bars on the rotor

z_s number of bars on the stator
rpm rotational velocity
τ^f time that a fiber remains in the refiner
A_f mean fiber straight-section longitudinal area
A_g straight-section groove area

The number of impacts (N_m) is a function of the number and length of the disc bars, the rotational speed, the fibers retention time in the refiner and the probability of the fiber contacting the disc bars during the passage through the refiner (A_f/A_g). The simplest way is to present the equation with directly proportional variables (Eq. 9.14) for which the number of bars in the rotor and stator are determined by Eq. 9.13 [595].

$$z_r \text{ or } z_s = \frac{\left(r_1 + \frac{r_2}{2}\right)}{(B+G)} = \frac{\pi\, DD}{(B+G)} \tag{9.13}$$

$$N_m = z_r.z_s.rpm.\tau^f.\left(\frac{A_f}{A_g}\right) \tag{9.14}$$

Where:

N_m number of impacts
n_r number of bars on the rotor
n_s number of bars on the stator
r_1 internal radius of the refiner working area, *m*
r_2 external radius of the refiner working area, *m*
B bar width, *m*
G average groove width, *m*
DD disc diameter, *m*

It is well established that increasing the number of impacts (or intercrossing of bars) improves fiber hydration with little effect on the reduction in average fiber length. Refining energy is transferred to the fibers only when the corners of the bars intersect. Therefore, the impact frequency is influenced by the configuration of the discs [710, 1051, 1117].

The retention time in the refiner, in turn, considers the internal volume of the refining area, which corresponds to the sum of the volume between the discs plus the total volume of the disc channels and the suspension flow through the refiner (Eq. 9.15) [595].

$$\tau^f = \frac{r_2 - r_1}{\cos\varnothing}\frac{\pi\langle \boldsymbol{r}\rangle(2\boldsymbol{g} + \boldsymbol{T})}{q} \tag{9.15}$$

Where:

G average channel width, *m*
q suspension flow through the refiner, (mass/time)
*r*4 inner radius of the working area, *m*
r_2 outer radius of the working area, *m*
T disc spacing, *m*
Ø angle of inclination of the bars
τ_f time that a fiber remains in the refiner

The probability of the fiber being impacted by the bars (A_f/A_g coefficient) is calculated as follows [595]:

$$\frac{A_f}{A_g} = \frac{d.l}{D.G} \tag{9.16}$$

Where:

d fiber diameter, *m*
l average fiber length, *m*
G average groove width, *m*
D average groove depth, *m*

Now, substituting 32 and 33 in E31, we have, for calculating the number of impacts (N), Eq. 9.17 presented below [595].

$$N_m = 4\pi^3.r^3 \frac{2+\left(\frac{D}{T}\right)}{G\left(G+B\right)^2} \frac{r_2 - r_1}{\cos\varnothing} \frac{rpm.d.l}{q} \tag{9.17}$$

Where:

N_m number of impacts
r radius of disc workspace
B width of bars
D average depth of the channels, *m*
G average channel width, *m*
T disc spacing, *m*
q suspension flow through the refiner, (mass/time)
r_1 inner radius of the working area, *m*
r_2 outer radius of the working area, *m*
Ø angle of inclination of the bars
rpm refiner rotational speed
d fiber diameter, *m*
l average fiber length, *m*

Thus, Leider and Nissan [1072] confirmed the proposal by Van Stiphout [1180] on the dependence that the probability of the fibers being impacted has on the characteristics of the equipment. In other words, regardless of size, configuration, and rotational speed, refiners treat fibers in the same way if these two factors (number of impacts and severity of impacts) are the same.

Knowing the number of impacts (N_m) and the specific energy applied (E_{net}), it is possible to determine the energy per impact (I_e), where (M) is the amount of refined pulp [595, 1072]. Energy per impact can also be defined as the severity of the impact.

$$I_e = \frac{E_{net}.M}{N_m} \tag{9.18}$$

9.8 A Refining Action Occurs on Suspended Fiber Flocs

A fundamental concept to understand the refining process is that it occurs on suspended flocs and not on individualized fibers. This concept was initially introduced in the late 1960s, but only gained increasing acceptance from the 1980s onward, and from then on it was widely accepted and used [88, 217, 234, 664, 742, 836, 838, 867–872, 874, 875, 877, 879, 887, 895, 989, 1018, 1051, 1057, 1074, 1112, 1115, 1121, 1128–1130, 1157–1160, 1170, 1226, 1285, 1373].

Questions involving floc dynamics, that is, how they are formed, how they move relative to each other, how they resist suspension turbulence, how they react to compression and shear, etc., are very important to a better understanding of the refining process.

Considering the floc as the unit receiving the impacts, the way we visualize the energy transferred to the pulp is changed. If we consider the impact on a fiber, all the impact energy will be absorbed by that fiber. However, if we consider the impact on a floc, all the energy of each impact will be distributed to the fibers inside the floc.

A more comprehensive form is the concept of net tangential force per bar crossing zone (F_t^{net}) or net normal per bar crossing area. The use of the net tangential force would contribute to obtain a better understanding and optimization of the pulp refining process either by varying the average length of the fibers or by the global effect of refining on the fibers (Schopper Riegler degree) [269].

$$F_t^{net} = \frac{P_{net}}{RN.L_{C2}.\sin\left(\varnothing_S + \varnothing_R\right)} \tag{9.19}$$

Where:

RN = number of revolutions
L_{c2} = equivalent total bar length of the refiner plate (Eq. 9.20)
$\varnothing_r$ = bar angle of the rotor

$\emptyset_s$ = bar angle of the stator
α_{AS} = average bar angles of stator (ØS)
α_{AR} = average bar angles of rotor (ØR)

$$L_{C2} = \frac{4\pi^2\left(r_1^3 - r_2^3\right)}{3\left(\alpha_R + b_R\right)\left(\alpha_S + b_S\right)} \tag{9.20}$$

Where:

r_1 = internal radius of the disc working zone
r_2 = external radius of the disc working zone
$\alpha_{R\,2}$ = average crossing angle – rotor
b_R = number of bars – rotor
α_S = average crossing angle – stator

9.9 Number and Effective Energy by Crossing Points

Also reporting that the edge-specific load model does not consider all the geometric parameters of the discs, Joris (1986) [1069] presented an extension of this model considering the missing geometric parameters. The parameter average number of crossing points of the bars was introduced in place of the parameter length of the bar edges. The crossing point is the exact point where the suspended flocs are taken by the refining bars.

The average number of crossing points is calculated based on the inner and outer diameters of the refining working areas, the widths of the bars and grooves, and the average radial angle of the bars in both rotor and stator discs, which varies depending on the relative position of the rotor to the stator discs. As the energy transfer to the fibers is carried out through these crossing points, the refining energy and the way in which this energy is transmitted will depend on this relative position of the rotor in relation to the stator [816, 1069].

The model was further optimized [710, 1079] with the introduction of effective energy per crossing point (ε) concept, that considers, simultaneously, the two parameters of the specific edge load (net refining energy and refining intensity) allowing the papermaker to choose the most suitable discs configuration in order to meet the desired results while minimizing operating costs [1079].

Assuming that the refining effects occur from impacts generated at the crossing points, the number of crossing points ($n_i(t)$) should provide an indication of the applied treatment. The number of crossing points (ni(t)) can be analytically calculated according to Eq. 9.21 [1079].

$$n_i(t) = \frac{S_i(t)\sin\gamma_i(t)}{LL'} \text{ where } i = 1,2,\ldots,p \tag{9.21}$$

The instantaneous number of crossing points is (N(t)) expressed by Eq. 9.22. Its average value, over a period (T) of the cut angle, which physically matters, is a function of nine geometric parameters (Eq. 9.23) and allows us to define the function of the characteristic cutting angle (γ*) (Eq. 9.22) in a disc refiner.

$$N(t) = \sum_{i=1}^{p} n_i(t) \quad (9.22)$$

$$N_{m=} \frac{1}{T} \int_0^{tN} (t)\, dt = \int r_1, r_2, W_r W_s, \theta_r, \theta_s, \text{Orient}_{s,}\ \text{Orient}_r, \beta \quad (9.23)$$

$$|\sin \gamma *| = \frac{2 N_m W_r W_s}{\beta \left(r_2^2 - r_1^2 \right)} \quad (9.24)$$

Where:

γ^* characteristic cutting angle
$\O_r$ rotor disc cutting angle
$\O_s$ stator disc cutting angle
W_r bar plus groove width – rotor
W_s bar plus groove - width – stator
β discs sectorization angle

The characteristic cutting angle (γ^*) is independent of the rotation speed. It represents the theoretically constant angle formed to obtain the same "cutting effect" as that caused by the studied geometry.

The effective energy per crossing point (ε) is proposed to replace the specific edge load (SEL), being used together with the average number of crossing points to characterize the refining. The energy per crossing point is calculated according to Eq. 9.25 [1079].

$$E_{eff} = \frac{P_{net}}{N_m \left(\frac{2\pi}{\beta} \right) \Omega} \quad (9.25)$$

Where:

E_{eff} crossing point effective energy, *J*
P_{net} refiner net energy, *W*
($2\pi/\beta$) number of sectors in the stator disc
Ω refiner rotational speed (revolution per second)
β disc sectorization angle

9.10 The C-Factor

This model, proposed by Kerekes [1073, 1351] in the early 1990s, was developed based on the model of the number and intensity of impacts (see Sect. 9.7), but introducing the estimative of the probability of impacts estimative and the cyclic deformation imposed by the refining elements. The model includes parameters of the disc configuration and fiber characteristics in addition to the pulp consistency. It was subject of several optimizations and later details [511, 583, 595,650, 710, 1020, 1027, 1035, 1036, 1073, 1076, 1077, 1080, 1227, 1315].

C-factor, expressed in (m/g), is a measure of the refiner's ability to impact the fibers. It tells us how impacts are received by the fibers. The model is considered, by many authors, as the most rigorous and adequately adjusted for refining [88, 402, 511, 595, 710, 880, 1021, 1073, 1076, 1077, 1105, 1107, 1352, 1353].

The model considers the average number of impacts (N_m) and the specific intensity of each impact (I_e) expressed as energy per impact per fiber (S), which is the average energy spent on a fiber during a load cycle (impact) divided by the fiber mass. The refining net specific energy (E_{net}) is related to the net power (P_{net}) applied and to the pulp flow (V) through the refiner, being defined as presented in Eq. 9.26 [511, 558, 595, 650, 710, 880, 1073, 1076, 1077, 1307, 1315].

The starting point is similar to other theories. The net refining energy per mass unit can be calculated from the number of impacts (N_m) multiplied by the energy of each impact (I_e). Two refining actions are equivalent only when the parameters (N_m) and (I_e) are equal, being fundamental, in the theory of C-Factor to be able to calculate these two parameters individually to allow, in each case, analysis of their preferential effect on refining.

$$E_{\text{net}} = N_m . I_e \tag{9.26}$$

Where:

E_{net} specific net refining energy
$N_m$specific number of impacts
I_e energy transferred by each impact

However, the same specific refining energy can be obtained with different combinations between the number of impacts and their intensities (Fig. 9.1). The two refining actions are equivalent only when the parameters (Nm) and (Ie) are equal.

Variations in intensity lead to different results. For example, with a greater number of low-intensity impacts, we favor fiber fibrillation, while with a lower number of high-intensity impacts, we generate more fiber length reduction action [511, 595].

Thus, it is essential to be able to calculate these two parameters (N_m and I_e) individually to allow, in each case, the analysis of their preferential effect on refining. The main idea of the C-factor model is to introduce a third variable.

The number of impacts (N_m) is calculated from C-factor (FC) and the mass flow rate through the refiner (q_m) by Eq. 9.28, while the intensity of each impact (I_e) is

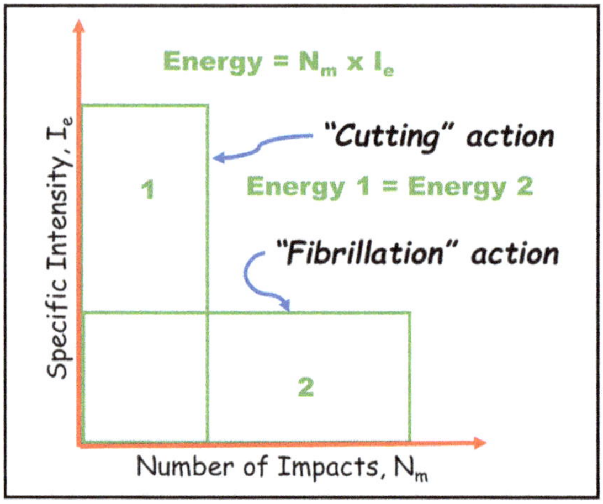

Fig. 9.1 The basic actions in refining. Cutting action for higher intensity and fibrillation action for shorter intensity. (Adapted from Refs. [589, 595, 1358])

calculated by Eq. 9.29. Equation 9.26 corresponds to the calculation of the specific refining energy (E_{net}) [595, 1155, 1227, 1307].

$$E_{net} = (FC / q_m)(P_{net} / FC) \tag{9.27}$$

being

$$N_m = (FC / q_m) \tag{9.28}$$

and

$$I_e = P_{net} / FC \tag{9.29}$$

Where:

N_m average number of impacts
FC C-factor
E_{net} specific refining energy
q_m mass flow rate through the refiner
P_{net} specific refining energy.

Due to the constructive differences between the conical and disc refiners, two different models were formulated to determine the C-factor, one for each type of refiner [1073].

9.10.1 C-Factor for Conical Refiners

For conical refiners, the mathematical model for C-factor is given by Eq. 9.30:

$$C = 4\pi^2 \rho C_f l n_s n_r \omega\left[T + \left(n_s D_s G_s\right) + n_r G_r D_r\right]\left(1 + tan\theta_s + tan\theta_r\right)$$
$$\left(r_1^2 L + L^2 + L^2 r_1 sin\theta + \frac{L^3}{3} sin^2\theta\right)\frac{1}{4w\left(l + D + t\right)} \tag{9.30}$$

Where:

C_f pulp consistency, fraction
D average depth of the channels, m
D_r depth of rotor channels, m
D_s depth of the stator channels, m
G_r width of rotor channels, m
G_s width of stator channels m
l average fiber length, m
L length of the refining zone, m
n_r number of bars/unit arc length of rotor, m-1
n_s number of bars/unit arc length of stator, m-1
r_1 inner radius of the working area, m
T disc spacing, m
w fiber coarseness, kg/m
ρ density of water, kg/m^3
$Ø_r$ bar angle of the rotor
$Ø_s$ bar angle of the stator
ω angular rotational velocity, $s = 1$
Θ angle of conical refiner, rad

In most conical refiners, the rotor configuration is similar to that of the stator, with the same bar widths and the same width and depth of the channels, while the spacing between the cones being much smaller than the depth of the channels. So, simplifying Eq. 9.28, we get Eq. 9.31 [5, 951, 1073, 1087].

$$C = 8\pi^2 GD\rho C_f l n^3 \omega\left(1 + 2\tan\theta\right)$$

$$\left(r_1^2 L + L^2 r_1 \sin\theta + \frac{L^3}{3}\sin^2\theta\right)\frac{1}{w\left(l + D\right)} \tag{9.31}$$

Where

C_f pulp consistency, fraction
D channel depth, m
G channel width, m
l average fiber lenght, m
L length of the refining zone, m.
n number of bars/unit arc lenght m^{-1}
r_1 internal radius of the refining zone, m

w fiber coarseness, kg/m
ρ water density, kg/m^3
ϕ bar angle
ω angular rotational velocity, s^{-1}

9.10.2 C-Factor – Disc Refiners

For disc refiners, the mathematical model for C-factor is given by Eq. 9.32:

$$C_{disc} = 4\pi^2 l.C_f.z_r z_s.w.rpm.\left(1+\left(\tan\langle \varnothing_s\rangle + \tan\langle \varnothing_r\rangle\right)\right)\frac{\left(T + z_r D_r G_r + z_s D_s G_s\right)}{w.\left(l + D + T\right)}\frac{\left(r_2^3 - r_1^3\right)}{3} \quad (9.32)$$

Where:

C_{disc} C-factor for a disc refiner
l fiber length
C_f pulp consistency
z_r number of bars in the rotor disc
z_s number of bars in the stator disc
rpm rotational velocity of the disc
$\varnothing_r$ bar angle of the rotor disc
$\varnothing_s$ bar angle of the stator disc
T distance between discs
D_r channel depth – rotor disc
D_s channel depth – stator disc
G_r channel width – rotor disc
G_s channel width – stator disc
r_1 disc refining area – internal radii
r_2 disc refining area – external radii
w fiber coarseness

As for the majority of disc refiners, the disc configuration is the same for both discs ($z^*r = z^*s$, $D_r = D_s$, $G_r = G_s$, and $\Theta_r = \Theta_s$), and Eq. 9.31 can be simplified as follows [595, 1073, 1087]:

$$C_{disc} = 8\pi^2.l.z^3.rpm.D.G.\left(1 + 2\tan\varnothing\right)\frac{\left(r_2^3 - r_1^3\right)}{3.w.\left(1+G\right)} \quad (9.33)$$

Where:

C_{disc} C-factor for a disc refiner
l fiber length

C_f pulp consistency
z number of bars in discs
rpm rotational velocity of the disc
D channel depth
G channel width
$Ø$ bar angle of the discs
r_1 disc refining area – internal radius
r_2 disc refining area – external radius
w fiber coarseness

The C-factor represents only impacts applied by the load between crossing bars. Minor impacts can occur as a result of contact between fibers, which, although consuming a relatively small amount of energy per cycle (impact), can, in large numbers, contribute to the refining action, such as that in high-consistency refining [1077].

The C-factor is based on the average number of impacts (strain cycles) (N_m), and the intensity of each impact (I_e) is expressed as energy per impact per fiber, which represents the energy dissipated in a single impact. Since energy is the product of force and distance, the intensity per impact provides a basis for estimating the force exerted if the distance over which it is applied is known [650, 798, 1032, 1036, 1082–1085, 1227].

At this point it is important to detail the refining intensity a little more.

The intensity determined by the specific edge load (SEL) refers to a "machine intensity" or "equipment intensity." It defines the energy transmitted per unit length of the bar edges. A machine intensity relates to *how energy is applied by the equipment*.

The specific intensity (I_e), in turn, defines the amount of energy transferred to each unit of refined pulp mass. It is, therefore, a mass intensity and is related to *how energy is received by the fibers* in suspension.

Machine intensity (SEL) and impact energy (I_e) (mass intensity) are related [595, 650, 1073, 1076, 1077]. The machine intensity (specific edge load (SEL)) can be expressed in terms of disc and refiner variables allowing comparison with impact energy per unit of mass (I_e), as determined by the C-factor. The model, for disc refiners, is presented below [1077].

$$I_e = \frac{1}{\rho C_f} \frac{1+\frac{1}{D}}{1} rpm \frac{1+\frac{B}{G}}{2\left(\cos^2 + 2\cos\phi\sin\phi\right)} \mathrm{SEL} \tag{9.34}$$

Where:

I^e specific intensity (energy/dry mass impact)
ρ fiber suspension density, f
C_f consistency of the fiber suspension (fraction)
l weighted average length of fibers (m)
D depth of the channels, m

rpm rotational speed of rotor disc, rev/sec
B width of the disc bars, m
G width of the disc channels, m
Ø angle of inclination of the bars

9.11 Specific Surface Load

This model presented by Lumiainen, in the early 1990s, can be considered as an optimization of the specific edge load model. It assumes that refining energy is transferred not only by bar edges crossing but also when the bars pass one another [365, 392, 1074, 1164].

The energy applied by bar surfaces onto the fibers depends on the distance between the discs, being greater as the smaller this distance is [589, 668, 1075, 1159, 1164, 1263]. This energy transference occurs in three distinct steps [659, 668, 1074, 1075, 1080].

- At first, the fibers captured by the bar's edges receive a strong impact directly on a small area of their surface or length. This momentary "cutting action" occurs when the leading edge of the rotor bar is in a corner position with respect to the leading edge of the stator.
- Next, the edges of the rotor and stator bars press the fibers against the bar surfaces. This causes a shear impact on the fibers as they move across the bar's surfaces. This shearing action occurs until the edge of the moving bar covers the entire surface of the opposite bar.
- At the same time, the fibers are compressed between the surfaces of the bars and the fibers slide over the surfaces, continuing the shear action.

The distribution of energy applied between the three phases above is strongly dependent on the disc configuration and will directly affect the operating results. In practice, the energy transferred to the fibers corresponds to the sum of the transfers in the three stages (edge to edge, edge to surface, and surface to surface), which justifies representing the total effective energy of each impact as the result of the intensity and impact length assessment [178, 668, 1074, 1080].

The total length of each impact depends on the width of the bars. In this way, not only the length of the bar edges but also the bar width will define the bar surface.

In the model, the amount of refining, expressed in kWh/t, is described by both the number and the energy of each impact.

The nature of refining, expressed in Ws/m^2, which characterizes the refining intensity, is described by the length of the impacts according to Eqs. 9.35 and 9.33 [220, 402, 655, 668, 669, 710, 765, 1073, 1074, 1080, 1105, 1183, 1352].

$$\mathrm{SSL} = \frac{\mathrm{SEL}}{\mathrm{BF}} \tag{9.35}$$

$$\mathrm{BF} = \left(\frac{\mathrm{B_r.B_s}}{2}\right)\left(\frac{1}{\cos\left(\frac{\mathrm{AA}}{2}\right)}\right) \tag{9.36}$$

Where:

SSL specific surface load.
SEL specific edge load.
BF impact length.
B_r bar width of rotor disc.
B_s bar width of the stator disc.
AA average bars intersecting angle.

It is natural that different types of fibers have different capacities to withstand the applied energy (surface load) [589, 668, 1075, 1164, 1263].

The distribution of energy applied between the three phases above is strongly dependent on the configuration of the discs and will directly affect the operating results of the process. In practice, the energy transferred to the fibers corresponds to the sum of the transfers in the three stages (edge to edge, edge to surface, and surface to surface), which justifies representing the total effective energy of each impact as the result of the intensity and length impact assessment [178, 668, 1074, 1080].

The total length of each impact depends on the width of the bars. In this way, not only the length of the corners of the bars but also their thickness define the surface of the bars.

In the model, the amount of refining, expressed in kWh/t, is described by the number of impacts and the energy of each impact. The nature of refining, expressed in Ws/m2, which characterizes the refining intensity, is described by the length of the impacts according to Eqs. 9.35 and 9.33 [220, 402, 655, 669, 668, 710, 765, 1073, 1074, 1080, 1105, 1183, 1352].

$$N_m = \frac{L_s}{q_m} \tag{9.37}$$

A better characterization of refining is attributed to the model by considering the geometry of the discs, especially its bar's width [178, 659, 668, 1074, 1087, 1102, 1164, 1352, 1355]. However, like the other models already discussed, it does not consider whether the fiber flocs cover the entire surface of the bars or not, nor whether the depth of the channels or the wear of the leading edges of the bars.

9.12 Modified Edge Load

One of the main criticisms of the specific load model is that it does not consider the configuration of the discs when determining the disc bar edge length, which is a fact if the measurement of this length is not made directly, that is, be considered only as the difference between the inner and outer radii of the working area of the discs multiplied by the number of bars in it. A recommended way of measuring the length of the edges of the bars is presented in Appendix II.

The bar's angle of inclination, for example, has a significant effect on the bars edge length. A large angle bar crosses more bars in the opposite disc than one with a smaller angle. Discs with a 60° angle have about three times as many crossing points as those with a 20° angle [583].

Based on the theory of the specific edge load (SEL), Meltzer and co-workers [583, 1066] presented, in 1994, some optimization considering the disc configuration (bars inclination angle, width of the bars, and width of the channels) modifying the initial model giving rise to what came to be referred in the literature as the modified specific edge load (MEL) model.

Considering L as the difference between the external and internal radii of the working area of the discs and Θ as the angle of inclination of bars, the projected length of the bar ($C = L.\tan \Theta$) defines the efficiency of the bar crossing. Together with the bar's width and channels width, a measure of the number of crossing points between bars is thus defined.

As shown in the sequence, the length of edges per second (L_s) is now complemented with (tanα) and projected as extended edge length per second (L_{sy}).

$$L_s = \left(\frac{z_r z_s L_n}{60} \right) \tag{9.38}$$

$$L_{s\gamma} = \left(L_{s\,stator} tan\text{Ø}_{stator} \right) + \left(L_{s\,rotor} tan\text{Ø}_{rotor} \right)$$

or, in the case of the same configuration for both rotor and stator discs,

$$L_{s\gamma} = L_s 2tan\text{Ø} \tag{9.39}$$

Where:

Ls bar edge length per second.
z_r number of bars in the rotor disc,
z_s number of bars in the stator disc,
Ø inclination bar angle of discs.
L_n distance from the internal to the external radii of the refining zone.
$L_{s\gamma}$ extended bar edge length.

The “new” disc edges length takes into account the bar inclination allowing what was considered as an exact calculation of the real refining intensity. A new expression is then derived for specific edge load (Eq. 9.53), which can be designated as extended specific edge load (SELγ), where the suffix (γ) indicates extension by the bar inclination angle [583, 1066].

$$\mathrm{SEL}_{\gamma} = \left(\frac{P_{\mathrm{net}}}{L_s}\right)\left(\frac{1}{2\tan\varnothing}\right) = \mathrm{SEL}\frac{1}{2\tan\varnothing} \tag{9.40}$$

After including the angle of inclination to the expression of the specific edge load (SEL), an equivalent solution was adopted [583, 1066] for the width of the channels, which transport the fiber suspension through the discs and are fundamental for the mass exchanges between rotor and stator, that is, for fiber treatment [583].

The influence of the bar width (*B*) is closely related to that of the channel width (*G*) [583, 710, 1066]. For this reason, the quotient [*B*/(*B* + *G*)] was defined considering the effects of the surface of the bars and the probability of the fibers being exposed to the refining zone. The length of the extended edge per second ($L_{s\gamma}$) is supplemented with the term [*B*/(*B* + *G*)] and designated as the modified length of bars per second ($L_{s\gamma}$BG).

$$L_{s\gamma BG} = L_{s\,\mathrm{stator}}\tan\varnothing_s\frac{B_s}{(B_s+G_s)} + L_{s\,\mathrm{rotor}}\tan\varnothing_r\frac{B_r}{(B_r+G_r)} \tag{9.41}$$

Or, in the case of the same configuration for rotor and stator discs,

$$L_{s\gamma BG} = L_s 2\tan\varnothing\frac{B}{(B+G)} \tag{9.42}$$

Thus, we have a new refining intensity value that considers both the width and the angle of inclination of the bars. This new model is known as modified edge load (MEL) (Eq. 9.43).

$$\mathrm{MEL} = \mathrm{SEL}\left(\frac{1}{2\tan\theta}\right)\left(\frac{B+G}{B}\right) \tag{9.43}$$

The model would make it possible to predict the refining result as a function of the refining degree and compare refiners of different sizes and configurations, allowing a more precise control of the process. Furthermore, as all parameters are known, it can be easily applied by operators, and the knowledge accumulated so far using the specific edge load (SEL) model can be fully exploited, keeping the usual units (J/m or Ws/m) [583, 1066, 1078, 1356].

9.13 Forces in Refining

The specific edge load, the specific surface load, and the modified specific load refining models do not consider the fibers characteristics and are based on the average power dissipated by the refining elements.

The fiber treatment frequency, effective energy per crossover points, and C-factor models treat refining as a fatigue process characterizing the refining action by the average number of strain cycles (impacts) applied to the fibers and by the energy dissipated in each deformation cycle. However, this average energy per strain cycle does not provide much information about the mechanical stresses that the fibers undergo, depending on their orientation and location within the floc [1021].

The C-factor model provides the average number of impacts (strain cycles) (Nm) and a specific intensity of impacts (Ie). The latter is expressed as energy per impact per fiber, which represents the energy dissipated in an impact cycle.

Since energy is the product of force and distance, the impact-specific intensity provides a basis for estimating force in the distance over which it is applied. The length of bars intercrosses on which the forces are applied determines the energy consumption of refining [650,798, 1032, 1036, 1082–1085, 1227, 1403].

As the refining action consists of repeated application of forces on fibers, a logical way to characterize the refining action is to divide the specific energy into two distinct factors: the number of impacts and the specific energy of each impact [650, 867, 1035, 1036, 1073, 1077, 1085, 1127, 1155, 1357, 1403].

The specific edge load (SEL) can be factored into the product of two variables that govern the energy expended by a moving body. This force is the force (F) multiplied by the distance of motion over which the force is exerted (s). Thus, SEL = F.s [1082, 1358, 1403]. In essence, this factorization is analogous to the refining specific energy factorization illustrated in Fig. 9.2.

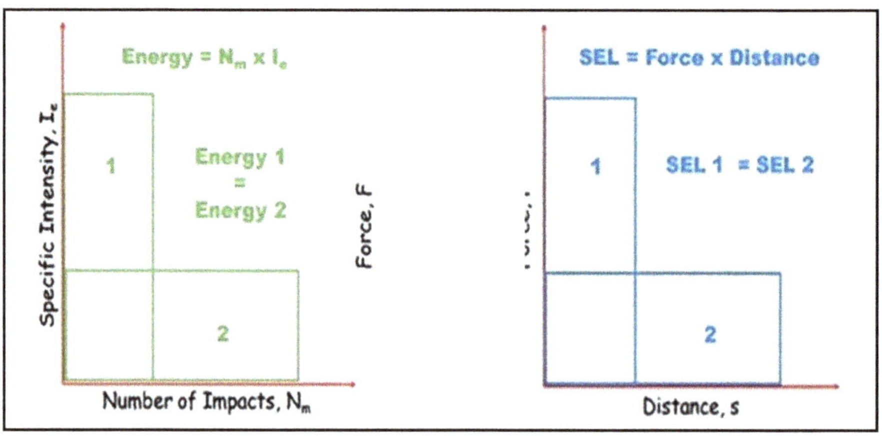

Fig. 9.2 Specific edge load (*SEL*) as the product of bar force and sliding distance over which force is applied, analogous to energy factored into number and specific intensity of impacts. (Adapted from Refs. [589, 595, 1358])

In both, beaters and refiners, the fibers are treated in the presence of water, under pressure and shear action between surfaces in constant motion. The suspended fiber flocs are subjected to forces with intensity and direction that vary continuously with time between the crossing bars. The cyclic load must overcome the cell wall compressive strength and flexural stress to cause bond breakage with consequent cell wall delamination [1123, 1127, 1224, 1225, 1243].

Although energy is the most traditional variable to characterize the refining action, force, not energy, is the causal variable of changes in fiber properties. In this way, the forces are the link between the bar disc and the fibers during refining.

The most recent models of refining are focused on refining characterization using forces. Energy is just a consequence of how forces are applied [1403]. To understand these "theories," it is necessary to know little better how forces are applied and how the pulp in suspension is distributed in the internal spaces of the refiner.

The distance between the discs (gap) controls the applied force [876, 1033, 1087, 1159, 1248, 1249]. This gap is on the order of 0.1 to 0.2 mm, which is as large as the thickness of several individual fibers evidencing that the fiber flocs are trapped between the stator and rotor bars during refining. Power-based intensities, known as machine intensities, do not account for this spacing between discs.

There is a critical distance, which is the one, from the discs approaching, in which the refining action begins. It is the distance at which the net refining power begins to be applied to the fibers initiating the flocs deformation between the bars. Energy is work, that is, the product of power and time or force and distance (Fig. 9.2). Thus, the mechanical refining energy applied on fibers is a product of an applied force multiplied by its active distance [650, 798, 1032, 1036, 1082–1085, 1227].

The magnitude of the applied forces is a function of the disc configuration and refiner rotational speed, the net energy applied (distance between the discs), the distance to the center of the discs (due to the difference in linear speed, the forces at the discs periphery are greater than those close to the inner radius), and the impacts intensity, while the extent of the fiber modifications depends on the finer characteristics, that is, on how the fiber reacts to the applied forces [650, 859, 866, 1033, 1035, 1036, 1084, 1085].

Refining begins with the fiber dehydration in the moment when the flocs are captured by the bar's edges. The flocs are compressed at the bar edges and strongly sheared between the bar surfaces, in a combination of compressive and shearing forces. As the applied pressure exceeds the structural elastic limit of the cell walls a physical change occurs in the fibers as the bars pass one another, with the fiber flocs pressed between adjacent bars surfaces and the cell wall, of at least part of the fibers, is permanently modified. Finally, the release of mechanical pressure allows the reabsorption of water by the fibers and the ruptured fibers, after which they are dispersed back into the suspension flowing through the refiner [456, 436, 452, 454, 546, 614, 650, 687, 688, 772, 816, 839, 1016, 1017, 1032, 1034, 1036, 1083–1087, 1089, 1091, 1095].

Lundin et al [1117] characterized the refining action on fibers considering the maximum force applied to the fibers (F_{max}), or tension (S), and the number of times this maximum force is applied. These two parameters were related to specific refining energy (SEC) and refining intensity (SEL), as shown below:

$$N = fg_0 \frac{l}{d^2} \frac{\text{SEC}}{\text{SEL}} \tag{9.44}$$

$$F_{\text{max}} = \frac{2.d.\text{SEL}}{l} \tag{9.45}$$

Where:

N number of impacts per fiber.
F_{max} maximum impact shear force.
f fraction of the bar that traps fibers,
g_o gap (disc distance) equivalent to initial net refining energy input,
l fiber length,
d fiber width,
w fiber coarseness,
SEC net specific energy.
SEL specific edge load.

Three distinct forces are imposed by the refining bars on the fiber flocs: a *corner force* (F_c) exerted on the bars edges; a *normal force* (F_n) due to the floc compression between the bars; and a shear *(or friction) force* (F_s) produced by the movement of bar surfaces over the flocs. Depending on the position and duration of the applied forces, they result in various irreversible physical deformations in the fiber structure (Fig. 9.3). [299, 315, 456, 436, 511, 577, 589, 650, 668, 710, 866, 1015–1018, 1020–1036, 1072, 1076, 1227]. . The fiber mat between discs forms a support

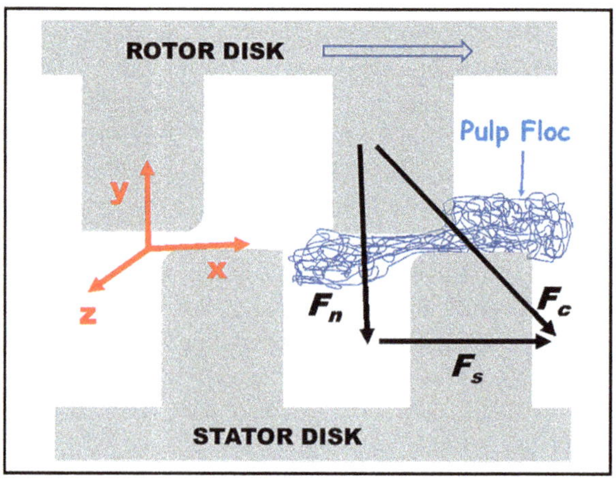

Fig. 9.3 Forces on pulp flocs during refining

blanket through which force is transmitted. The reaction force of the compressed fiber mat produces the torque measured as refiner energy input.

The force of the bar multiplied by the distance of movement of the bar during the application of this force represents the energy consumed during the bar crossing.

Impact energy is the work done by the forces acting on the fibers in the bar movement direction [1227].

9.13.1 Normal Force

The torque applied by the refiner motor is directly proportional to the normal force acting to push a refiner rotor against a stator. This means that, with a fixed motor speed, the motor power is proportional to the normal force [816].

The normal force holds the fibers between the surfaces of the bars being directly related to the fiber flocs compression between the stator and rotor discs. Tangential and normal forces are exerted. However, if the spacing between the discs is large enough that no refining effect occurs (limit between no load power and the beginning of net power), the normal reaction force can be written [456, 589, 595, 1077, 1087] as follows:

$$\mathrm{FR}^0 = \mathrm{AMF}^0 \varepsilon \pi \left(r_2^2 - r_1^2 \right) \tag{9.46}$$

$$\varepsilon = \frac{B_s B_r}{\left(B_s + G_s\right)\left(B_r + G_r\right)} \tag{9.47}$$

Where:

FR normal reaction force.
AMF average mechanical force.
r_2 external radius of the disc working zone,
r_1 internal radius of the disc working zone,
ε ratio of the inter-crossing area versus the global area,
B_s bar width of stator disc.
B_r bar width of rotor disc.
G_s groove width of stator disc.
G_r groove width of rotor disc.
With superscript ^{0} meaning no refining effect.

In any situation, refining or no refining, it is assumed that the hydraulic pressure is not a time-dependent variable being maintained in the spaces between the discs regardless of the distance between them. In this way, the normal reaction force (*FR*) does not vary and is not affected by the degree of pulp refining [1087].

Another way of estimating the normal force (F_n), now from the specific edge load (SEL) is presented in Eq. 9.48 [650].

$$F_n = \frac{\sqrt{\pi}\ d.C_{f.}\ I^{3/2}\ \sqrt{Gk}}{0.012T}\left(\frac{SEL}{\mu_s.s.g_f.f}\right)^{0.7} \tag{9.48}$$

Where:

F_n normal force on fibers (average forces on fibers).
d_0 fiber diameter without compression, m,
C_f pulp consistency.
l average fiber length, m,
T spacing between discs.
G channel depth.
K fraction of the channel in which fibers are captured.
SEL specific edge load.
μ_s coefficient of friction,
s distance from the course of the bar during which this force is exerted,
g_f distance of the width of the bar covered by the fibers, m,
f fraction of the length of the bar occupied by the fibers,

In addition to the force compressing the pulp floc between the bars, two other primary forces act on the fiber mat: a corner force from the bar edges and a sliding friction force on the bar surfaces, causing important changes in fiber characteristics [348, 511, 668, 866, 1026, 1033, 1087, 1227, 1240].

9.13.2 Corner Force

Corner force is generated by the edges of the bars compressing the flakes in the "*y*" direction (Fig. 9.3) Corner force causes the flocs densification and expansion in the "*x-z*" plane. The reorganization of fibers brings fibers to attach to the short side oriented in the direction of compression [1021].

The corner force generates a short duration, highly localized and highly concentrated tension, which is greater the sharper the leading edge of the bar is. Depending on the impact intensity, the fiber can suffer internal delamination, and the damage, or defect, caused in the cell wall can be propagated with the refining sequence. The tension forces applied cyclically in the longitudinal axis direction of the fibers are responsible for the average length reduction of the fibers [456, 589, 595, 1035, 1036, 1074, 1077, 1087, 1246].

Most of the refining energy is consumed at the leading edges of the bars, where the mechanical pressure on the fibers reaches high values, resulting in a blanket of compressed fibers [178, 456, 589, 595, 1029, 1035, 1036, 1074, 1077, 1087, 1121, 1126, 1246, 1361]. The consistency in the space between the discs can reach 50 to 60% as water is expelled from the cell wall structure at a high rate.

9.13.3 Shear Force (Sliding Friction Force)

The shear force corresponds to the frictional force and hydrodynamic friction between the surfaces of the bars and the fiber flocs. It originates in the tangential component of the corner force and in the friction of the fibers by the bars surface during the passage of the flocs between the bars.

These processes result in shear fatigue, causing delamination of the outermost cell wall layers, which is one of the primary refining goals, as fiber with a large surface area bonding with other fibers will form a more consolidated and stronger paper [456, 589, 595, 909, 1024, 1036, 1077, 1082].

The shear force (F_s) is related to the normal force (F_n) through the coefficient of friction (μ_s). The corner force (F_c) also depends on the normal force (F_n) through a complex function with a series of variables, one of the most important being the radius of curvature of the bar edge (ac) ("bar sharpness") [1030, 1036, 1077, 1082].

$$F_s = \mu_s . F_n \tag{9.49}$$

$$F_c = a_c \, F_n^{5/3} \tag{9.50}$$

The coefficient of friction (μ_s) is the shear force (Fs) divided by the normal force (F_n). Knowing the μ_s values a_c and I_e, it is possible to determine the F_n values and, from that, estimate F_c and F_s. Then, the refining intensity can be characterized by the forces in the fibers and, consequently, by the stresses imposed on the fibers [650, 1033, 1036, 1055, 1077, 1082, 1083, 1126, 1237].

9.13.4 Machine Intensity and Specific Intensity

The action of refining on fibers is traditionally related to the refining intensity characterizing the stress suffered by the fiber's cell walls. Each impact transferred to the suspended pulp has its own energy and, therefore, the basic treatment, that is, the refining effect, can be characterized by the impact intensity distribution [1026, 1036, 1072, 1073, 1109].

The process is not analyzed from the pulp, from the fibers that receive the treatment, but in relation to how the refiner applies these impacts. Therefore, the number of impacts and the energy per impact must be calculated in relation to a refiner reference, for example, the bar edges length and the intensity defined as, for example, specific edge load that is a well-known and used model [1109].

The specific edge load (SEL), as well as the specific surface load (SSL) and the modified specific edge load (MEL), corresponds to *machine intensities,* being applicable for both high and low consistencies. They represent the energy expenditure at the crossings of the bars without reference to how this energy is distributed to the suspended fibers [650, 1032, 1033].

In case of low-consistency refining where the refiners are completely filled with the pulp suspension, the machine intensity has a better relationship with the pulp properties evolution compared to high-consistency refiners. This occurs because fiber suspensions are more homogeneous because they are biphasic, that is, formed by two constituents (water and fibers), while in high-consistency refining, there are three-phase suspensions (water, fibers, and air or steam) generating irregular fiber distribution [1032, 1033].

The changes promoting by refining are caused by the energy received by the fibers, therefore, the energy amount transferred by impact, per unit of fiber mass. This intensity is the *specific refining intensity* (I_e) which is obtained by dividing the specific refining energy by the number of applied impacts [558, 1032, 1073, 1242, 1293]. The number of impacts (Nm) determination has already been presented in Sects. 9.7, 9.9, and 9.10. It is a fundamental parameter in fiber treatment frequency models, average number of crossings and effective energy per crossing point, and, also, in the C-factor.

In cases of low-intensity impacts, the fibers tend to fibrillate internally and externally, whereas high-energy impacts tend to reduce the average length of the fibers [511].

For industrial refiners, and using the C-Factor, the specific intensity (I_e) was determined to be about 10^{-5} to 10^{-6} J/impact/fiber. From a theoretical point of view, only 5.4 to 8.6 kJ/kg/impact is needed to modify a fiber, which can be an important reference in refining, including eucalyptus pulps [798, 1072, 1085, 1227].

In industrial practice, the specific edge load (SEL) and the specific energy per impact (I_e) have been used at values slightly higher than theoretical estimates, especially for eucalyptus pulps [1085]. This finding makes room for optimizing the refining of eucalyptus pulp with the use of lower refining intensities.

The specific intensity (I_e) is used to determine the forces applied in refining. The corner force (F_c) exists while the bar edge is in contact with the fiber and, therefore, the distance it acts is approximately the length of the fiber (l). In the case of the surface force (F_s), the force originates from the bar surface and, therefore, must act over the entire width of the bar (B). Thus, the impact specific energy, or specific intensity (I_e), must be expressed as the sum of these two forces [1032, 1033, 1077].

$$I_e = F_c I + F_s B \tag{9.66}$$

The force per bar length, obtained by a simple calculation from the machine intensity (SEL), is a useful parameter to describe the threshold force levels below which the levels of reduction in the average fiber length can be avoided [1082]. For the simplest case, where the angle of inclination of the bars is zero [650, 1362], we have

$$F_s = \frac{\text{SEL}}{s} \tag{9.51}$$

Where:

F_s shear force, in N/m.
SEL specific edge load.
s bar length on which this force is exerted,

When the width of the bars B (mm) is greater than the average fiber length l (mm), $s = l + 5$ Cf l, and in the opposite case, $s = W + 5$ Cf l, where Cf is the pulp consistency [687, 1403]. When s is the width of the bar, F is similar to the specific surface load (SSL).

The bar normal force (NB) can be related to the shear force by an "equivalent friction coefficient" (μ_E) which combines the corner force with the friction force, resulting in [650]

$$F_n = \frac{\text{SEL}}{\mu_E \, s} \tag{9.52}$$

Where:

F_n normal bar force.
SEL specific edge load.
μ_E equivalent coefficient of friction,
s distance from the course of the bar during which this force is exerted,

Forces F_B and N^B are average forces applied by bars per unit length of bar. However, the main interest is the force exerted on the pulp mass and on the individual fibers. As a vector, force is transmitted from bars to flocs and from fiber to fiber through columns of load bearing fibers at intervals during bar crossing events. The load bearing fiber can occupy only a fraction (z) of a bar length. Consequently, the force on the pulp mass per unit length of the bar can be expressed as (FF) and (NF), are: [1032, 1033].

$$F_s = \frac{\text{SEL}}{z_r s} \tag{9.53}$$

$$F_n = \frac{\text{SEL}}{\mu_E . z_f . s} \tag{9.54}$$

Where:

F_s shear force on pulp mass per unit length of bar.
F_n normal force on pulp mass per unit bar length.
SEL specific edge load.
μ_E equivalent coefficient of friction,
s distance from the course of the bar during which this force is exerted,
z_f fraction of the length of the bar occupied by the fibers,

Our interest is to determine the normal pressure (σ_N) and the shear stress (σ_S) on the fibers captured in the spaces between the bars, that is, the force per unit area. These can be determined from the previous equations using the bar width value covered by the captured fibers (*b*) as follows [1032, 1033]:

$$\sigma_N = \frac{\text{SEL}}{\mu_E.B.z_{f.}s} \tag{9.55}$$

$$\sigma_S = \frac{\text{SEL}}{B.z_f.s} \tag{9.56}$$

Where:

σ_S shear stress in the mass of pulp captured between the bars,
σ_N normal pressure in the mass of pulp captured between the bars,
SEL specific edge load.
μ_E equivalent coefficient of friction,
s distance from the course of the bar during which this force is exerted,
z_f fraction of the length of the bar occupied by the fibers,
B width of disc bars.

9.14 Refining as a Fatigue Process

From the 1990s onward, refining began to be considered as a fatigue process and began to be analyzed not only considering how energy is transmitted to the fibers but also how the fibers receive this energy – in other words, "from the fiber point of view."

Fatigue phenomena are observed in all materials subjected to cyclic loads of sufficient magnitude to introduce an irreversible structural change, that is, fatigue. In addition to the load per cycle, the number of cycles is also an important parameter in the process.

The decomposition of the refining action into two variables (intensity and quantity) corresponds to characterizing it as a fatigue process allowing studies about pulp properties development as a function of the number of compression cycles, as well as the determination of the forces applied to the fibers considering the geometry of the bars and the distance between the discs [470, 511, 589, 595,650, 680, 710, 839, 1015, 1026, 1027, 1035, 1036, 1073, 1077, 1087, 1109, 1154, 1213, 1227, 1274, 1363–1368].

The concept of refining as a fatigue process considers that faults (defects, cracks, or damage) formed at the internal interfaces of the cell wall are a result of repeated stresses that produce new surfaces that eventually propagate along the fiber surface. The process is directly linked to the energy spent by the bar movement, that is, the refiner energy [1032, 1367].

The result of these cycles of impacts is the hydration of the cell wall, causing its swelling and the solubilization of constituent material. There is a reduction in the cell wall intrinsic strength, which become more and more flexible as they are refined.

Only one or a few charge cycles of appropriate intensity are required. The breaking strength of the fibers represents the upper limit for the impact intensity. If the intensity exceeds this value, the fibers will break rather than fibrillate. The number of cycles required to reach failure depends on the levels of internal stresses.

The impact intensity, for a cell wall deformation must exceed a certain minimum value (minimum fiber intensity) to induce an irreversible structural change (fatigue, or defect, or damage) [1032, 1077, 1088, 1107, 1109, 1145, 1154, 1368].

The force applied by impact must be between those necessary to overcome the yield point (minimum impact intensity) and the breaking point (maximum impact intensity) (Fig. 9.4).

This will cause irreversible changes, making the fiber more flexible, but without causing the fibers to rupture with the consequent lamellar fines generation that impair the drainage of the refined pulp.

Fatigue causes damage to the fiber's cell wall. This variable (DEF) is defined as the relationship between the effective surface subject to stresses (S_{eff}), which can present cavities, cracks or other defects, and the total surface (CWS) considered, being (n) the surface normal [589, 595].

The defect variable varies from 0 to 1, where DEFn = 0 corresponds to a sample without defects, $DEF_n = 1$ corresponds to the material rupture in two parts, and $0 < DEF_n < 1$ corresponds to the defect state of the material. So, by definition

$$DEF_n = \frac{\left(S_{eff} - CWS\right)}{CWS} \tag{9.57}$$

In regions where defects already exist, the applied stresses become localized and concentrated, so that the effective stress (σ_{eff}) can be written as:

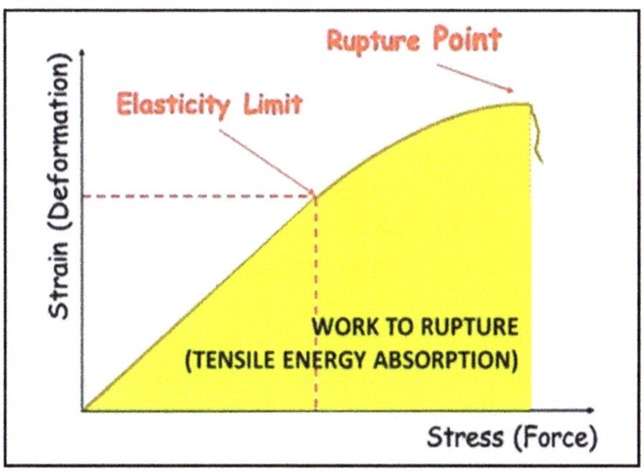

Fig. 9.4 Range for minimum and maximum intensities to avoid fiber breakage in refining

$$\sigma^{eff} = \frac{\sigma}{(1-\text{DEF})} \tag{9.58}$$

The defect variable is not directly measurable. Its quantitative evaluation is established by another variable that represents the fatigue phenomenon. One of the possible ways is to correlate fatigue with the Young's modulus of the material [589, 595] and thus the effective stress can be rewritten as

$$\sigma^{eff} = \frac{\sigma}{(1-\text{DEF})} = E.\varepsilon^{eff} \tag{9.59}$$

and therefore:

$$\sigma = E(1-\text{DEF})\varepsilon^{eff} \tag{9.60}$$

The term E(1-DEF) can be interpreted as the refiner fiber Young's effective modulus. The inverted terms result in the expression of the defect variable (DEF), that is, the reduction in Young's modulus is related to the increase in the defect variable. In the refining process, this is what happens: the fibers become more conformable (lower stiffness due to lower Young's modulus) as they are refined [589, 595].

$$\text{DEF} = 1 - \frac{\sigma}{E.\varepsilon^{eff}} = 1 - \frac{E^{eff}}{E} \tag{9.61}$$

Regarding the number of cycles, the damage variable follows a progression law. In the case of refining, a direct relationship can be established from the analysis of some paper property according to the pulp refining degree. Martinez e Park [589, 595] used the paper breaking length, which is related to the tensile strength.

Numerically, this property increases proportionally with refining until it reaches a limit after which it begins to decline, which can be understood as follows: in unrefined pulp (ideally assuming "zero defect"), the individual fibers are stiffer, or less collapsed, due to a larger Young's modulus, with a low entanglement capacity, therefore, forming a smaller number of fiber-to-fiber bonds and, due to this, a smaller elongation.

At the limit where the average defect in the fibers reaches unity ($\text{DEF}_n = 1$) and where cell wall rupture and fiber length reduction occur, there is a numerical reversion on the paper breaking strength (tensile index) due to the smaller number of whole fibers in the refined pulp [589, 595].

Chapter 10
Refining Operating Variables

Refining is the mechanical treatment of fibers designed to confer flexibility, collapsibility, and the ability to connect with each other, with the aim of attributing strength and quality to the final paper. This is done by passing an aqueous suspension of fibers through a narrow gap between two surfaces equipped with bars and channels, one of which is fixed, called the stator, and the other moving at high speed, which is called the rotor.

In this way, energy is transmitted to the fibers, providing the necessary fibers modification to endow the paper with the desired optical, structural, and resistance properties, in addition to guaranteeing adequate behavior in the paper machine during the production, conversion, and use of the paper [300, 436, 595, 688, 816, 1040, 1101, 1192].

In paper production, refining is the only unite operation that modifies the structure and morphology of the fibers and, together with the selection of the raw material to be used, has the greatest influence on the quality of the finished product [816].

Currently, with larger paper machines reaching around 12 m width and operating at speeds of 2000 m/min, it is of the utmost importance that refining is fully known and controlled, so as not to adversely affect the paper machine productivity or exceed planned production costs [595, 816, 840].

In addition to the chemical and anatomical characteristics of the pulp, as well as the pulping method, specific energy, refining intensity (SEL), and disc configuration (inclination angle and width of the bars, shape of the bars edges, and depth and width of the channels) are traditionally reported as important variables in the process. Disc wear and operating conditions such as pH, consistency of the fiber suspension, and its volumetric flow must also be considered, in addition to the differential pressure in the refiner and the rotation speed of the discs [26, 56, 67, 68, 80, 82, 83, 96, 105, 108, 113, 126, 165, 178, 209, 217, 219, 266, 281, 298, 299, 308, 341, 402, 437, 443, 454, 456, 466, 468–470, 496, 502, 511, 539, 555, 588, 618, 625, 650, 653, 659–673, 803, 836, 838, 866, 868, 1114, 1117, 1137, 1145, 1159, 1161, 1170, 1183, 1184, 1194, 1199, 1213–1216, 1227, 1253, 1295, 1307–1320].

© The Author(s), under exclusive license to Springer Nature Switzerland AG 2024

V. Manfredi, *Eucalyptus Kraft Pulp Refining*,
https://doi.org/10.1007/978-3-031-47285-5_10

These operating variables and their effects on refining results will be discussed next.

10.1 Refining Energy

The imposition of cyclic strains on the fibers is the common denominator of refining. The magnitude of each cycle can be characterized by various types and intensity parameters, and the total work done on the fibers is usually represented as “specific energy,” that is, by the amount of energy applied for each fibers mass unit.

The process has high electrical energy demand [747, 1036, 1092]. In the case of double-disc refiners, the most used today, about 80–95% of the operating costs of refining are related to the energy consumed [815, 816, 1015, 1086, 1092].

In energy terms, industrial refiners are characterized by their specific energy consumption (SEC), normally expressed in kWh/t, which is the energy applied per one ton of refined pulp. In practice, the main control variable used is the net power applied, which is mainly regulated by the spacing between the discs (gap), the flow rate, and the consistency of the fiber suspension.

In the case of old Hollander beater or laboratory beaters, such as the Jokro or Valley ones, the amount of treatment is determined by the grinding time, while in the PFI laboratory mills the determination is made by the number of rotor revolutions which, indirectly, also reflects a measure of time.

In industrial refiners, the refining load, or power, is applied as the discs approach. The friction between the fibers increases markedly with the increase of pressure on the pulp caused by the distance reduction between the discs (gap). Mechanical energy is transmitted from the main shaft to the refining zone and then completely converted into fiber modification energy and heat [877, 1089]. Thus, not all the energy applied in refining is transferred to the fibers during the short time in which the leading edges of the bars cross each other [668, 1089, 1140, 1159, 1317, 1371].

The power transferred to the fibers is called “net power,” or useful power, while the power dissipated by electrical and magnetic losses in the motor; mechanical losses in the main motor and refiner bearings and hydrodynamic losses, among others, are called “no-load power” or “empty power.”

10.1.1 No-Load Power

The empty power of a refiner is required to run, with the pulp suspended in the working consistency and discs apart, without any treatment taking place on the fibers. It corresponds to the energy needed to overcome the hydraulic, the pumping, and mechanical losses of the refiner’s operation having no refining effect on the fibers.

Based on different disc models, manufacturing, or processing precision, the no-load power of a disc refiner operating at low consistency ranges from about 15–35%, even 50% in some specific cases, as those of laboratory refiners, of the total refiner power [67, 866, 1027, 1063, 1067, 1084, 1087, 1097, 1102, 1111, 1170, 1117, 1199, 1213, 1216, 1227, 1314–1320].

The no-load power is consumed with electrical and magnetic losses in the motor; mechanical losses in the main motor and refiner bearings; and hydrodynamic losses for water and suspended fibers (rupture of fiber networks). The no-load power of the refiner is a function of the mass flow, the consistency of the suspension, the rotation speed, the configuration of the discs (width of the bars, width and depth of the channels, and angle of bars inclination), and the level of wear of them. That is, it is a function of the turbulence caused by the pumping characteristics and the motor used, being composed of friction forces, pumping forces, and losses due to the efficiency of the motor [67, 178, 219, 354, 511, 630, 886, 988, 1063, 1067, 1096, 1097, 1111, 1187, 1235, 1236, 1321]. Hydraulic power, depending on the flow rate, is the dominant component and accounts for more than 90% of the no-load power [1063].

The total refining energy consumption is closely related to the distance or distance between the discs (gap). From the maximum distance between the discs and an approximation of them to the level of refining the fiber suspension, it is observed that the energy consumption curve can be divided into three sections, constant power section (without the application of power over fibers and the no-load power), power reduction section, and power increase section, which represents the segment of the power curve in which power is being applied to the fibers [1063, 1317].

The beginning of the capture of pulp flocs by the bars, that is, the point at which the refiner begins to consume refining power, is closely related to the characteristics of the fibers, in particular length and coarseness, as these control fiber flocculation. The size of the flocs influences the distance between the discs as well as the ability to pulp flocs capture. In low consistency refining (2.0–5.5%), the effect of pulp flocculation tendency is greater than the effect of consistency variation.

In this way, different pulps, as they have different flocculation tendencies, will have different distances between the discs, reflecting different empty power for the same refining operating conditions [178]. However, not only the type of pulp, but also its level of refining, influences the value of the refiner's empty power.

This is because the flocs are reduced in size during refining, and smaller distances between the discs are required in relation to unrefined pulp. This reduction is bigger for long fiber pulps and can reach 5–10% of the initial value. From a practical point of view, this means that in series operation, a constant total power will in all refiners means a slight increase in the net power applied from the first to the last refiners arranged in series [459, 839, 1156].

The no-load power is strongly influenced by both disc diameter and refiner rotational speed. As the peripheral speed of the disc is increased by the rotational speed or by the diameter of the disc, the level of suspension turbulence in the refiner is increased and the no-load power is also significantly increased [67, 219, 365, 1017, 1039, 1063, 1067, 1096, 1111, 1156, 1187, 1317]. For disc refiners, the no-load power varies with the cube of the rotational speed and with the fifth power of the disc diameter [219, 401, 511, 880, 886, 1015, 1017, 1063–1067]. The relationship between the empty power of the refiner and its size and rotation speed is presented by the equation below.

$$P_{\text{nol}} = k.(rpm)^3.(DD)^5 \tag{10.1}$$

where:

P_{nol} no-load power (Hp or kW)
rpm rotational speed, *rpm*
DD disc diameter
k proportionality constant

The "useful power" defined as the difference between the total applied power and the no-load power is responsible for the changes in the fiber structure during refining. Thus, to accurately characterize the refining process,

To improve the control and efficiency of the process and also the quality of the results obtained, it is necessary to accurately determine the powers involved. Full power is easy to obtain as it is indicated on the controllers on the refiner's control panel. The no-load power must be measured, and the useful power is calculated from these.

Particularly in low-intensity refining, where the proportion of no-load power is high, the application of an accurate procedure for estimating it is particularly important. Its rigorous and periodic determination is necessary to obtain the true net power and, in addition, to compare low intensity refiners [736, 936, 1156, 1253, 1315].

Consistent methods for this measurement are essential for the correct determination of the refining intensity. Although there are citations in the literature of no-load power measurements with refiners operating only with water [156, 354, 1150], this determination must be made with the fiber suspension flowing, at the same working consistency, through the refiner; this is the correct and most usual practice [178, 630, 1063, 1224, 1317, 1369]. In fact, there is a special gap where the total power reaches a minimum value, that is, no refining action takes place on the fibers. This minimum value corresponds to the refiner's no-load power.

Under industrial conditions, no-load power is normally estimated only at the commissioning of the refiner installation or occasionally thereafter, depending on the interests of the plant [1253]. This can lead to less accurate estimates of effective power and net specific energy if the no-load power has changed and is not corrected in the control algorithm. This may be a reason for the low comparability of refining results from different refiners.

In practice, no-load power is generally considered to be constant, and refiner manufacturers often present no-load (kW) ratings fixing the refiner and its operating speed [1097]. However, if measured regularly, the refiner's no-load power can act as an indicator of the wear condition of the discs and equipment's moving parts. High no-load power is indicative of wear and should be used as a "yellow light" for maintenance personnel [67, 1133, 1253].

10.1.2 Useful Power (Net Power)

The "useful power," or "net refining power," must be used to link the action of refining to its effects, that is, to the structural changes in the fibers. The power effectively transferred to the fibers is calculated as the difference between the total power and the no-load power. Thus, the determination of net power must be accurate as it is an important parameter for adequate process control [178, 630, 1063, 1087, 1156, 1168, 1224, 1317, 1369, 1372].

The two most important refining parameters are refining specific energy and refining intensity (SEL). The amount of energy applied per unit of pulp mass, that is, the specific energy of refining, corresponds to the energy used in viscoelastic (reversible) and plastic (irreversible) modifications in the fibers. The refining intensity characterizes the way in which this energy is transmitted to the fibers, that is, the intensity at which the refining is conducted [178, 454, 511, 1082, 1307].

10.1.3 Specific Refining Energy

As the discs are approached, they reach a critical clearance limit at which the fiber mat begins to absorb the applied load starting the refining treatment. This critical point in the distance between the discs is strongly influenced by the flocculation tendency of the fibers and depends on the production (flow rate and consistency of the fiber suspension), the applied refining power, and the degree of fiber refining.

Associated with the suspension flow (m^3/h) and the refining consistency, which characterize the refiner's production, the specific power gives us the amount of energy used by per mass unite of pulp (kWh/t), which quantifies the action done on the fibers in relation to the number of cyclic impacts received by the suspended fibers [105, 153, 178, 219, 456, 511, 829, 884, 937, 1032, 1035, 1076, 1077, 1091, 1145, 1215, 1316, 1318].

More specific energy results in increased swelling (fiber hydration), which contributes to higher paper density and better tensile strength. However, if the increase is excessive, it will cause a reduction in the average finer length and a considerable increase in fines amount, which may even lead to a reduction in the strength of the paper [1015, 1372, 1370] and, certainly, a reduction on paper drainage rates.

The refining-specific energy, or refining time in the case of laboratory beaters, and the type of pulp to be refined have always been recognized as important process variables. Thus, we will start our approach on the operating variables of refining considering these two variables.

Figure 10.1 illustrates the results obtained in the refining of four different samples of commercial bleached eucalyptus kraft pulps (BEKP) produced by three traditional suppliers. The treatments were carried out in a pilot plant whose installation description, as well as the adopted experimental procedures, is described in Appendix III.

The results confirm that the pulp properties evolve with the increase of the specific refining energy. For each considered property, the evolution pattern is the same for the four pulps considered, differing only in the absolute values obtained (Fig. 10.1).

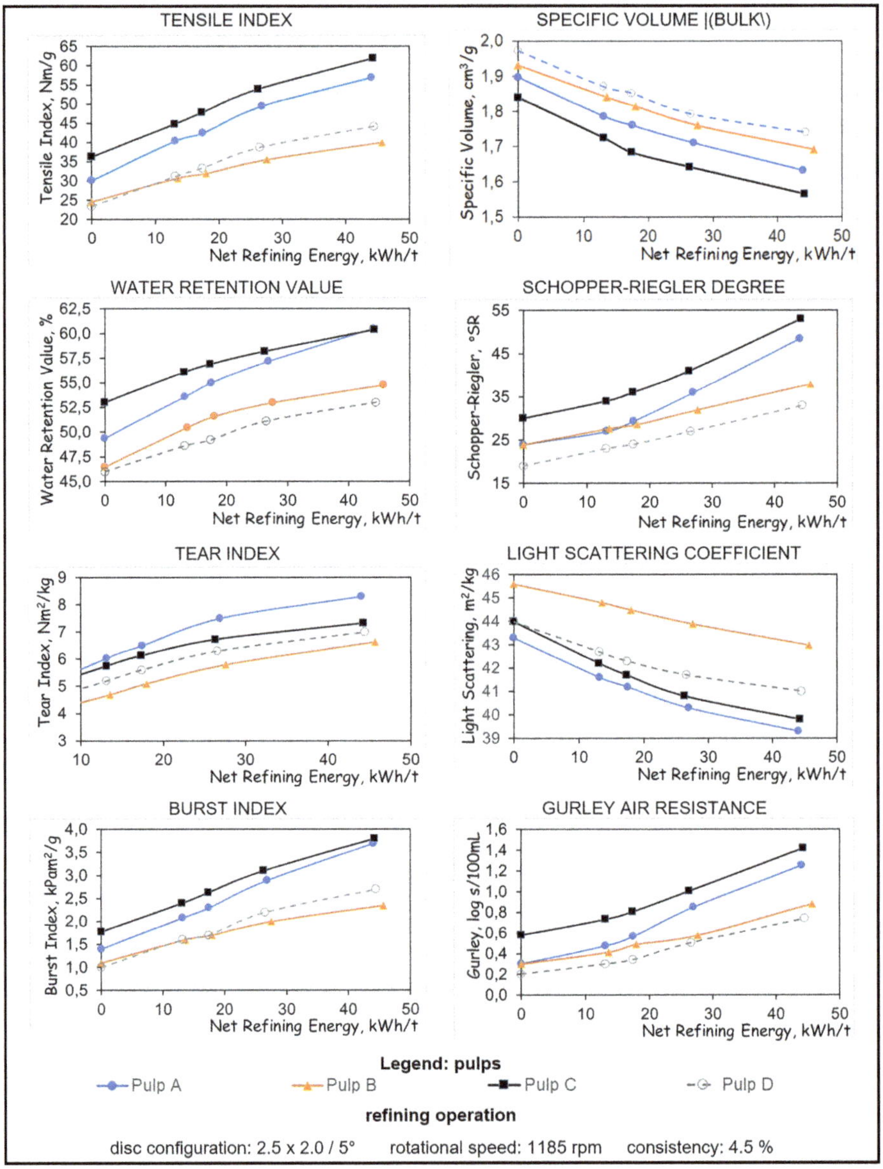

Fig. 10.1 Pulp properties evolution with refining. Effect of the pulp and specific refining energy. Comparison at the same net refining energy

It is evident, from the results of water retention index (WRV) in all samples tested, that the primary effect of refining is the pulp hydration, which makes the fibers more collapsed, with greater surface for connection between the fibers. This was also confirmed by the evolution of the specific volume and the tensile strength, respectively. In this way, we have shown the two basic effects of fiber hydration, that is, that refining compacts and consolidates the paper structure.

The four samples presented different energy demands for the development of the considered properties, indicating that it is coherent to state that for refining energy consumption, we should use pulps that already present, in their unrefined state, properties close to the desired values in the finished product.

Figures 10.2 and 10.3 illustrate the evolution of the paper properties considering the evolution of the paper structure consolidation as defined by the tensile strength (Fig. 10.2) and paper structure compaction as defined by the paper bulk, or specific volume (Fig. 10.3).

For all samples tested, the evolution of both paper compaction and consolidation gives us a linear pattern showing the synergistic effects between the two main effects of refining. This is expected because the paper structure compaction comes from the fiber cell wall collapse, which increases the fiber flexibility by improving the surfaces for connections between fibers.

Linear behavior is also observed, in all tested samples, for the opacity evolution either in the interaction with the tensile strength or with the specific volume (Figs. 10.2 and 10.3). Opacity is a function of the unbound area between the fibers. Then the light scattering should decrease with increasing bonding capacity between fibers due to the greater relative binding area (RBA) and increase with the specific volume, due to the increase of the relative bonding area between the fibers (Fig. 10.3).

Fiber collapse is caused by internal and superficial delamination of the cell walls, reducing its transverse modulus of elasticity (lower cell wall intrinsic resistance). The superficial delamination also releases fines, most of which are fibrillar.

The Elmendorf tear and tensile indices are important indicators of these changes in the fibers as the tear strength is strongly influenced by the intrinsic strength of the fiber's cell wall, while the tensile strength is strongly influenced by the fibers bonding capacity.

At the beginning of refining, and for all pulp samples tested, the Elmendorf tear strength increased probably reflecting a greater bonding capacity between the fibers in relation to the simultaneous reduction of cell wall intrinsic resistance. With the refining sequence and, therefore, with a decreasing intrinsic strength, the tear strength values tend to stabilize despite the increase in the fibers bonding capacity (Fig. 10.2). Probably, if the refining curves had been enlarged with greater specific energy application, we would observe a reduction in tear rates due to greater weakening of cell walls, despite greater fibers binding capacities.

The Elmendorf tear strength evolution with the paper structure compaction (Fig. 10.1) also shows a linearity because both properties are dependent on the same basic refining action, that is, the reduction of the cell wall elasticity modulus.

The tear-tensile relationship is traditionally reported as a predictor of paper properties and, in association with the elongation-tensile relationship, is used as an

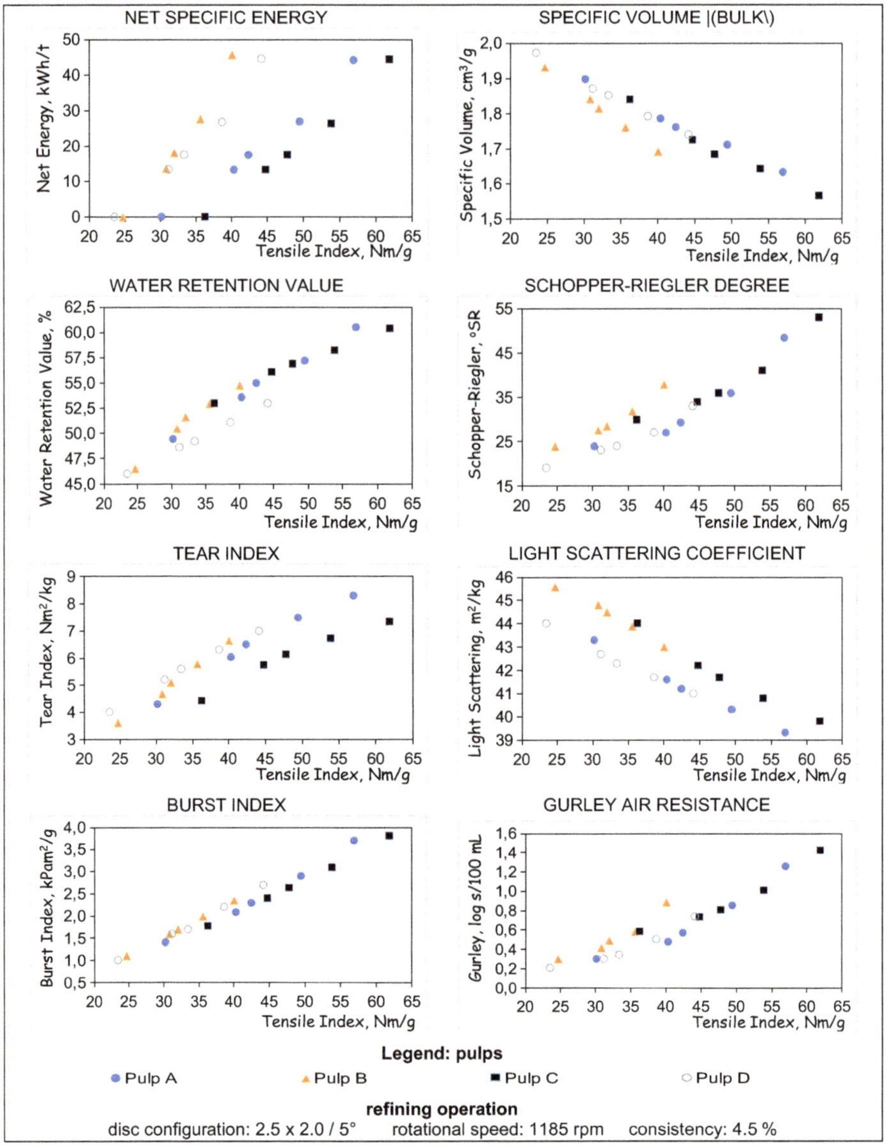

Fig. 10.2 Pulp properties relationships. Effect of the pulp and specific refining energy. Comparison at the same paper consolidation (tensile index)

indicator of the paper wet strength (paper machine runnability) [153, 154, 204, 208, 218, 249, 252, 254, 268, 270]. Nonetheless, there is no consensus about this because the Elmendorf tear resistance measures the resistance in two plans while the machine runnability occurs in one unique plan.

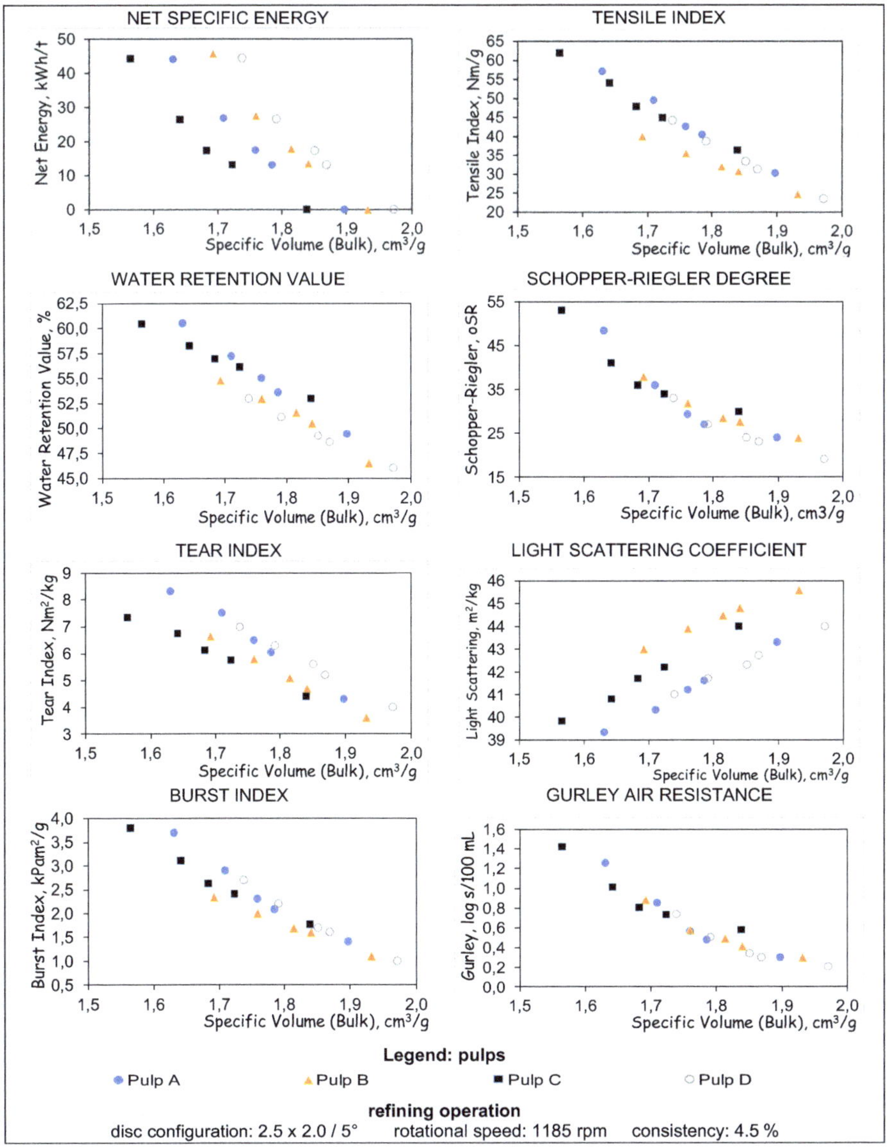

Fig. 10.3 Pulp properties relationships. Effect of the pulp and specific refining energy: Comparison at the same paper compaction (specific volume or bulk)

10.2 Pulp Type (Fibrous Material)

Although all cellulosic pulps are basically composed of the same materials (cellulose, hemicelluloses, lignin, inorganics, and extractives), different pulps have different refining responses, even when apparently similar, as seen in Fig. 10.1.

The ease with which pulps are refined and the variation in their properties during refining depend on the structure of the fiber cell wall and its chemical composition. Fibers from different wood species, or obtained by different pulping processes, respond differently to variations in refining operating conditions [26, 56, 68, 80–83, 96, 105, 108, 113, 126, 165, 209, 217, 219, 262, 266, 289, 298, 299, 308, 341, 437, 443, 466, 468, 470, 493, 496, 502, 511, 539, 555, 625, 650, 838, 868, 1308–1313].

The pulp to be refined is the main operating variable involved in the process. The pulp to be used depends on the paper to be produced. Generally, differences in fiber and quality of unrefined pulps are maintained throughout the process [82, 209, 618, 1137].

The intrinsic fiber properties that influence pulp quality are fiber length, fiber coarseness, fiber width, wet fiber flexibility, fiber wall thickness, and its collapsing degree. In summary, its flocculation tendency and floc's characteristics [42, 43, 49, 56, 58, 78, 80, 83–85, 87, 96, 105, 108, 115, 117, 132, 160–166, 343, 389, 397, 437, 467, 478, 493, 529, 531, 534, 555, 593, 618, 634–658, 660, 662, 667–669, 672, 676, 682, 683, 685, 801, 836–839, 868–877, 1338].

The stronger the fibers, the greater the need for refining energy. Long-fiber pulps require higher refining intensities than short fiber pulps. The pulping process used and the fibers drying history are important in this evaluation as they determine how the fibers will respond to deformation when subjected to the refining forces [26, 178, 454, 816, 875, 1024, 1096].

Dried pulps consume more energy to refine as they lost their swelling capacity when subjected to drying. Dry chemical pulp loses almost 30% of its water absorption capacity compared to never dry pulp, whose high conformability and fibrillation translate into lower refining energy demand, reducing refining costs [124, 454, 938, 1114].

The selection process of a pulp suppliers starts with the comparative evaluation of commercial samples carried out by most companies by means of laboratory beating followed by the formation of test sheets for the paper properties determination. The most used mills are PFI, Jokro, and Valley.

In beating, the main interest is to determine the beating time, or the number of revolutions, as an estimate of the refining energy demand of the pulp being evaluated to reach the target values of drainage resistance and paper properties of interest in the final paper. The most used tests of resistance to drainage (refining grade) are the Schopper-Riegler and the Canadian Standard Freeness, while the most important properties in the refined pulp are the tensile and tear strengths, specific volume (bulk), opacity, and porosity.

Companies specify the quality of the raw material in terms of the performance designed for their machines, as well as the quality specified in the final product. The key is to select the fiber, or fiber blends, that considers the papermaker needs as well as the requirements of the finished product [406, 618].

As each company has its peculiarities within its industrial design (chippers, digesters, washers, bleaching line, evaporation and recovery systems for liquor, refiners, paper web dryer, etc.), the bottlenecks (or restrictions) of production are the factors that determine the raw material specifications. For example, if the paper mill has low drying capacity the specification will be for pulps with excellent wet end drainage and press dewatering [406, 115].

Comparisons of pulp properties interactions in the refined pulp at a given specific volume (bulk) or tensile strength are of more interest to the papermaker than comparisons at the same drainage resistance or specific refining energy (kWh/t, beating time, or PFI revolutions) [52, 1220].

10.3 Effect of Treatment Type (Beating and Disc Refining)

"Beating" and "refining" are widely used terms and refer to different equipment used in the pulp treatment. The fibers mechanical treatment mechanism is quite the same for both cases [765, 877], occurring through structural modifications of the fibers. However, the type of equipment used significantly influences the way in which such modifications occur.

Beating is the term used for the treatment carried out in a beater, where the pulp in suspension is loaded into the equipment and discharged after the treatment, which is carried out in batches cycles. Refiners are continuous flow equipment. The pulp suspension is pumped through rotating and fixed discs separated by a very narrow spacing. In both equipment, the fibers are treated by cycles of repetitive impacts [456, 511, 1016, 1063].

Disc refiners and conical refiners, which differ in operating aspects, are the most common configurations for low consistency refiners. When operated under the same refining intensity (SEL) and consistency, the qualitative results are similar [221, 765, 871, 983, 1096, 1097, 1100, 1372–1374].

Each type of equipment has its inherent advantages and the type of pulp treatment obtained is not characteristic of the used geometry but rather the operating parameters used [387].

Due to the characteristics of the equipment, the widespread use of refiners in industrial facilities is justified, to the detriment of beaters, which have been more used in laboratory activities. In this case, the most known examples are the Valley, Jokro, and PFI mills, the latter being the most common in laboratory experiments [709, 1040, 1375, 1376].

The pulp potential to make paper is generally evaluated after treatment simulating industrial refining, followed by evaluation of key parameters that comprise the pulp adaptation potential related to machine behavior or quality of the finished product. This evaluation is, as a rule, performed in a laboratory where milling, as it is a standard method, presents results that should reflect both the effects of the fiber obtaining process and the structural changes in the fibers during the mechanical treatment [62, 178, 692].

Traditionally, it is recommended that the results of these laboratory evaluations be used, avoiding direct extrapolation to the industrial reality. The justificative is that beaters and refiners operate using different fiber treatment mechanisms, generating different levels of fiber flexibility and fines.

In beaters, compression treatment prevails which effectively generates internal flexibility and fiber straightening [63, 221, 300, 387, 709, 747, 1375–1377] and less fines generation.

10.3.1 PFI Mill

The PFI mill is the largest used in the laboratories as it uses a small amount of sample and provides a uniform treatment of the pulp, generating a lower content of fines and promoting high levels of fibers hydration without significantly shortening the fibers [161, 747, 1040, 1375, 1378, 1410].

In the PFI mill, as in the Jokro mill, the pulp is compressed between the inner wall of the chamber and the cylindrical roller with bars. The pulp is treated at 10% consistency by means of compressive shear forces. The roller with bars (rotor) and the treatment chamber, provided with a smooth inner surface, rotate around a vertical axis at different speeds (Fig. 10.4) [764, 709, 1040, 1141, 1333]. The fiber treatment level is controlled by the number of rotor revolutions. Depending on the type of pulp to be evaluated, the treatment levels vary from 1000 to 15,000 revolutions, being higher for the long fiber pulps.

The rotor has 33 bars, and each bar is 50 mm long and 5 mm wide. The bars are arranged radially, parallel to the rotor axis. The diameter of the rotor is 200 mm, measured across the bars, and the depth of the cavities between the bars is 30 mm. The rotor is driven by a motor of approximately 1 kW and the rotation frequency is $24.3 \pm 0.5\ s^{-1}$ when there is no pressure. The number of rotor revolutions is indicated by a counter. The beating chamber, with an internal diameter of 250 mm, is powered by a 400 W motor. The casing speed must be adjusted so that a difference in peripheral speed of 6.0 ± 0.2 m/s between the refining elements is obtained when under zero load and a rotor rotation frequency of $24.3 \pm 0.5\ s^{-1}$.

The refining pressure is obtained through a load applied by a lever, which presses the rotor against the wall of the casing. The refiner has a screw, with which the

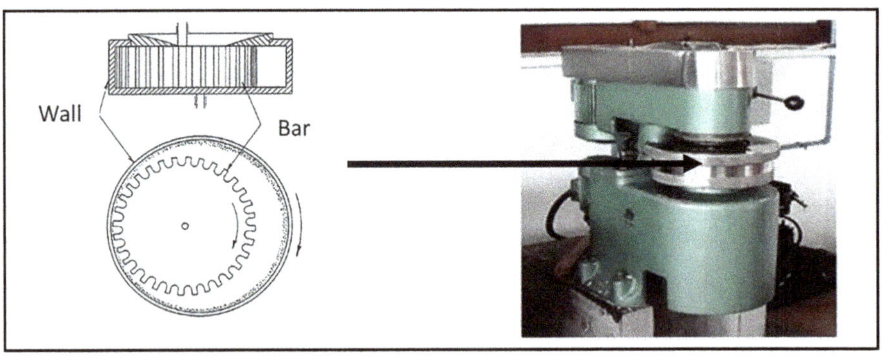

Fig. 10.4 Main element of a PFI mill. (Adapted from Ref. [1280])

clearance between the impeller and the casing is adjusted when the bars are sharpened or conditioned. The refining elements are made of stainless steel.

In the PFI mill, the specific refining intensity is approximately 1 kJ/kg/impact [1107].

10.3.2 The Jokro Mill

The Jokro mill uses the principle of centrifugal force. Like the PFI mill, it consists of a roller with several bars that rotates inside a chamber. The depth of the grooves is lower than in the PFI mill. There are six chambers with rollers that rotate around their own axes, like planets that, at the same time, orbit a central point. The pulp is compressed between the inner wall of the chamber and the cylindrical roller with bars. This principle ensures a consistent grinding load (Fig. 10.5). The milling time is the variable that quantifies the treatment applied [456].

10.3.3 Valley Mill (Hollander Beater)

This mill, which is one of the best known and used by pulp and paper laboratories, is a reproduction of the Hollander beater and, as such, consists of an open container in which the pulp suspension in low consistency (1.57%) is kept. in circulation. The pulp is treated between a beater roller equipped with bars and a base with a loading lever arm. The treatment time, depending on the type of fiber being milled and the quality required after treatment, varies from 0.5 to 1.5 h, being longer for softwood pulps [1377].

One of the most common criticisms of the Valley mill is that it does not maintain a constant condition with continuous use, therefore, not producing reproducible results. Furthermore, their results would not correspond to the results obtained in an industrial refiner, not being, therefore, able to evaluate the quality of the fibers in the same way as in an industrial installation (Fig. 10.6).

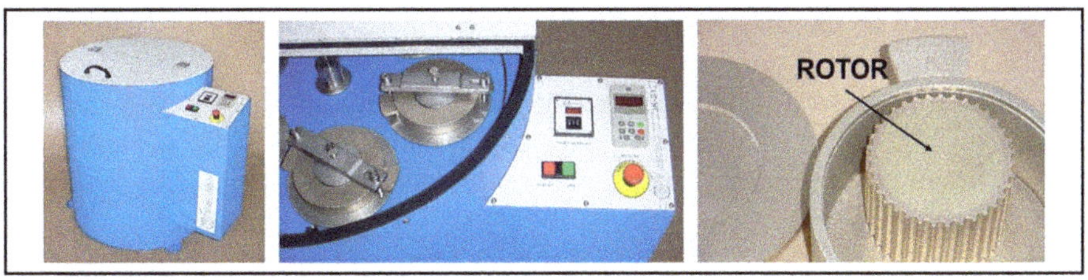

Fig. 10.5 Jokro Mill [530]

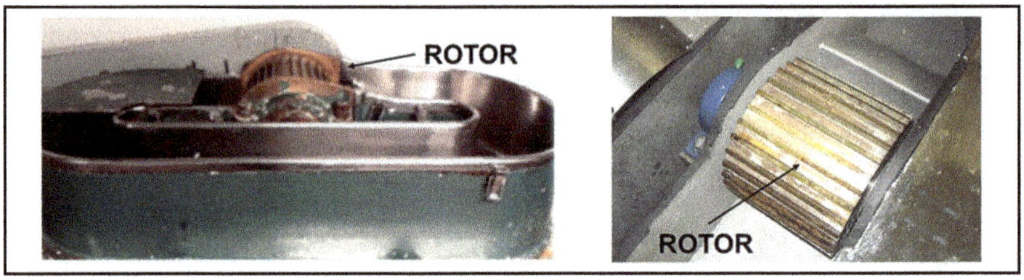

Fig. 10.6 Laboratory Valley beater. (Adapted from Refs. [530, 911])

Compared to the PFI mill, it requires more treatment time and more pulp sample. In addition, it promotes the shortening of the fibers, generating more fines, than the PFI mill [456, 691, 1150].

10.3.4 Comparing the Equipment for Pulp Treatment

As previously mentioned, it is traditionally recommended that the results of pulp evaluations in the laboratory are used with caution avoiding direct extrapolation to industrial reality. What other differences would there be between the results of mills and refiners, in addition to the way of quantifying the effect of the treatments (milling time, number of revolutions, and specific energy applied)?

Would the laboratory results be relevant for the pulp evaluation for industrial use, or to assist in the solution of operating problems or, still, in the generation of data to optimize the industrial operation?

To quantify the differences between the two treatments (beaters and disc refiner), the same commercial bleached eucalyptus kraft pulp (BEKP) was treated in Jokro, PFI, and Valley mills, according to standardized procedures for each equipment, in addition to being refined in pilot refining installation equipped with 3.0 mm × 3.0 mm/5° discs, 12 inches in diameter, operating at 1640 rpm, consistency 4.5%, and refining intensity (SEL) of 0.5 Ws/m. The treated pulps were evaluated according to the standard sheet formation procedure and tests (Appendix III).

Each equipment has a different way of treatment quantification as a function of the refining energy, which prevents the comparison of results with the same applied energy. However, the effects of each equipment can be analyzed using the fiber modifications as a reference.

Figure 10.7 illustrates the evolution of the properties considered with the consolidation of the paper structure (comparisons at the same tensile strength) and Fig. 10.8 with the compaction of the paper structure (comparison at the same specific volume).

In relation to the refiner, the PFI mill applies a greater proportion of compressive to shear forces and operates at extremely low intensities. The result is more internal fibrillation and less external fibrillation, with less fiber shortening and less fine

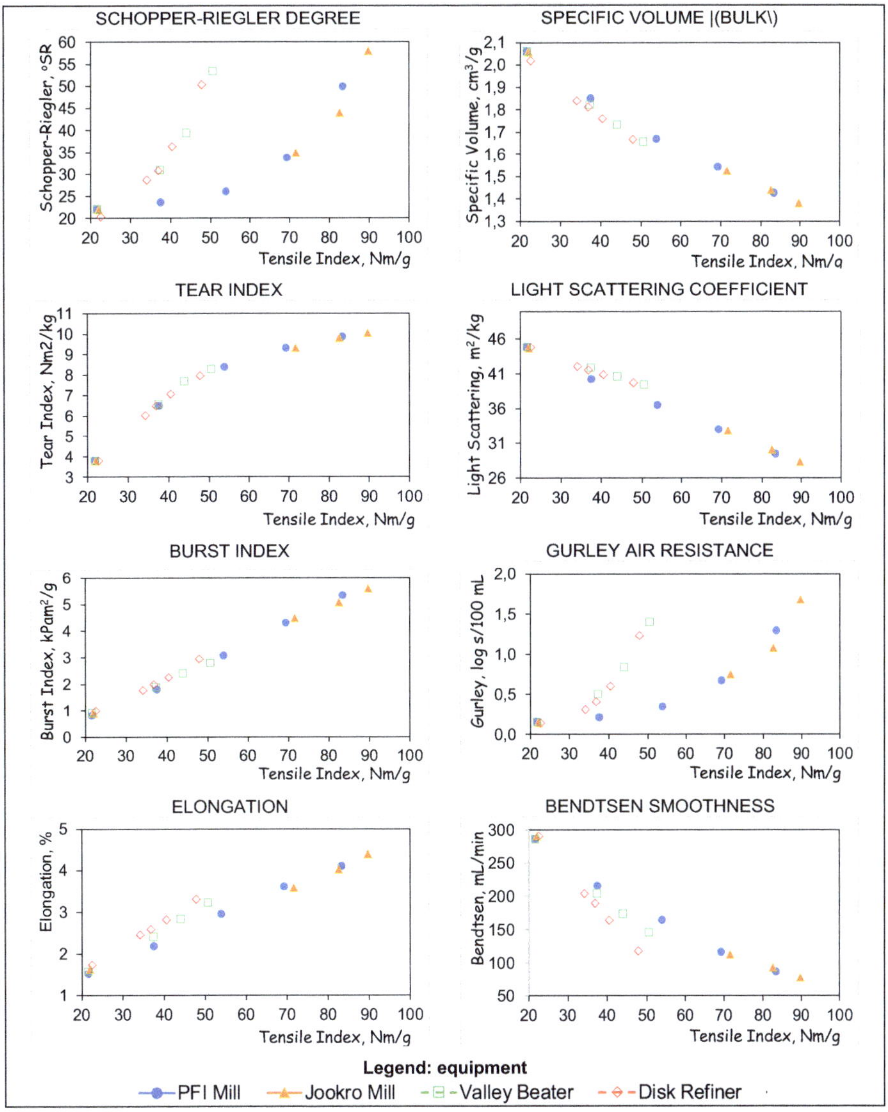

Fig. 10.7 Effect of the beating or refining device. Pulp properties relationships. Comparison at the same paper consolidation (tensile index)

formation [161, 300, 381, 764, 709, 830, 1040, 1049, 1150, 1377, 1403]. The Jokro mill has constructive and operating characteristics similar the PFI mill; therefore, the same trend of response to refining is expected.

Comparing the results from Valley mill and the refiner, the same trend of variation of pulp properties with both consolidation and compaction of the paper

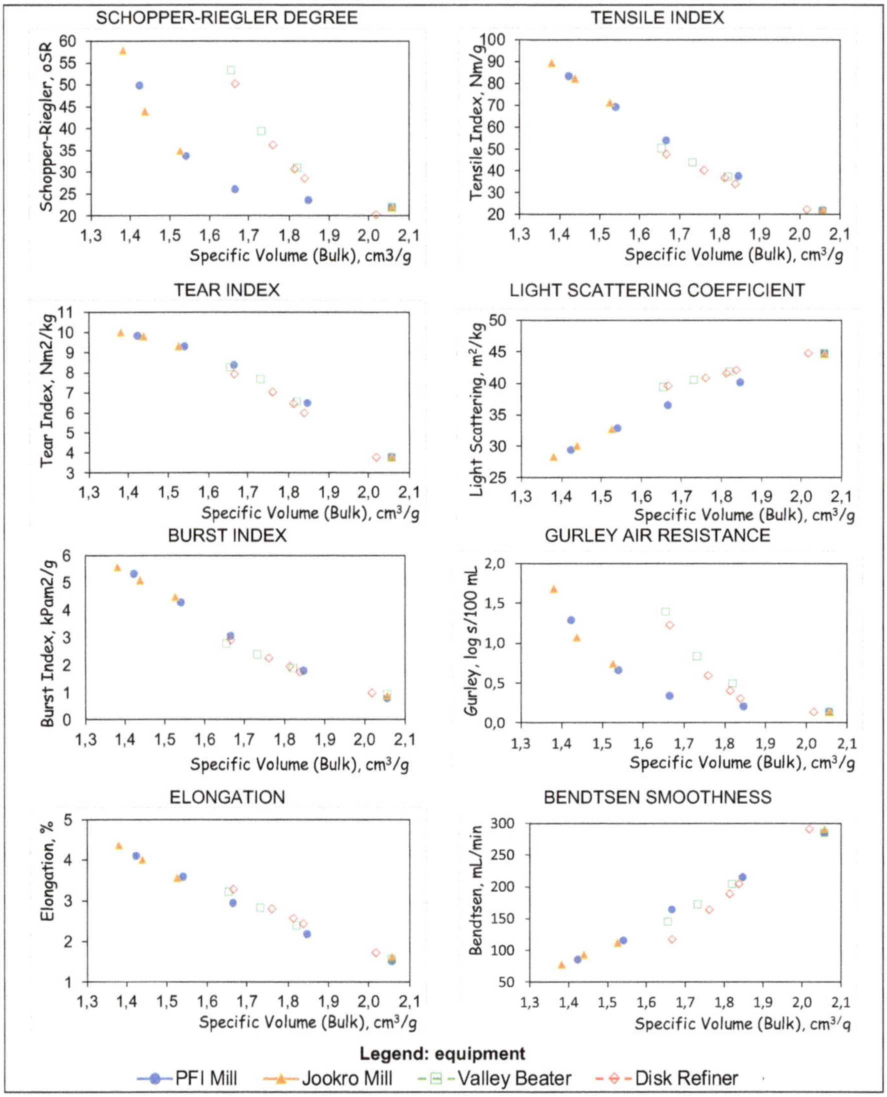

Fig. 10.8 Effect of the beating or refining device. Pulp properties relationships, comparison at the same paper (specific volume or bulk), and compaction

structure is observed. The Valley mill, similar the refiner, treats the fibers through impacts. These results confirmed information from the literature [456, 691, 1150].

When the results are compared at the same paper structure consolidation (Fig. 10.7), an important aspect is observed for the drainage resistance (Schopper-Riegler degrees) and porosity (air resistance–Gurley) evolution, with higher values being obtained in the treatments in refiner and in the Valley mill.

These two properties are dependent on the type and content of fines present in the slurry. The results indicate that, at same fiber bonding capacity, there is a greater fines generation in this two equipment.

The higher fines generation is also evidenced by the greater surface smoothness, opacity, and elongation in pulps treated by the Valley mill and disc refiner. However, for these properties the differences are not as accentuated as for the drainage degree or porosity.

These paper properties evolution tendencies indicate that lamellar fines generation, which cause the clogging of the internal channels of the paper, is favored in both Valley mill and disc refiner.

Pulp properties comparisons at the same paper structure compaction, which are illustrated by Fig. 10.8, lead to similar conclusions. Thus, we have that the equipment that promotes the treatment through repetitive impacts, in this case the Valley mill and the disc refiner, promotes greater formation of fines, which is expected because it is a direct consequence of the impacts on fibers.

On the other hand, the Jokro and PFI mills, with similar constructive characteristics, but different ways of quantifying the treatment (time and number of revolutions, respectively), promote the compressive treatment, with little or no direct impact on the fibers, favoring fibrillation of cell walls.

Knowing the differences between the equipment and how these differences impact final results is important for conducting studies to improve industrial systems, preliminary evaluation of commercial pulps, and even the formulation of theoretical models for refining.

Many studies have been developed considering the evolution of refining as the evolution of the "degree of refining" measured as a restriction to the pulp suspension drainage. Depending on the equipment used in the treatment, this approach may lead to discrepant results and, therefore, must be avoided.

The results of the equipment comparison indicate we should be careful when relating the properties of pulps treated in mills with those treated in a refiner. However, although the results obtained in mills do not directly reproduce the results of industrial refiners, it is possible to evaluate pulps in laboratory having a reasonable estimate of their results in industrial processes [1141, 1150, 1377].

10.4 Spacing Between Discs (Gap)

In a refiner, the distance between the discs corresponds to both the energy being applied by the equipment and the refining. The smaller the distance greater is the applied power and greater are the forces generated, being a direct indication of the total deformation to which the fiber flocs are submitted [160, 222, 610, 692, 773, 839, 879, 1027, 1036, 1039, 1084, 1100, 1109, 1117, 1156, 1159, 1166, 1223, 1224, 1227, 1238, 1253, 1380].

The size of the fiber flocs, depending on the pulp type, varies from 1 to 6 mm, a value that is quite high when compared to the average space between the discs, commonly in the range 0.1 to 0.2 mm [1087, 1379].

This distance is determined by several factors, such as the type of pulp, the configuration of the discs, the pulp consistency, the applied energy, and the refining intensity (SEL).

The type of pulp influences the process because fibers obtained from different woods, or by different pulping processes, have different flocculation tendencies and also different floc sizes. Fibers with greater flocculation capacity will support greater distances between discs [178, 650, 866, 1017, 1033, 1035, 1036, 1057, 1084, 1085, 1090, 1100, 1129, 1160, 1163, 1166, 1227, 1232].

A shorter fiber will wrap less around neighboring fibers thus forming a weaker network. In this case, the shear forces generated when passing the bars are likely to cause these flocs to break more easily than those of longer fibers [839, 1166].

Reducing the consistency decreases the relative thickness of the fiber mat between the discs, increasing the forces on fiber flocs. The direct effect is an increase in the fiber shortening, reducing the efficiency of the refining process [1117, 1163, 1238, 1242, 1317].

In addition to the parameters related to fiber flocculation (fiber morphology, surface chemistry, etc.), other factors, such as number of crossing bars, channel geometry, turbulence, and flow velocity, also can influence the number of flocs present in the working area of the discs influencing the distance between the discs [839].

Keeping the other operating variables constant, different pulps will support, for the same refining power applied, different distances between the discs because they present different flocculation tendencies forming flocs with different sizes.

In published articles about forces in refining, the spacing between discs is always considered as a primary parameter. There is a critical gap between the discs at which the floc will withstand the applied force without changing. At critical clearance, the flocs are assumed to be fully compressed (near zero void volume).

This critical distance is the limiting distance between the no-load power and the start of application of the net refining power. As the clearance is further reduced, due to the floc size reduction during refining, the applied power increases until, at some even smaller clearance, a maximum load is reached that causes the flocs to break. Power cannot be increased and attempts to reduce clearance only cause friction between the discs [178, 798, 839, 1027, 1033, 1100, 1227].

As the distance between the discs is related to the floc size, the critical distance (limit between the empty and useful power of the refiner) can be considered as a measure of the initial floc diameter, that is, the diameter of the flocs, compressed only to the point of being able to maintain a minimal load.

Thus, the strength of the fiber mat plays an important role in the refiner load capacity. The closer the discs are, smaller is the number of fibers between them and greater is the applied forces. These forces are a source of stress and deformation and, once the fibers are stressed, they tend to undergo delamination, disrupting the cell wall structure and fibrillating internally and externally [459, 837, 839, 879, 886, 1035, 1036, 1159, 1168, 1381].

10.5 Mass Flow Through the Refiner

The pulp suspension flow rate through the refiner is an important operating parameter because, associated with the consistency, it defines the refiner's production capacity and, therefore, the net specific refining of refining. This energy is pointed out in the literature as the most important factor defining the operating refining efficiency [459, 839, 1097, 1100, 1111, 1138, 1374].

At the same consistency and energy applied, the variation in flow reflects three important refining parameters: the production capacity, the specific energy applied, and the retention time in the refiner, which is directly related to the number of impacts received by the suspended fibers.

Situations combining low flow and high energy reduce the thickness of the mat of fibers, which can result in lower strength properties as well as lead to excessive wear of the bars [1097, 1100, 1111, 1374].

In addition to the net power available, which limits the amount of refining that can be done, a refiner's capacity is also limited by its volumetric flow capacity. This volumetric flow capacity is determined by disc diameter, by its operating speed, by the hydraulic section (ratio of the channel width to the sum of the bar and channel width), and by bar angle of the refiner's discs [401].

10.6 Refining Intensity (SEL)

The imposition of cyclic strains on the fibers is the common denominator of the refining treatment. The total work done on the fibers is usually represented as specific energy while the magnitude of each cycle, which is related to fiber deformations and paper quality improvement, is defined as the refining intensity (Eq. 10.2). The specific edge load (SEL) is a viable tool to define the refining intensity because it considers, simultaneously, the three basic refining parameters (production, energy, and intensity) [48, 73, 105, 153, 178, 219, 511, 765, 772, 829, 866, 937, 1016, 1032, 1035, 1036, 1063, 1076, 1077, 1091, 1097, 1109, 1142, 1145, 1215, 1224, 1316, 1318, 1410].

$$\mathrm{SEL} = \frac{\left(P_{tot} - P_{nol}\right)}{\left(\dfrac{\mathrm{BEL} \times rpm}{60}\right)} \tag{10.2}$$

where

SEL refining intensity, Ws/m or J/m
P_{tot} refiner total power, kW
P_{nol} refiner no-load power, kW
BEL disc bar edge length, km/rpm
rpm refiner rotational speed, rpm

The treatment intensity (SEL) is one of the primary parameters of refining, being widely used, in association with the specific refining energy (SEC) for the process characterization, including in scientific research.

It is a consensus that high intensities result in greater fines and fibers average length reduction, being recommended for the refining of long fiber pulps, while there is a predominance of cell wall fibrillation, which increases fiber flexibility, in the refining at low intensities. In lower intensities, the paper became more compacted and more consolidated and lower refining energy is required for the same final paper quality. Furthermore, at the same refining intensity, all refiners could be compared with each other [68, 73, 179, 209, 217, 219, 222, 249, 299, 393, 402, 452, 470, 493, 640, 709, 710, 747, 765, 815, 816, 836, 872, 875, 877, 1059, 1066–1119].

Long fiber pulps withstand higher intensities due to their physical characteristics that give their fibers a good flocculation tendency [816, 1085, 1100, 1102]. On the other hand, short fiber pulps, with shorter lengths and thinner cell walls, require a gentler treatment.

Identifying the refining intensity, which maximizes the results of the process, is fundamental for the definition of the refining operating conditions, aiming at both the energy consumption optimization of and the quality control of the refined pulp.

Next, experimental results referring to the evaluation of the effect of refining intensity (SEL) obtained in a pilot refining plant as described in Appendix III are presented.

Figure 10.9 illustrates the pulp properties evolution during refining using a 2.5 mm × 3.5 mm/5 ° disc configuration, operating at 1185 rpm and 4.5% consistency. The pulp used was the bleached eucalyptus kraft and the analyzed intensities ranged from 0.3 to 1.25 Ws/m.

The results obtained confirm the literature data. At the same specific refining energy, the treatments carried out under lower refining intensity allowed the best use of the energy in the development of the considered pulp and paper properties.

The net specific energy does not, in isolation, provide all refining information. Refining intensity (SEL) must also be considered. To the development of strength properties, pulps respond better to a larger number of soft impacts rather than a reduced amount of high intensity impacts. This is more evident for eucalyptus pulp due to its greater difficulty in forming coherent networks than some long and flexible fibers, which explains the excellent paper formation [68, 73, 209, 217, 219, 249, 299, 319, 393, 402, 452, 470, 493, 511, 595, 640, 667, 747, 765, 816, 837, 838, 865, 866, 868, 872, 875, 877, 1015, 1033, 1055, 1119].

Lower refining intensities lead to lower specific energy consumption to obtain the same desired paper properties, whether they depend on fiber collapsing and flexibilization that increase the contact area between fibers (Fig. 10.9).

Figures 10.10 and 10.11 illustrate the development of the properties of the refined pulp according to paper structure consolidation (tensile index) and compaction (specific volume), that is, they illustrate the effects of the refining intensity (SEL) in the quality of the refined pulp.

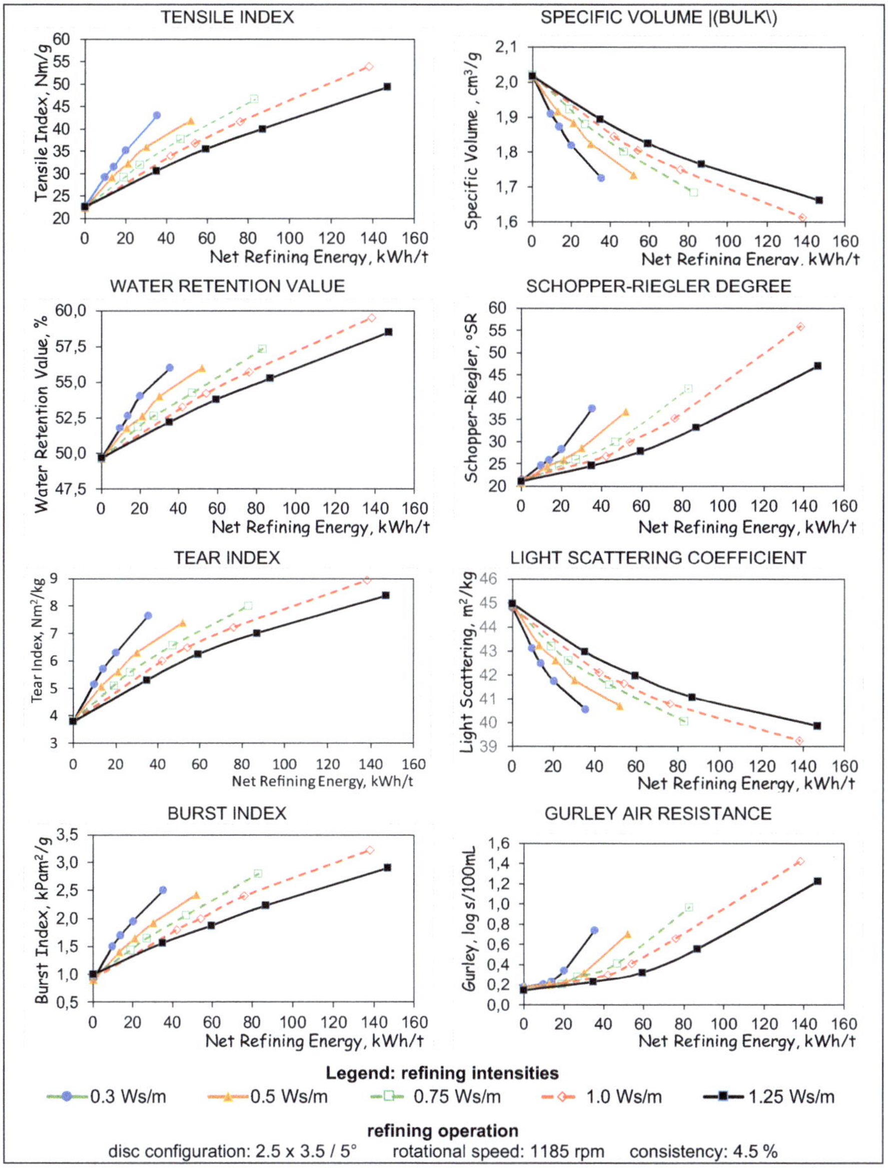

Fig. 10.9 Pulp properties evolution. Effect of the refining intensity (SEL). (Adapted from Ref. [1116])

It is observed that although the specific energy consumption varies with the refining intensity, the interaction between the properties was not affected. Therefore, the reduction of the refining intensity leads to the reduction of energy consumption, maintaining the final quality of the paper. This shows that the quality of the refined

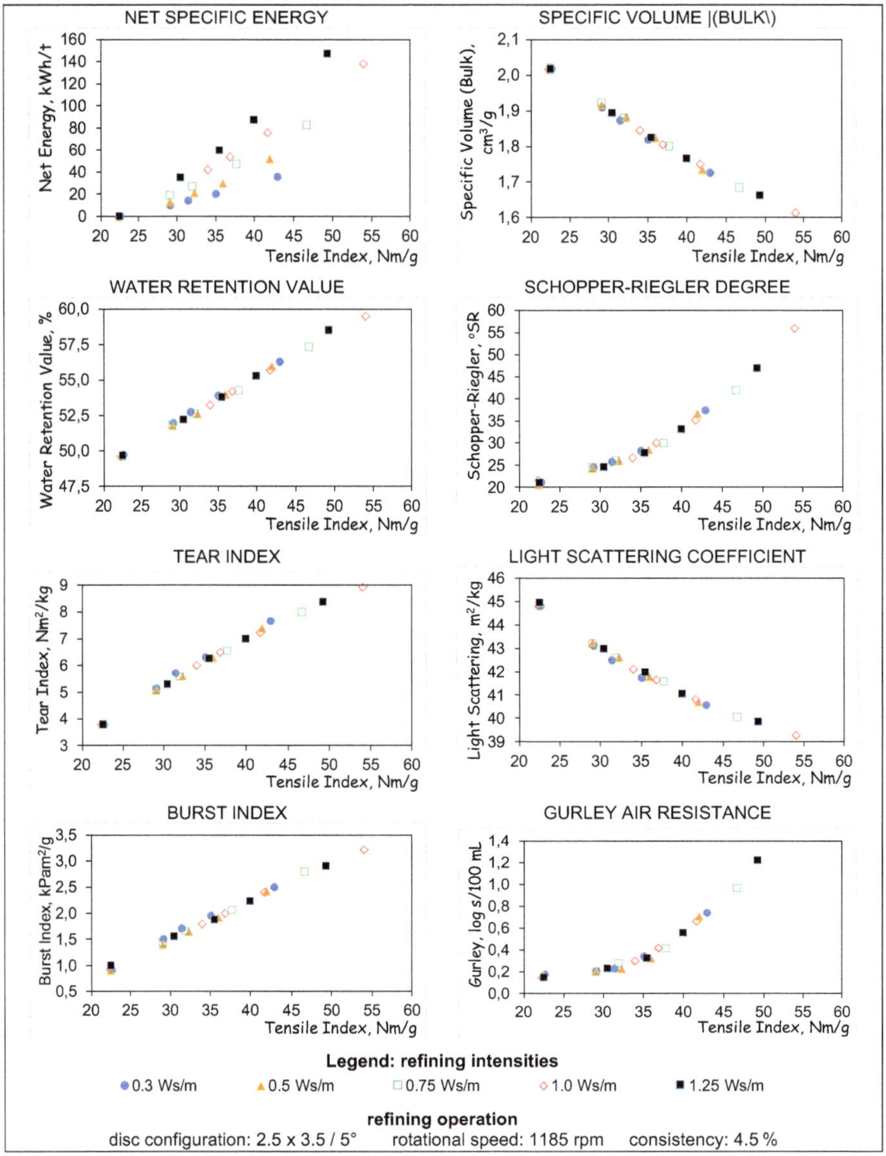

Fig. 10.10 Pulp properties relationships. Effect of the refining intensity (SEL): comparison at the same paper consolidation (tensile index). (Adapted from Ref. [1116])

pulp, that is, the interaction between the properties of the paper, depends on the pulp used.

The reduction in refining energy demand, keeping the quality of the refined pulp unchanged, seems to be a general rule for eucalyptus kraft pulp when refining intensity is reduced. This was also demonstrated by other authors [333, 815, 838, 865,

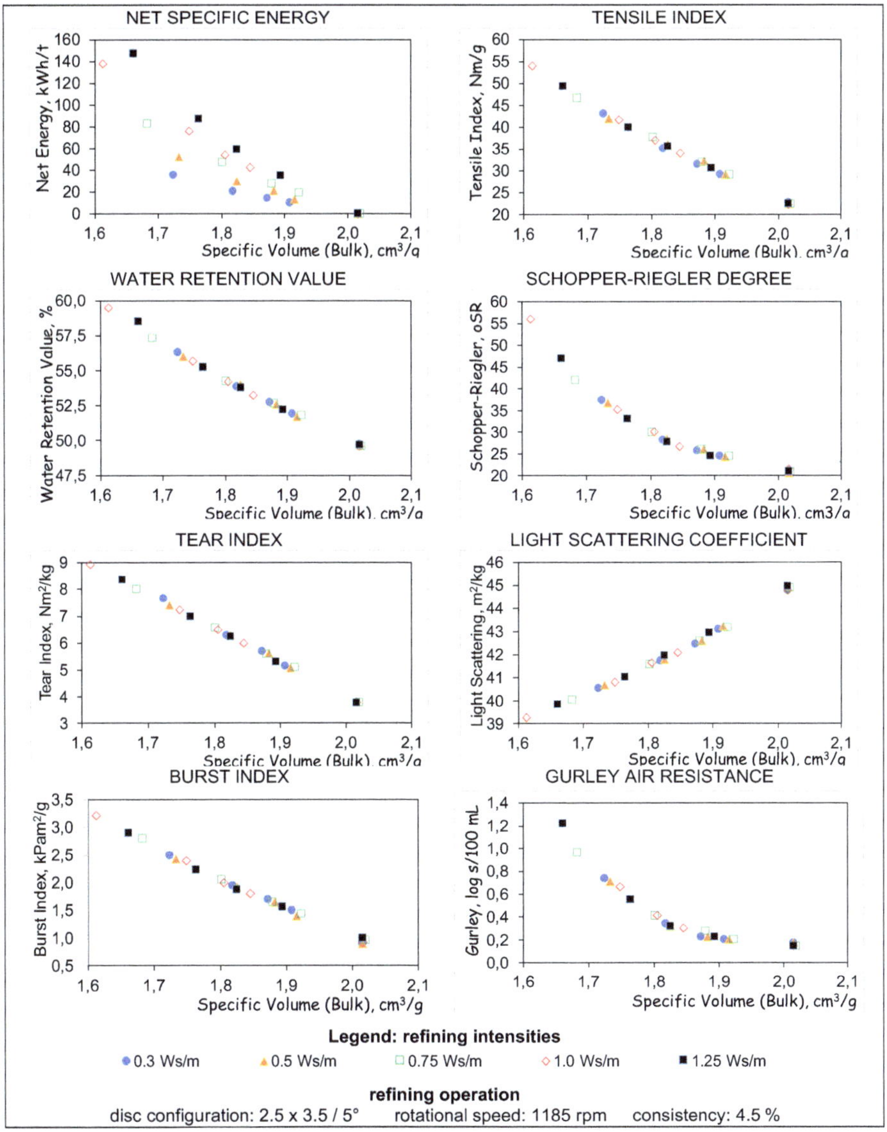

Fig. 10.11 Pulp properties relationships. Effect of the refining intensity (SEL): comparison at the same paper compaction (specific volume/bulk). (Adapted from Ref. [1116])

868, 1059, 1100, 1114, 1116, 1118, 1412] as well as for other short fibers [410, 866, 1101, 1102, 1116, 1255, 1291], for bamboo pulp [1275], for individual long fibers [244, 837, 1140, 1143, 1384], or in blends with short fibers [436, 866, 1217], plus recycled fibers from white chips [815, 816] or from OCC corrugated board [1385].

Thus, to optimize operating costs due to refining energy, we must conduct refining at the lowest intensity (SEL) as possible, regardless of the type of pulp to be refined.

To reduce the refining intensity, we must increase, as much as we can, the refiner operating bar edge length (km/s). There are three possibilities:

- Increasing the diameter of the discs. This alternative, which implies the change of refiners, is the most expensive and should only be considered if the refining facility rebuilding is already being considered.
- Increasing the refiner rotation speed. This alternative is accompanied by an increase in the refiner's no-load power, which also generates additional refining energy costs.
- Changing the discs for new ones having higher bar edge, which is the alternative most used by papermakers.

When changing the discs is the chosen alternative, increasing the bars edge length is obtained, as a rule, by increasing the number of bars in disc by reducing the width of the bars and channels. In this case, care must be taken to ensure that the width of the channels is not reduced to the point of hampering the pulp flow through the discs working area, which would increase the refiner no-load power.

Due to the strength limitations of the materials used in the discs manufacture, in the beginning the refining intensities for the short fiber pulps were in the range of 0.6–0.8 Ws/m. With the evolution of more appropriate metal alloys, the intensity can be reduced to levels of 0.2–0.3 Ws/m and, more recently, with new disc manufacturing techniques, it can be reduced to even lower values (0.1 Ws/m) indicated as the theoretical levels of intensity, which, without changing the hydraulic capacity of the refiners, can modify the short fibers beyond their elastic limit without reaching their breaking limit [393, 410, 651, 765, 816, 839, 1028, 1073, 1093, 1241, 1386].

Even lower intensities can be achieved. A disc with a configuration of 1.3 mm (bar width) × 1.6 mm (channel width) was used [410], as a means of achieving an "ultra-low" intensity of 0.05 Ws/m, in the refining of eucalyptus kraft pulp in a pilot plant. The potential gain of energy and the development of pulp quality showed that the fibers can develop their potential through the extensive treatment of the fiber wall, removing folds and undulations without fiber length reduction or high fines generation [410].

Typical results with ultra-low intensity refining discs have shown gains of 5–10% in strength properties and energy savings on the order of 10–15% with short fiber pulps [1388].

The ultra-low refining intensity (0.1 Ws/m) has been industrially tested in the production of writing and printing papers with 100% kraft pulp of eucalyptus. A 30-inch diameter refiner equipped with disc configuration 1.6 mm (bar width) × 2.0 mm (channel width) was used [410]. The results, after an evaluation period of more than one year, showed a refining specific energy reduction in the order of 50% with improvement of paper properties and also dimensional stability [410].

For short fiber pulps, the refining performance can be improved by using ultra-low refining intensity (SEL). The generally observed benefits are [110, 410, 679, 816, 1255, 1291, 1386]:

- Significant reduction in energy requirements to achieve the target desired pulp and paper properties
- Significant increase in disc life
- Homogeneous treatment of the cell wall without fiber "cuts"
- Better vessel picking resistance
- Development of the full potential of fiber properties in a similar way to grinding in the laboratory, such as in PFI mills

10.7 Disc Configuration

Structural changes in the fibers are controlled not only by the specific energy applied to the fibers, but also by the refining intensity which is influenced by the disc configuration defined by their diameter, number of bars, width of bars, bars inclination angle, and the width and depth of the channels.

The disc configuration directly affects the operating efficiency of the refiners. It is a widespread concept among papermakers that for each type of pulp, there is a more favorable disc configuration. For example, the long and thick fibers of softwood pulps require discs with wider bars and channels which, under other refining operating conditions, lead to high refining intensities. Short fiber pulps require lower refining intensities [67, 68, 125, 209, 221, 222, 614, 710, 742, 865, 870, 875, 946, 1022, 1052, 1067, 1068, 1106, 1111–1113, 1117, 1286, 1319, 1388–1391].

Refining discs are typically characterized by their bar edge length (BEL) that corresponds to the length of the bar leading edges, usually specified in km/rev. When used in a given refiner, the (BEL) is multiplied by the disc rotation speed (ω) resulting in the operating edge length (OBL), OBL = BEL·ω.

When using discs with high operating edge length, fiber hydration is greater than with discs with low operating edge length, leading to lower energy consumption [765, 244]. This occurs because, keeping the other operating variables constant, the disc with the greater operating length of edges implies less refining intensity.

10.7.1 Detailing the Disc Configuration

The configuration of a refining disc is the set of items that characterize the disc, differentiating it from the others. The most considered items are the width of the bars (B), the bars inclination angle (also known as the "cutting angle" ($\emptyset$), the width (G) and the depth (D) of the channels, as well as the bars edge length of (BEL). These configuring items are illustrated in Fig. 10.12.

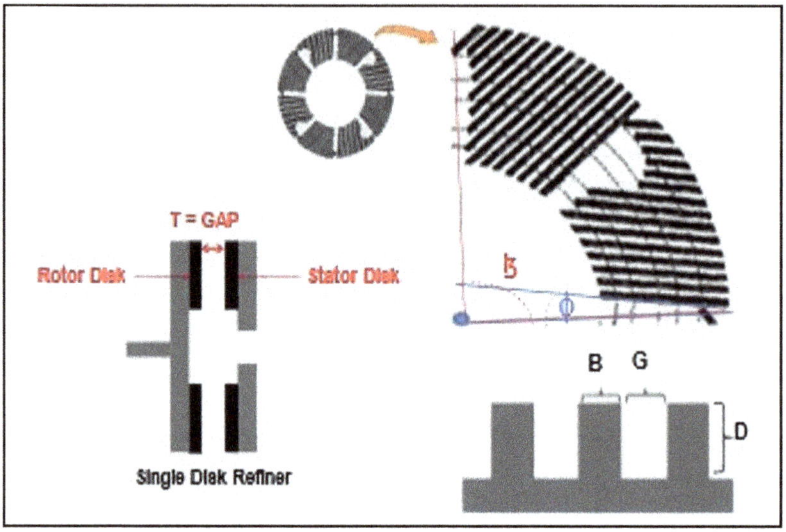

Fig. 10.12 Characteristic features of a refiner plate

The bar inclination angle (*Ø*) is the average angle of the bars with the radius of the disc, while the sectorization angle (*β*) is formed by the lines that join the outer ends of the first bars of each disc segment, being measured at the center of the refiner disc. The sectorization angle (*β*) and the segmentation of the discs in different working areas (Fig. 10.12) can also be described in the disc configuration.

As for the segmentation of the work area, there are two possibilities, those of one stage, in Fig. 10.13a, and those of several stages, as in Fig. 10.13b. Each stage is characterized by the dimensions of its disc bars and channels.

The disc configuration or its "barcode" is usually expressed as below

$$2.0\,\text{mm} \times 2.5\,\text{mm} \times 4.0\,\text{mm} / 5°, 0.98\,\text{km} / \text{revolution}$$

where the first number is the bar width, the second number is the channels' width while the third number is depth of the channels. After the bar, we have the bar inclination angle followed by the disc bar edges length. Three units of measurement are traditionally used: 1/16″, inches and mm [1015].

10.7.2 Bar Edges Length

An important practical issue is the proper measurement of bar edge length (BEL). Among the measurement methods, evidently the most accurate is the direct measurement of the bars, one by one, which is also the most laborious method [1405].

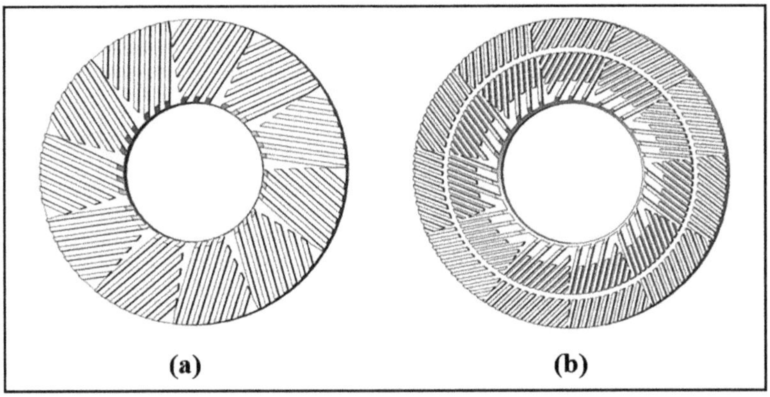

Fig. 10.13 Segmentation of the discs work area. With one stage (**a**) or with several stages (**b**) [1162]

A simple way to calculate the bar edges length is presented by Eq. 80 [1319]. Another way to measure this length is by sectoring the disc, as shown in Appendix II. There is also a specific method, widely used in the literature, formulated by TAPPI (TAPPI Standard TIP 0508-05.

$$\mathrm{BEL} = \frac{(2\pi)^2}{\cos\varnothing}\left(\frac{r_2^3 - r_1^3}{3(B+G)^3}\right) \tag{10.3}$$

where

BEL bar edge length
$\varnothing$ angle of inclination of the disc bars
r_1 internal radius of the disc working zone
r_2 external radius of the disc working zone
B bars width
G groove width

Under the same set of operational conditions, a longer bars edge length results in a lower refining intensity (SEL), which leads to an improvement in process performance [614, 1405] allowing for lower energy requirement.

10.7.3 Operating Bars Edge Length

In its simple form, the specific edge load (SEL) is easy to calculate. However, when the full complexity of bar patterns and angles are taken into account, determining a representative SEL is a complex matter [1032, 1087].

The bars edges length (BEL) depends on the disc diameter, the bars width, and channels' width, as well as the bars angle of inclination. This length, measured in km/revolution, is not the most important factor to be considered, but the operating length (OEL) is. It considers the refiner rotation speed, therefore referring to the length of bars generated per unit of time available to capture fiber flocs in suspension. The operating bar edges length is the parameter used in calculating the refining intensity [614, 816, 1015, 1100, 1392].

The average number of impacts received by the fiber will be proportional to the operating edges length along which the pulp flocs can be captured. The use of discs with greater operating length provides a reduction in refining intensity, promoting greater fiber hydration, which leads to a reduction in specific refining energy consumption, improves the quality of the refined pulp, reduces fiber shortening, increases the disc's life, and, depending on the refining area configuration, can reduce the number of refiners in operation [1291, 1384]:

10.7.4 Disc Configuration – Bars Inclination Angle

The bar inclination angle, also referred by the literature as the "cutting angle," is an important parameter in the disc configuration. It significantly influences the bar edge length, as well as the interaction between the bars and the suspended flocs. Smaller angles reduce the edges length, increasing the refining intensity, which favors the reduction of the average fiber length, while smaller angles reduce the refining intensity, accentuating the fiber fibrillation [221, 583, 614, 687, 872, 875, 1067, 1096, 1097, 1372, 1111, 1133, 1392–1394].

Discs with greater "cutting angles," in addition to having greater pumping capacity of the suspension [1097, 1111], have bars that cross, or are crossed, by a greater number of contact points. A disc with an angle of 30° has about three times more crossing points as one with 10 ° angle [875, 1066, 1133]. A disc with a bar angle of 10° will show a variation in the interaction bars angle between 15° and 40°, which will lead to an unstable pulp flow in the refining zone [614, 1117].

Aiming to standardize this bars intersection angle, also known as effective "cutting angle," discs with curved bars were developed in order to maintain the same intersection angle throughout the entire disc working area of the refiner (Fig. 10.14).

Discs with curved bars have a better pumping rate than those with straight bars and treat the fibers more gently. They allow for lower refining intensities and, as a direct result, less reduction in average fiber length, less fines formation, and more intense fibrillation, favoring paper strengths [614, 815, 1162]. However, discs with straight bars are still the most used by papermakers.

Table 10.1 presents the configuration parameters of two discs of the same characteristics and size but differing on bars shape. On the left disc we have straight bars one, while the disc on the right is a curved bars disc [639]. With the curved bars disc, a greater bar edges length is obtained, which means that for the same net refining energy applied, we will have a lower refining intensity (SEL).

Fig. 10.14 Disc refiner equipped with curves bars disc [614]

Table 10.1 Bar parameters of the discs [639]

	Straight bar				Curved bar			
Plate Segment	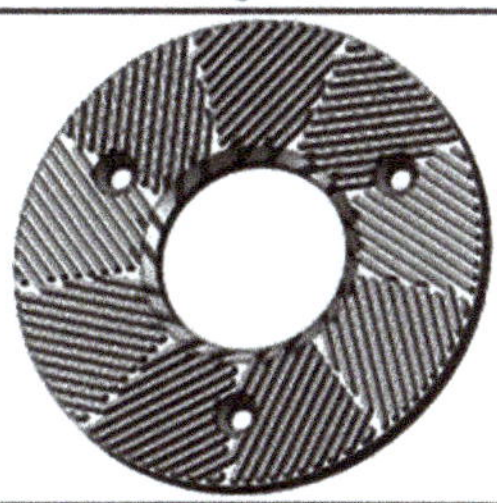				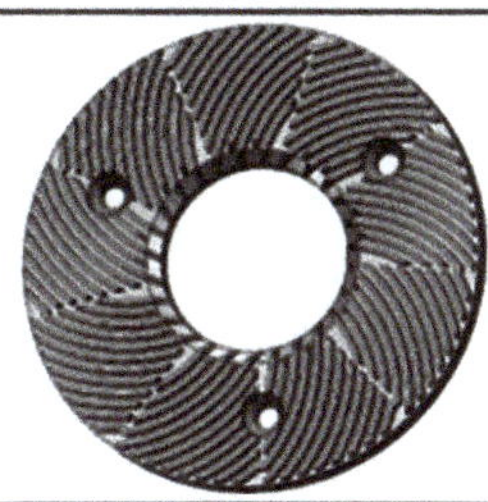			
BEL	276.55 m/rev				327.58 m/rev			
Common bar parameters	Bar width	Groove width	Bar height	α	Inner radius	Outer radius	Center angle of segment	Number of bar
	2 mm	3 mm	4 mm	42°	82.5 mm	203 mm	40°	117

10.7.5 Disc Configuration – Bars Width

The power applied to the refiner and the number of disc bars determine the type of treatment given to the fibers. The bars width defines the number of bars in the discs and, therefore, the bars edges length. Discs with a greater number of bars have a greater length, allowing, for the same refining specific energy, a greater number of smaller impacts than discs with wider bars [222, 710, 816, 1015, 1067, 1080, 1093, 1096, 1100, 1102, 1111, 1113].

However, simply adding bars to the discs reduces the channels' width negatively interfering on the suspension flow through the refiner, reducing the internal pressure differential, and impairing the capture of the flocs by the bars [710, 1130, 1166].

10.7.6 Disc Configuration – Width and Depth of Channels

The channels ensure the fiber suspension transport through the discs and are extremely important for exposing the fibers flocs to the intersecting bars, that is, for the capture of the flakes by the bars. The channels must be wide and deep so that the flocs have enough space to move [583, 663, 1051, 1066, 1084, 1087, 1117, 1129, 1151, 1156, 1165, 1167, 1170, 1233, 1356].

The short fiber pulps suspensions have a low apparent viscosity, with a lower rotation flow in the channels than long fiber pulps. The flocs size is fundamental in the development of the double flow in the discs, and in the discs with narrower channels, there is a smaller volume of free spaces. This is the reason why short fibers are refined using discs with narrower channels than those used for long fiber pulps [1165].

Thus, the channels' width is of fundamental importance for the refining action and must be selected according to the flocculation characteristics of the pulp to be used. Long fibers, which form larger flocs than short fibers, are more likely to be captured by the bar's edges than shorter fibers, which, with smaller flocs, tend to flow through the disc channels. Flocs with diameters similar to the channel's width show the fastest pulp properties development with the specific energy applied [216, 742, 836, 838, 866–868, 1127, 1238].

There is an optimal maximum channel width for each combination of fiber type and refining consistency. If most of the suspension flow is in the channels, little or no energy will be transferred to the fibers, which leads to poor quality development. If excessive reduction in channel width occurs, the pressure differential between the rotor and stator bars can be reduced in some areas of the discs, making it difficult to maintain tertiary flow [710, 1117].

The depth of the channels and the disc material are also important parameters, although they are not used in the determination of the refining intensity. Excessive channel depth results in more fibers passing straight through the refiner without being treated. Shallow channels expose more fibers to bars edges promoting the refining action. Nonetheless they restrict the pulp suspension flow and reduce the refiner's hydraulic capacity. The refiner hydrodynamic losses increase with deeper channels [63, 178, 459, 475, 511, 583, 1017, 1063, 1067, 1091, 1099, 1100, 1113, 1121, 1151, 1160, 1163, 1235, 1237].

Changes in the width and height of the bars have always been constrained by the strength and durability of the materials used in the disc production. The use of new materials and new manufacturing processes have allowed the production of discs with very thin bars and high volumetric capacity. The combination of strong narrow bars and rectangular channels allows greater hydraulic capacity than previous U-channel discs [1067, 1388].

10.7.7 Comparing Different Disc Configurations

The effect of disc configuration on the refining of commercial bleached eucalyptus kraft pulp (BEKP) was evaluated in a pilot refining installation (Appendix II) according to the methodology described by Appendix III. Seven different configurations (Table 10.2) were compared in refining curves at the same refiner rotation speed, consistency, and intensity [1116].

As the disc configuration is a variable extensively cited by the literature as a variable of great influence on the refining, the results obtained were surprising. Figure 10.15 illustrates the evolution of pulp properties with specific refining energy. It is observed that none of the properties considered, whether they are dependent on collapse or fiber bonding capacity, subject or not to the type and content of fines, presented variation that could be attributed to the used disc configuration.

When the pulp properties evolution is compared at same bonding capacity between the fibers (paper structure consolidation), the same result is observed for all configuration discs evaluated (Fig. 10.16), that is, no difference can be attributed to the disc configuration.

Figure 10.17 illustrates the pulp properties evolution with the paper structure compaction (specific volume). The same results are observed, that is, no difference caused by use of different disc configurations.

Equivalent results were also identified in other studies presented in the literature, that is, if the refining intensity (SEL) is kept constant, there will be no differences due to disc configuration, including for large "cutting angle" variation [217, 511, 838, 868, 1085, 1115, 1225, 1291]. That was not tested by the trial presented by Figs. 10.15, 10.16 and 10.17.

Thus, the effect of the disc configuration must be considered with caution as it reflects the operating length of the edges and, if the comparison occurs at the same specific refining energy (kWh/t), we will have different refining intensities, which explains different operating results. However, when discs, even with different configurations, are compared for the same pulp and refining intensity (SEL), the results will be the same.

We have the same pattern of properties evolution with the refining energy because the intensity was used, that is, discs with different configurations will give the same results when the same refining intensity is used. The similarity of the results when

Table 10.2 Characterization of the discs used in the evaluation

Design				Operational edge length, km/s		
Bar	Channel	Bar angle	Bar edge length, cm/rev	At 890 rpm	At 1185 rpm	At 1640 rpm
2	2.5	5	98,147	14.56	19.38	26.83
3	2	5	86,466	12.83	17.07	23.63
2.5	3.5	5	75,629	11.22	14.94	20.67
3	3	5	63,028	9.35	12.45	17.23
3	3	10	61,952	9.19	12.24	16.93
5	5	5	25,988	3.85	5.13	7.10

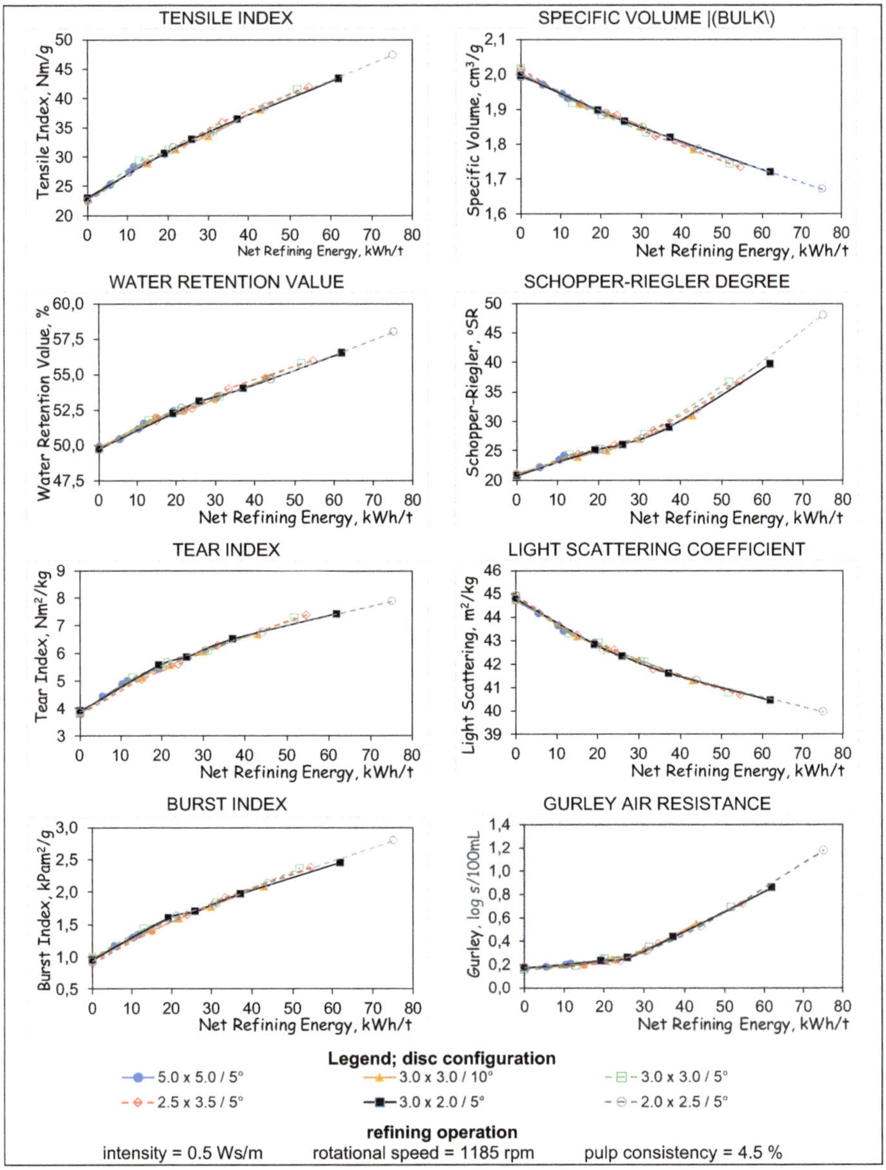

Fig. 10.15 Pulp properties evolution with the specific refining energy. Effect of disc configuration. (Adapted from Ref. [1116])

the properties are compared to the same bonding capacity or fiber collapse occurs because the same pulp was used in all refining curves.

The pulp properties relationship depends on the pulp characteristics as shown in Sect. 10.2 of this chapter.

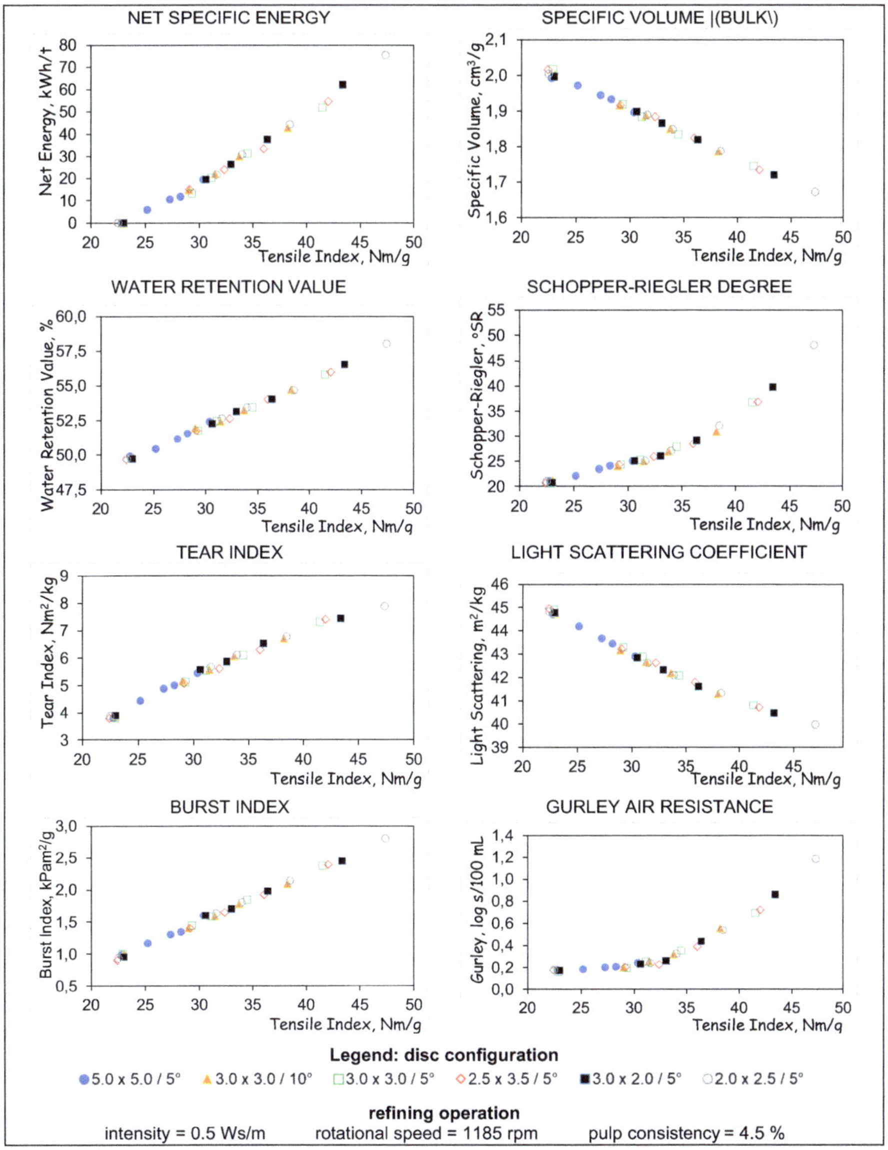

Fig. 10.16 Pulp properties relationships. Effect of disc configuration. Comparison at the same paper consolidation (tensile index). (Adapted from Ref. [1116])

These results apparently contradict industrial practice as it is known by papermakers that an efficient way to optimize refining energy consumption is changing discs to most appropriated configuration [67, 125, 209, 221, 222, 614, 710, 742, 865, 870, 875, 946, 1052, 1067, 1106, 1111–1113, 1117, 1286, 1319, 1387–1391].

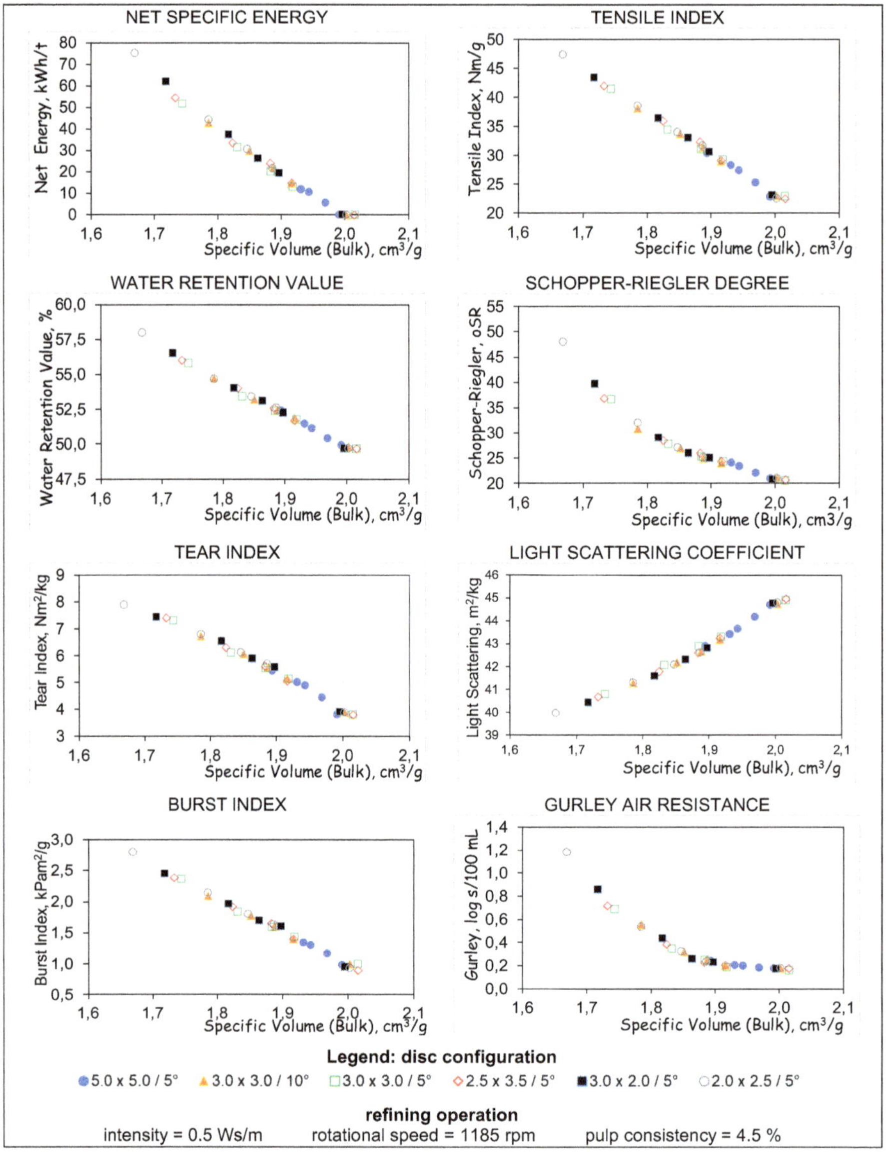

Fig. 10.17 Pulp properties relationships. Effect of disc configuration. Comparison at the same paper compaction (specific volume). (Adapted from Ref. [1116])

How to explain this difference between the obtained results and the "industrial practice"?

The answer is simple. In industrial practice, whenever discs are exchanged by discs with different bars width, for example, which is the most common practice, the new discs are operated at same way as the previous ones (net refining energy, pulp consistency and pulp flow). By dividing the same net refining applied energy by a longer edge bars length, we will have a lower refining intensity when the new discs are used.

The expected result is lower refining energy consumption to obtain the same properties in refined pulp. In the trial's results presented, all refining were run out under the same intensity although with different disc configurations. In the industrial practice, the disc changing leads a refining intensity changing too.

Would there be a more appropriate disc configuration? Here again the answer is simple. Yes!

Although the pulp properties evolutions with refining energy are the same for all disc configuration, there are differences in their production capacity (Table 10.3). The longer bar edges discs are those allowing the highest production levels.

Thus, the most suitable disc for operation, allowing you to reduce the energy requirement, maintaining the quality of the finished product, is the highest bar edges length, as it reduces refining intensity (SEL) with the highest production capacity.

Refining discs can be produced by different processes, the most common being the machined (milling) discs and the castings that gradually replaced the discs with welded bars. Two important aspects, but not considered by any of the theoretical models about refining, are the material of the discs and their operating wear.

The resistance of the materials used in the discs manufacture is a function of the amount of free chromium left in the material's structure after manufacturing and heat treatment. Larger amounts of free chromium result in corrosion resistance. Low-carbon stainless steels have the highest levels of corrosion resistance due to their high chromium/carbon ratio [67].

Often the only criterion used is the operating disc lifetime being selected the most abrasion-resistant material. Low carbon stainless steels, although less resistance to wear resistance, resist the rounding bar edges and produce a uniform pulp quality over the disc lifetime [67, 1108].

During refining, when very thin gaps (discs distance) are used, such as when high refining energy is applied, the flocs rupture may cause generating friction between

Table 10.3 Estimative of production capacity, by disc design, for a target tensile index

Disc configuration		Estimated production with the Target Tensile Index, kg/h		
Design	Operational length, km/s	30 Nm/g	35 Nm/g	40 Nm/g
5.0 × 5.0/5	5132.0	134.8	[a]	[a]
3.0 × 3.0/10°	12216.2	398.1	188.0	87.7
3.0 × 3.0/5°	12448.0	405.3	203.4	117.0
2.5 × 3.5/5°	14936.9	416.3	253.4	142.2
3 × 2.0/5°	17077.0	478.7	266.3	144.2
2.0 × 2.5/5°	19384.2	512.6	316.3	185.7

Based on Figs. 10.12 and 10.13
[a]No refiner capacity to meet the target value

discs causing damages on bars surfaces and attack bar edges [879, 1018, 1129, 1159, 1227].

This abrasive process typically causes radially uniform disregard that results in the loss of disc parallelism affecting the refiner's hydraulic behavior. The generation of secondary flows in the spaces between the discs, which are the flows responsible for the fiber flocs transference between neighboring bars and channels, also shows variations making the flocs capture process irregular. This contributes to the abrasive wear of the bars [67, 663, 1018, 1129, 1133].

A rough bar surface transfers larger shear and stress forces to the fibers that will be more easily "shortened." The corner force is affected by wear, which alters the curvature radius of the bar edges. When the attack edge shape profile has a radius less than 30–50 μm, the risk of severe fiber shortening is higher [178, 1126].

The bar attack edge wear resistance is more important than bars surface wear resistance. Cluttered edges bars reduce the efficiency of flocs capture, which becomes a less uniform process and reduces the refining efficiency due to higher power consumption to meet the target pulp properties [67, 1018, 1036, 1051, 1055, 1067, 1097, 1108, 1121, 1126–1135].

The wear level determines how the bar attack edge penetrates on the suspended flocs. In the same pulp consistency, sharper bars collect more fibers than worn edges and cause greater fiber 'shortening' [23, 178, 1011, 1253, 1313].

Refiner no-load power, if measured regularly, can act as a disagree state predictor because high no-load powers are indicative of wear bars [67, 1096, 1097, 1133].

10.8 Refiner Rotational Speed

The rotational speed determines the operating edges length and the refiner no-load power. Therefore, for the same net power, the rotational speed is related to refining intensity (SEL) [219, 365, 663, 829, 866, 1067, 1015, 1096, 1104, 1111, 1115].

Disc velocity is the determining factor in the impact frequency on suspended pulp flocs. Thus, for the same refining intensity, production capacity, and refining consistency, larger number of impacts will be transmitted to the pulp as greater the velocity is.

For the same specific net energy applied, higher rotational speed increases the refining capacity because it increases the operating edges length. However, this action has well-defined limits because rotational speed increasing also produces higher refiner no-load power, which increases in the cubic ratio of increased disc rotational speed [262, 493, 710, 829, 946, 1018, 1067, 1115, 1117, 1133, 1187].

To evaluate the effect of disc rotation speed on the reefing of bleached kraft eucalyptus pulp was done. The pilot plant installation described by the Appendix II was used, and the disc configuration was 3.0 mm × 3.0 mm/5 °, operating in 4.5% consistency. Three rotational speeds (890 rpm, 1185 rpm, and 1640 rpm) were compared. For each rotation velocity, two refining intensities (0.5 Ws/m and 1.0 Ws/m) were used [1116]. The results are illustrated by Figs. 10.18 and 10.19.

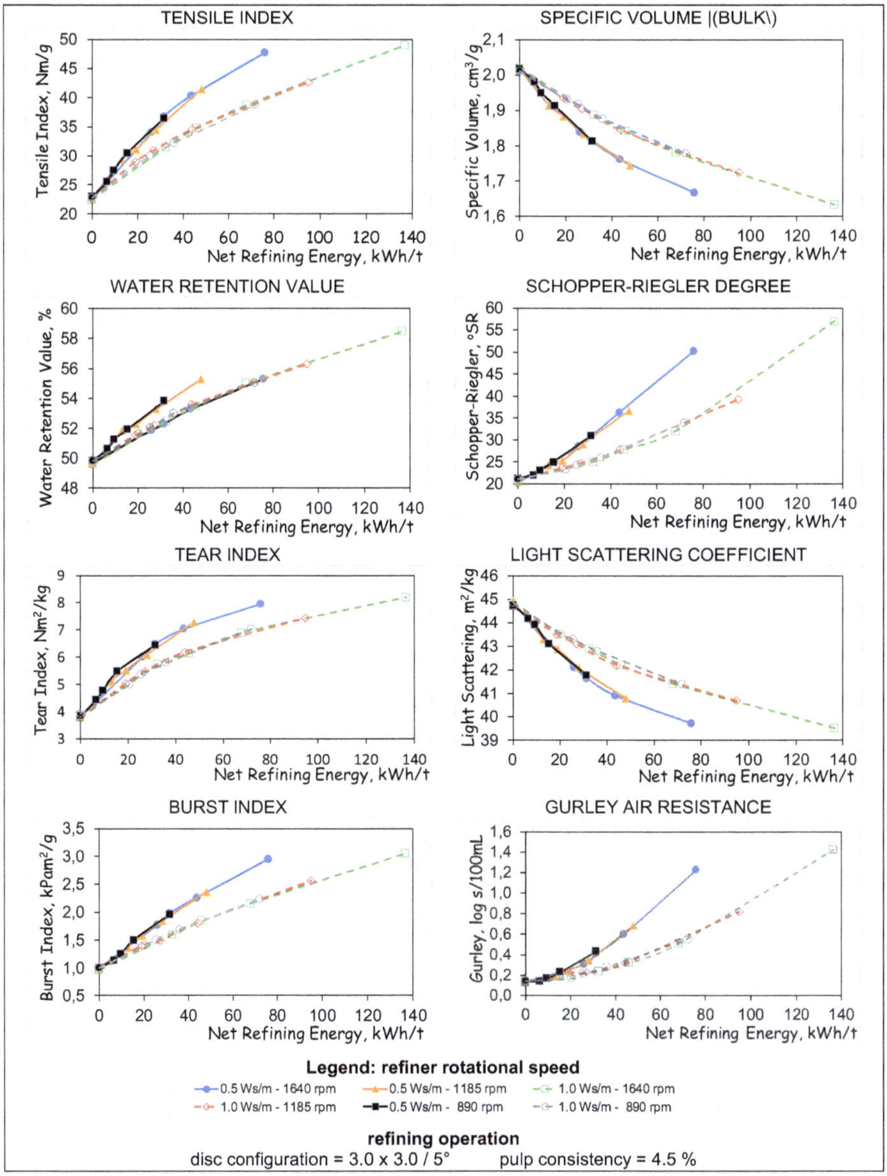

Fig. 10.18 Pulp properties evolution. Effect of the refiner rotational speed. (Adapted from Ref. [1116])

The pulp properties development against net refining energy and the refined pulp properties relationships were not influenced by the refiner rotational. The differences observed in Fig. 10.18 are due to refining intensity. Thus, keeping intensity constant, there is no difference in specific energy demand to meet the target pulp properties values (paper quality) (Figs. 10.19 and 10.20).

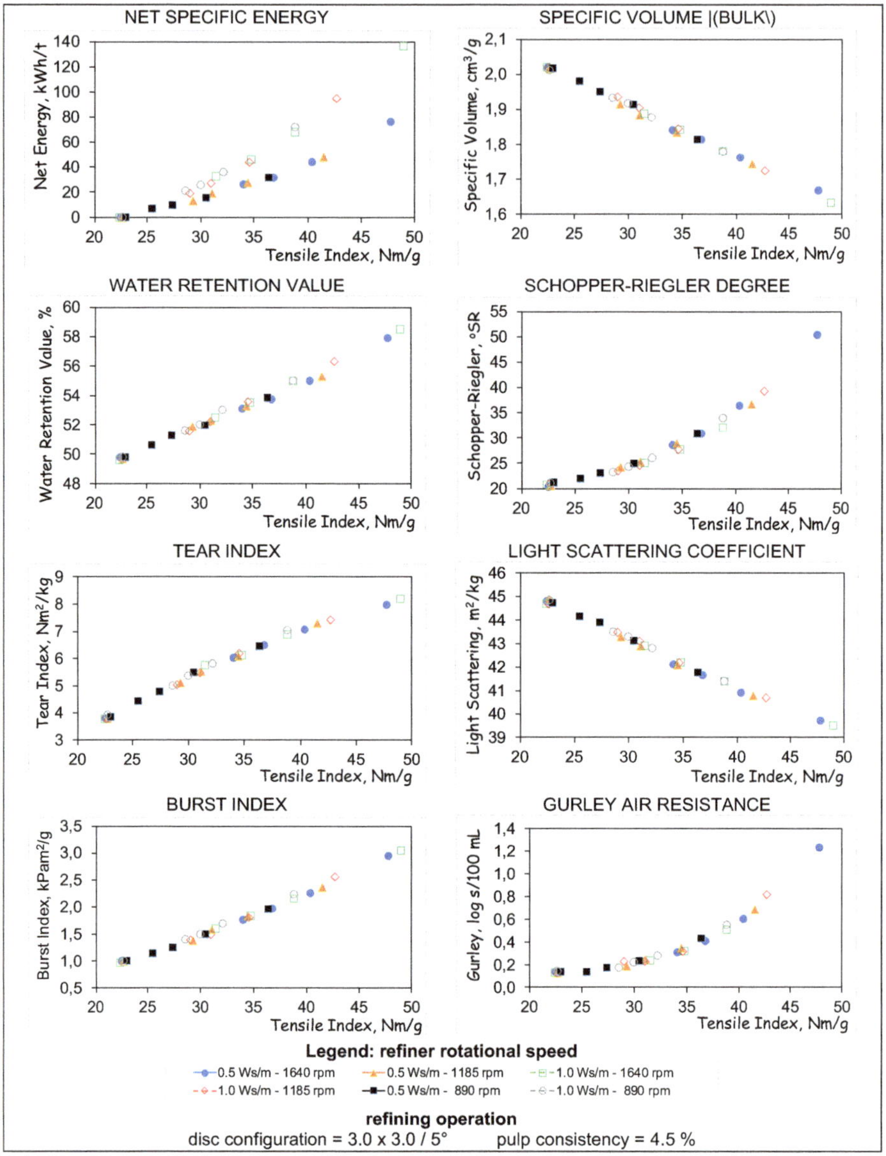

Fig. 10.19 Pulp properties relationships. Effect of the refiner rotational speed. Comparison at the same paper consolidation (tensile index). (Adapted from Ref. [1116])

However, the variation in rotation speed causes variation in the refiner's empty power. Rotation speed variation also influences the production capacity of refining installation and thus operating costs. Table 10.4 details the effects of disc speed on the total refining energy consumption and production capacity for pilot installation used to obtain the experimental data presented.

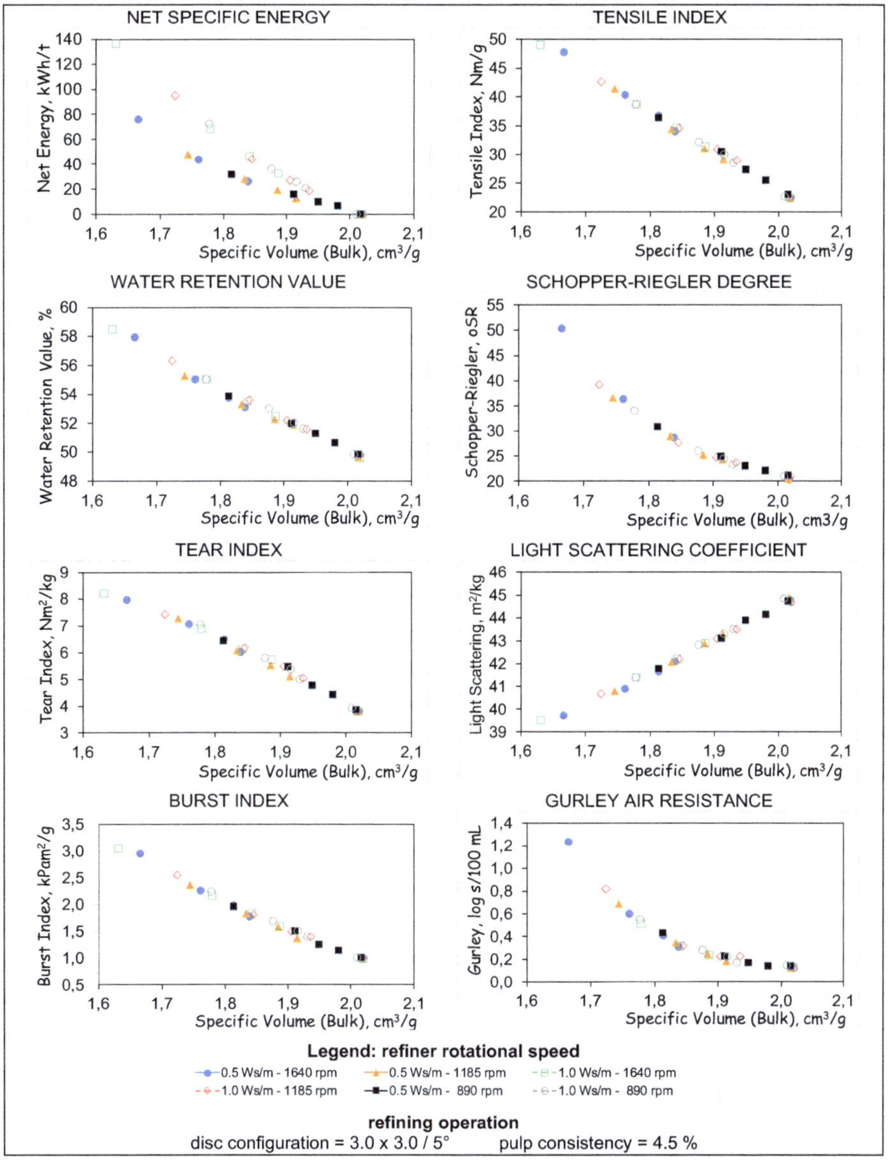

Fig. 10.20 Pulp properties relationships. Effect of the refiner rotational speed. Comparison at the same paper compaction (specific volume or bulk) (Adapted from Ref. [1116])

For the same production and net refining power, higher speeds mean greater refiner empty power contributing to increase the total energy demand in the refiners. This result agrees with previous information in the literature [262, 493, 710, 829, 946, 1018, 1067, 1115, 1117, 1133, 1187].

Table 10.4 Effect of disc rotational speed on total refining energy demand for same refining intensity (SEL)

Rotational velocity (rpm)	Specific net energy (kWh/t)	Pulp production (t/h)	Number of needed refiners	Power for each refiner (kW)			Total power (kW)	Total power reduction (%)
				No load	Util	Total		
1200	50	0.26	1	9.0	13.0	22.0	22.0	–
	100	0.13	1	9.0	13.0	22.0	22.0	–
1000	50	0.21	1.24	6.0	10.6	16.6	20.6	6.4
	100	0.11	1.24	6.0	10.6	16.6	20.6	6.4
600	50	0.13	2	2.5	6.4	8.9	17.8	19.1
	100	0.06	2	2.5	6.4	8.9	17.8	19.1

Reference values for refining pilot facility used for the study [1115]

10.9 Refining Consistency

Chemical pulp refining is predominantly done at low consistencies, between 3% and 6%, as low consistency systems are cheaper to buy, install, and operate. In this consistency range, the fiber slurries are easily pumped through the installation using standard pumps, and the quality of the refined pulp is suitable for most applications.

Refining consistency is an important parameter as it influences both the effects of refining on the fibers and the refiner's load capacity (production) and also its operating efficiency [710, 765, 766, 841, 867, 872, 1087, 1096, 1104, 1111, 1118, 1138, 1404]. All refiners have an optimal operating range for consistency. Operating outside this range, variations are usually observed both in the type (quality) and amount of treatment, as the pulp consistency directly influences the number of fibers that are captured in each bar intercrossing points. This directly affects the forces applied to the fibers [262, 511, 610, 841, 867, 879, 1051, 1121, 1138, 1291, 1317, 1390, 1398].

The consistency effect is greater for short fiber pulps than for long fiber ones. For short fibers, refining consistencies in the range of 4 and 6% are recommended, with the use of the maximum consistency possible in a sustainable way being desired [52, 84, 318, 393, 493, 1085, 1096, 1118].

High consistency refining has been used in specific applications, such as the production of paper for sacks, where the effect of high consistency is reflected in the excellent paper strength. High consistency often induces curling, cell wall displacement, and internal fiber fibrillation, resulting in greater sheet extensibility. Low consistency refining tends to cause more surface fibrillation and fiber straightening [312, 1106, 1271, 1328, 1329, 1359, 1399].

In the sequence, there are presented results from an evaluation of the refining consistency effects on the bleached kraft eucalyptus pulp treatment. The test was carried out in a pilot refining facility (Appendix III) equipped with a 12-inch diameter disc refiner, using discs with 3.0 mm × 3.0 mm/5° configuration, operating at 1640 rpm under 0.5 Ws/m intensity [1116].

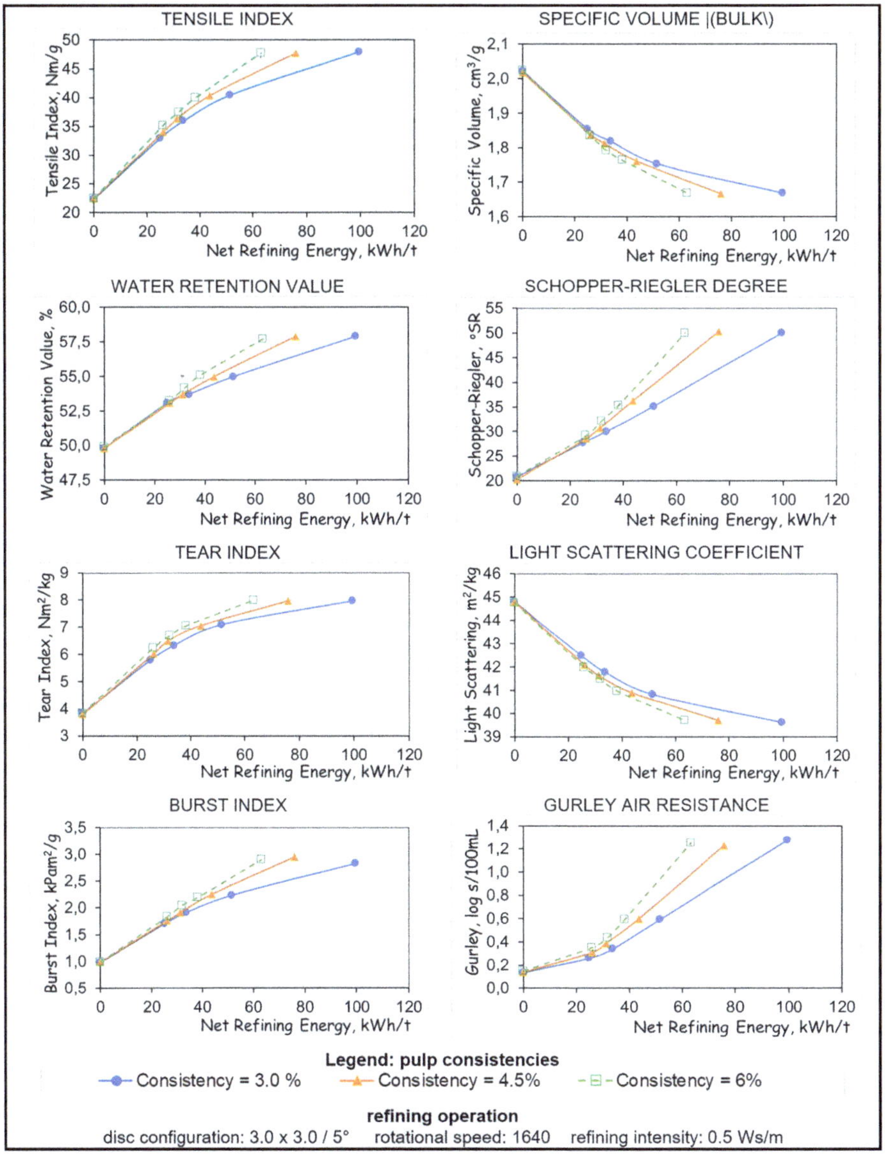

Fig. 10.21 Pulp properties evolution. Effect of pulp consistency. Comparison at same net refining energy. (Adapted from Ref. [1116])

The results, illustrated by Fig. 10.21, demonstrate the significant effect of consistency on net specific energy demand, especially when higher levels of net energy are required. When considering the same level of net specific energy, better results are obtained at the highest consistencies, confirming results from the literature [178, 217, 710, 765, 766, 841, 867, 868, 872, 879, 1125, 1085, 1087, 1096, 1099, 1100, 1104, 1106, 1111, 1115, 1118, 1138, 1187, 1291, 1307, 1317, 1390, 1404].

Higher consistency allowed smaller amount of net specific refining energy to reach the target values for refined pulp properties. According to the Specific Edge Load Theory, this result is not expected.

As observed in the evaluation of the others refining operating variables, also for consistency, no effects were observed on the refined pulps properties interaction (Figs. 10.22 and 10.23). In other words, although there were different net specific

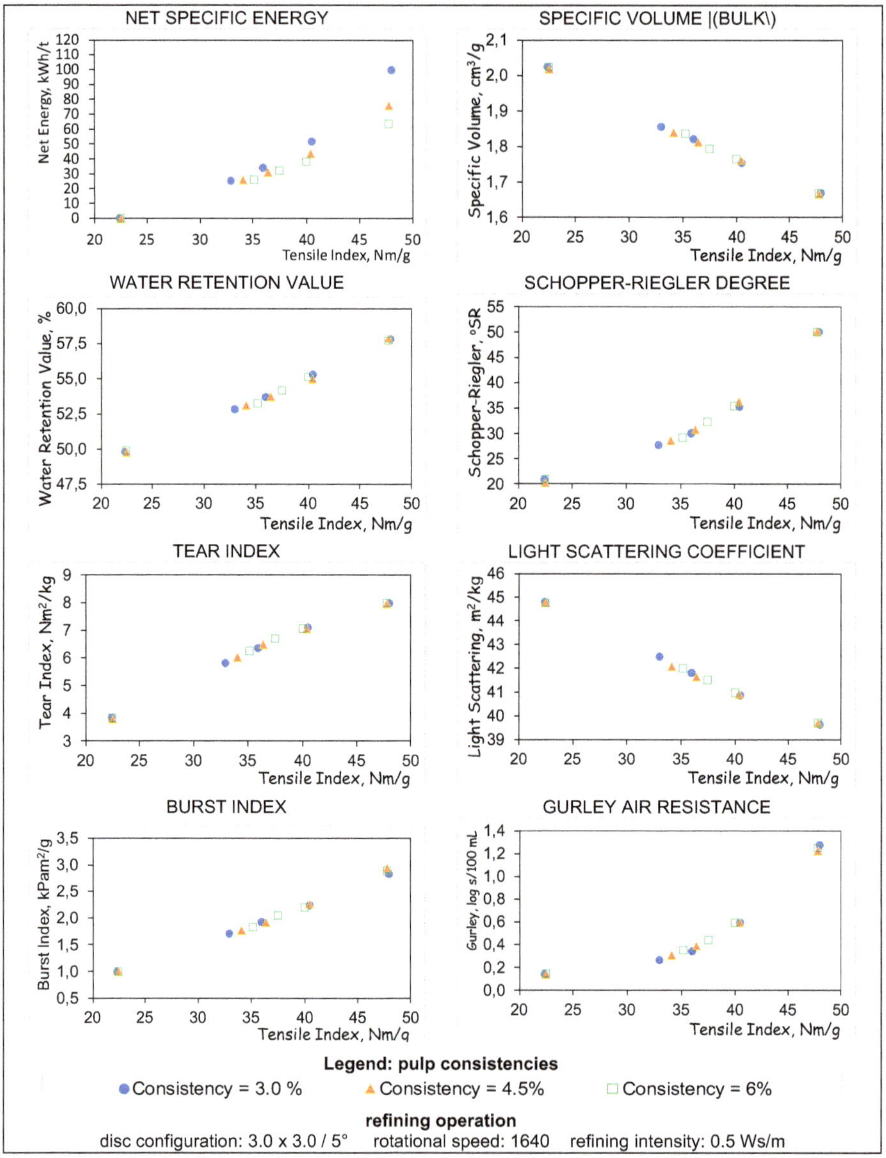

Fig. 10.22 Pulp properties relationships. Effect of pulp consistency. Comparison at the same paper consolidation (tensile index). (Adapted from Ref. [1116])

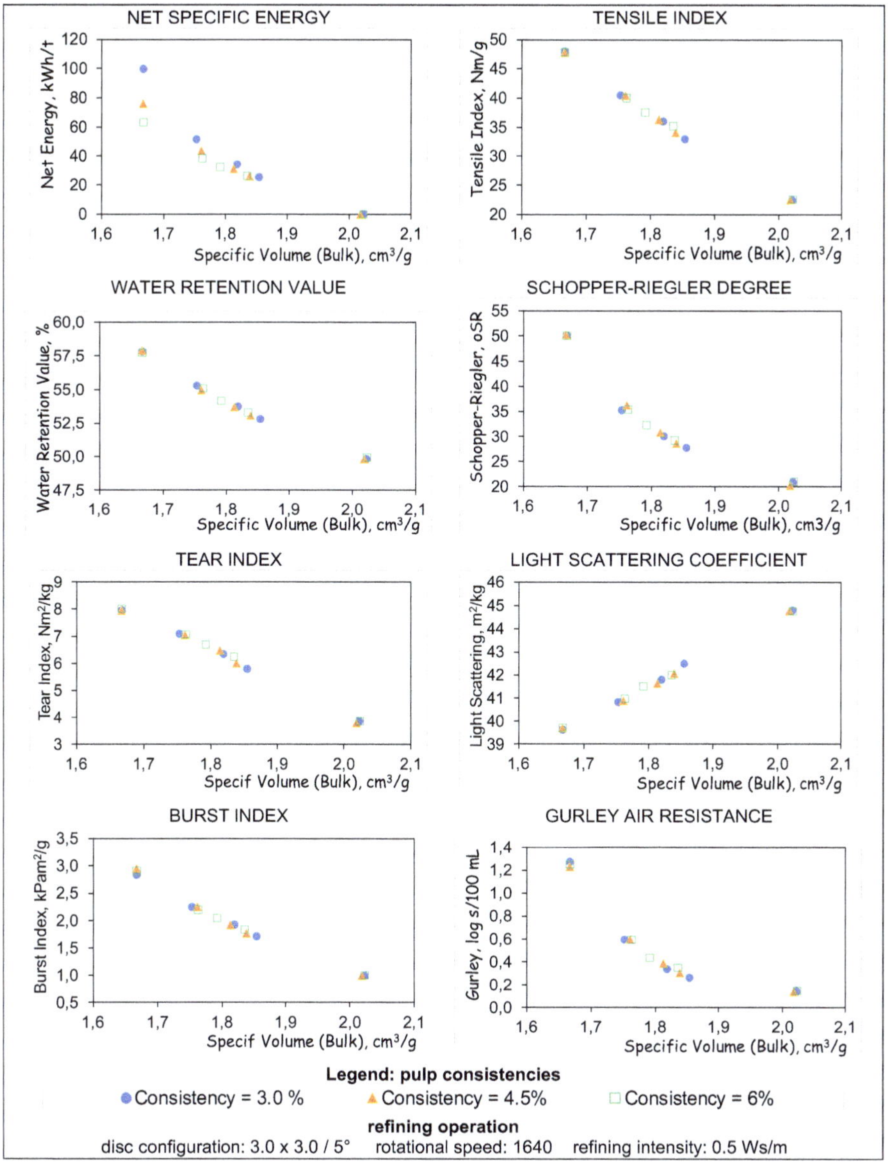

Fig. 10.23 Pulp properties relationships: Effect of pulp consistency. Comparison at the same paper compaction (specific volume or bulk). (Adapted from Ref. [1116])

refining energy consumptions to reach the target property, the refined pulp quality was not influenced by the refining consistency.

There are several explanations presented in the literature for the better refining performance under greater consistency. All are based on increased interactions (contacts) between fibers.

When refining is conducted at low consistency, the number of fibers in the space (volume) between the discs is reduced, generating energy loss in the water-disc friction, fiber cutting, and even more accentuated wear of the disc bars [262, 511, 610, 841, 867, 879, 1051, 1121, 1138, 1291, 1317, 1390, 1398].

In low-consistency refining, the increase in consistency allows a reduction in the specific energy demand to obtain the same refined pulp quality because it increases the fibers flocculation tendency, forming larger flocs, which makes it difficult to flow through the channels. By this way

- The bar surface shear action of the fibers is reduced, and the shear action is stimulated through the greater fiber-to-fiber interaction [765, 836, 838, 868, 1057, 1096, 1097, 1106, 1111, 1112, 1125, 1170, 1183, 1307, 1386, 1393].
- Increases the number of fiber-to-fiber impacts generating higher hydration levels [178, 765, 879, 1099, 1100, 1110, 1125, 1291, 1307, 1317, 1359, 1368, 1383].
- Increases the density of the fiber mat between the discs, favoring fiber fibrillation [879, 1047, 1099, 1100, 1110, 1359].
- Increases the thickness of the mat between the discs, allowing a more homogeneous treatment of the fiber flocs, minimizing the intensity of the impacts received by each floc [217, 511, 867, 879, 1015, 1051, 1121].
- Although all these explanations are apparently quite coherent, we believe that the better refining efficiency when increasing the consistency from 2% to 5.8% is due to the higher impact probability on the suspended flocs. The increase in consistency, keeping the other operating conditions constant, including the hourly production, as occurred in the evaluation carried out, leads to an increase in the retention time of the pulp in the refiner (Fig. 10.24).

To maintain the same hourly production, the lower consistency slurry flows faster through the refiner than the higher consistency one. In this way, the flocs in greater consistency suspension spend more time receiving impacts in the refiner. Receiving more impacts (energy), the fibers develop more properties at the same applied refining energy.

Several studies demonstrated the importance of consistency for both the pulp refined quality and the refining energy consumption [109, 217, 219, 354, 692, 765, 798, 817, 838, 868, 895, 1075, 1085, 1115, 1130, 1133, 1155, 1170, 1187, 1229, 1291, 1307, 1336, 1387, 1395–1397, 1404]. Thus, the retention time of the fibers inside the refiner must also be considered, which is the most likely explanation for the obtained results.

Retention time is determined by several factors, such as flow rate, pulp consistency, disc configuration, and even disc wear [67, 1085, 1096, 1117, 1133, 1291, 1397].

To optimize the refining energy consumption, not only eucalyptus kraft pulp but all others must be refined under an operationally stable condition, which allows the lowest intensity associated with the highest consistency as possible. This guarantees the best operating result to the papermakers.

It is important to avoid consistency variations as these cause variations in the refining energy, expressed in kWh/t, leading to variations in the refined pulp quality,

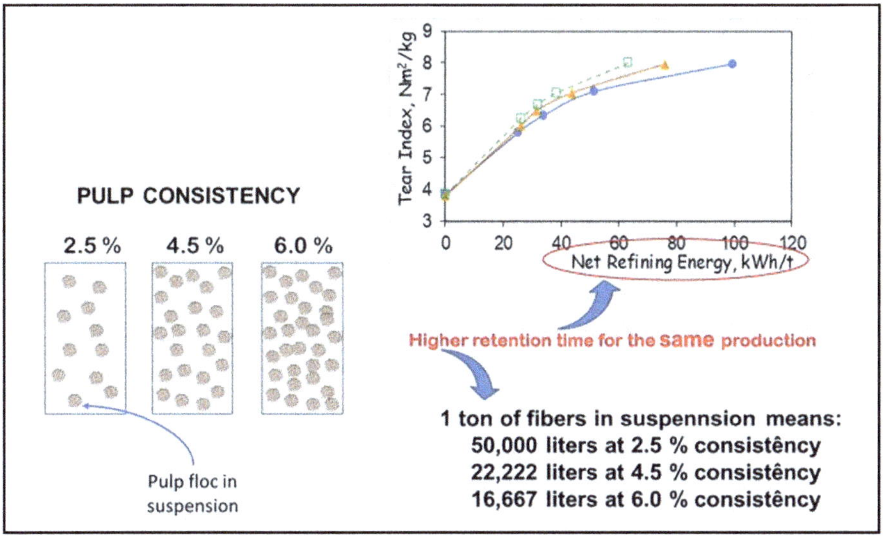

Fig. 10.24 Effect of refining consistency on retention time in the refiner. Greater consistency means longer retention time allowing an increase in the number of impacts on fibers despite the same refining intensity (SEL)

reducing the treatment uniformity. Keep the consistency as stable as possible, controlling the consistency variation in the range of ±0.1% (desirable) to ±0.2% (tolerable) [1015, 1085, 1118].

In Chap. 9, Sect. 9.6, we comment on the criticisms to the specific edge load (SEL) model. In general, such criticisms are presented because the model does not consider the disc configuration nor the fibers characteristics. Some of these criticisms, especially those related to the disc configuration, have evolved into new models or refining theories. Other refining models, especially those considering the forces acting on the refiner, consider the fiber characteristics, giving special importance to those related to the rheological behavior of the pulp suspension.

The results illustrated by Fig. 10.21 confirm the need to also characterize refining consistency as an operating parameter in theoretical refining models. Thus, from what has been discussed up to this point, three parameters are necessary for the proper refining characterization: refining energy, refining intensity (SEL), and refining consistency.

10.10 The Pulp to Be Refined

So far, we have identified that the energy input, intensity, and consistency at which the refining is conducted are essential variables to characterize the process. The edge specific load model proved to be effective in expressing the eucalyptus kraft pulp refining results, since the bar edges length of the discs is properly measured and the refining consistency is reported.

In Sect. 10.3, we highlight the importance of net specific energy in the pulp properties evolution. We did this by presenting refining results from a pilot installation, using four different commercial eucalyptus kraft pulps. They showed different results despite having been refined at the same intensity and consistency. From a practical point of view, this means that the most suitable pulp for the manufacture of a certain type of paper is the one that, in the unrefined stage, presents properties closer to those desired in the final product.

The pulp to be refined is also an important operating variable for the whole refining process characterization. It must be used in conjunction with refining energy, intensity (SEL), and consistency to characterize the process.

As a way of validating this proposal, we present below results of refining curves carried out in a pilot plant, for the same commercial eucalyptus kraft pulp, but refined under a miscellany of operating conditions. We varied the refining specific energy, the slurry consistency, the refining intensity, the disc configuration, and the refiner rotational speed. In summary, in the refining curves whose results are illustrated (Figs. 10.25, 10.26 and 10.27), only the used pulp was the same.

The results from pulp properties evolution with the net specific refining energy (Fig. 10.25) varied according to the operating conditions used. The specific energy consumption is reduced when the refining intensity decreases, confirming the previous results.

However, when the quality of the refined pulp is considered, no differences are observed in the pulp properties interactions, either in the comparisons to same paper structure consolidation (Fig. 10.26) nor in the comparisons at same paper structure compaction (Fig. 10.27).

Thus, for the same pulp, the quality of the refined pulp, that is, the set of its properties, will always be the same, regardless of the set of operating variables used in refining. For the same fibrous raw material, the interaction between the refined pulp properties (the paper quality) will be a kind and "fingerprint" of that fibrous material, regardless of the way it was refined.

This is a significant result as it validates the performance of pulp evaluations in the laboratory as indicative of their industrial behavior. In other words, the quality of the paper depends exclusively on the raw material and the pulping method used.

The results illustrated by Figs. 10.25, 10.26 and 10.27 show the need to define the type of pulp as an operating parameter in theoretical refining models. A simple way to justify this need is presented in Fig. 7.1 in Chap. 7, Sect. 7.1, which illustrates the great variability in the anatomical characteristics of the fibers and other anatomical elements of wood in general, the different pulps resulting in different refining behavior, mainly due to their different flocculation tendencies.

Thus, from what has been discussed up to this point, four pieces of information are necessary for the proper characterization of refining: refining energy, refining intensity (SEL), pulp suspension consistency, and the pulp to be refined.

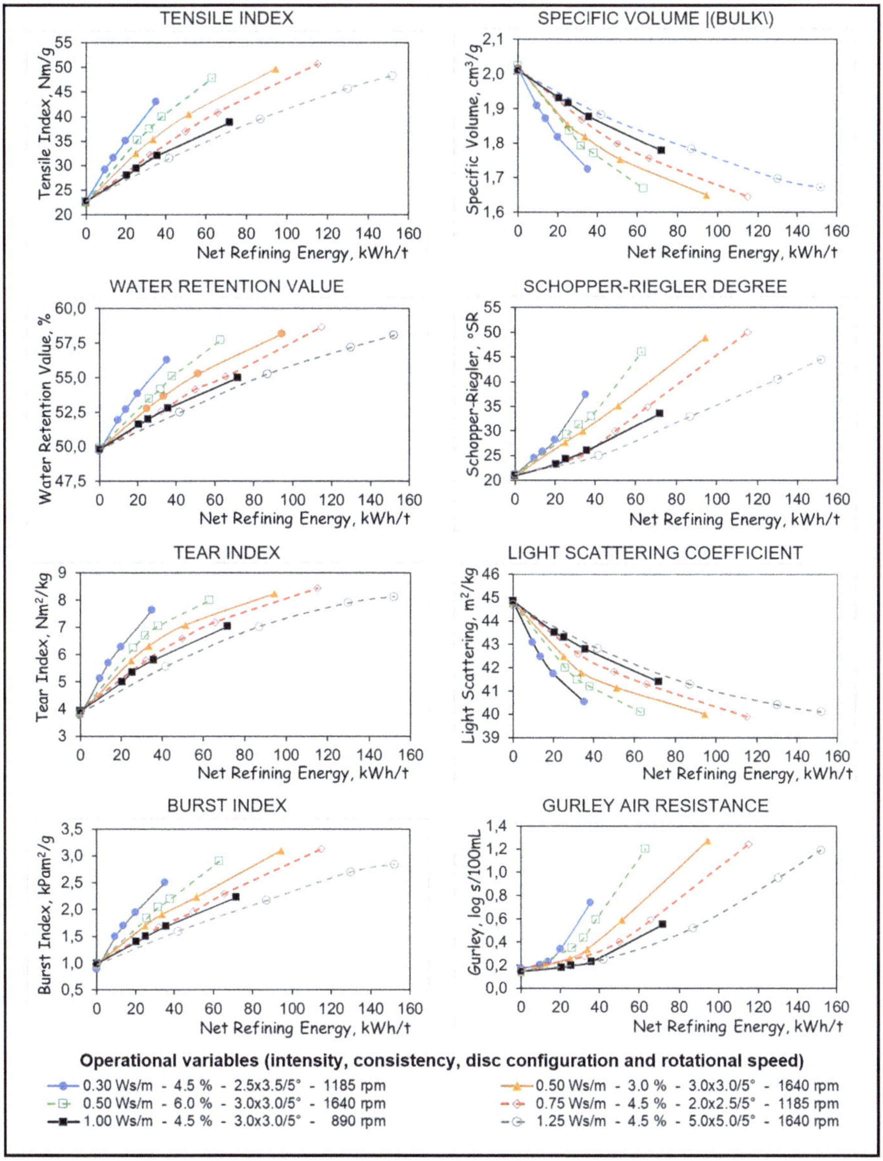

Fig. 10.25 Pulp properties evolution during the refining for same commercial eucalypt kraft pulp under several operating refining conditions

10.11 Pulp Suspension pH

Since the refining action occurs by hydration of the fibers cell wall, the refining efficiency is dependent on the fiber hydration capacity (fiber swelling). So, theoretically, the more swollen (hydrated) the unrefined fibers are, the lower is the refining energy requirement to obtain the target properties in the refined pulp [704, 1106].

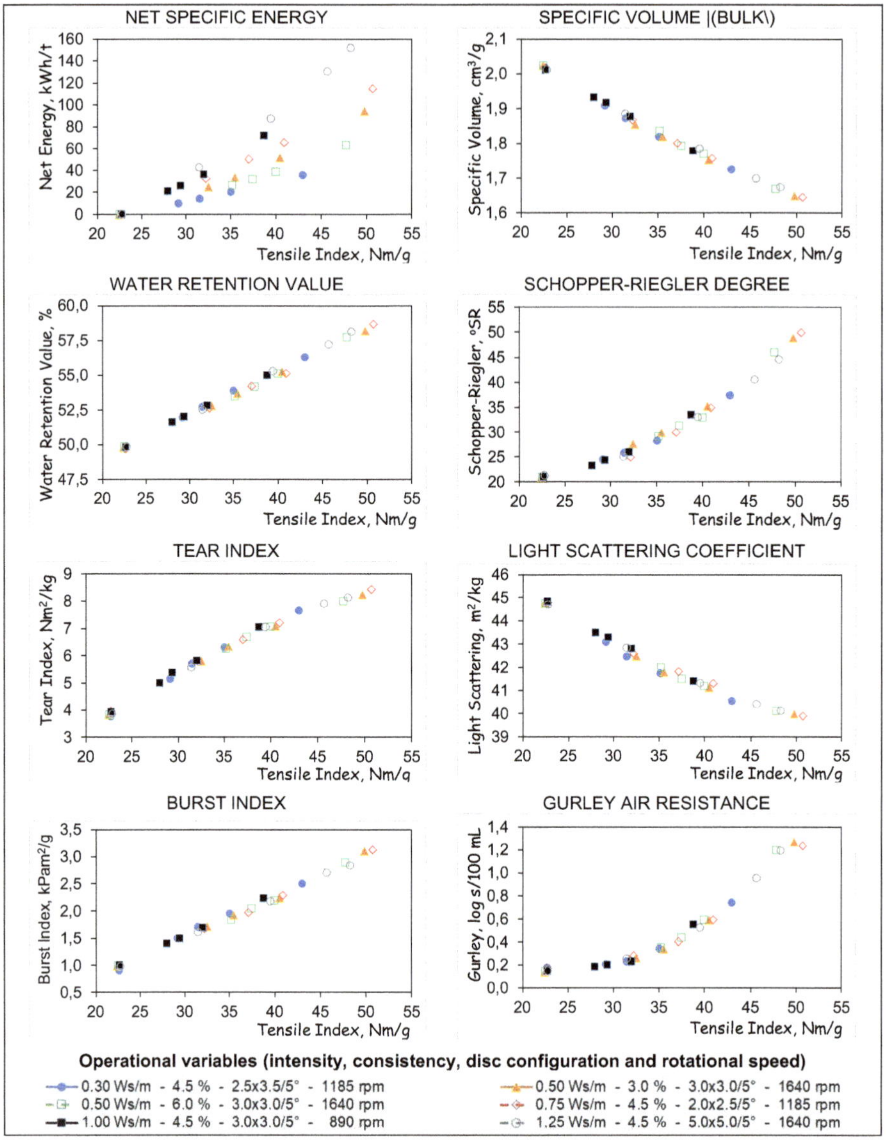

Fig. 10.26 Pulp properties evolution during the refining for same commercial eucalypt kraft pulp under several operating refining conditions. Properties at the same paper consolidation (tensile index)

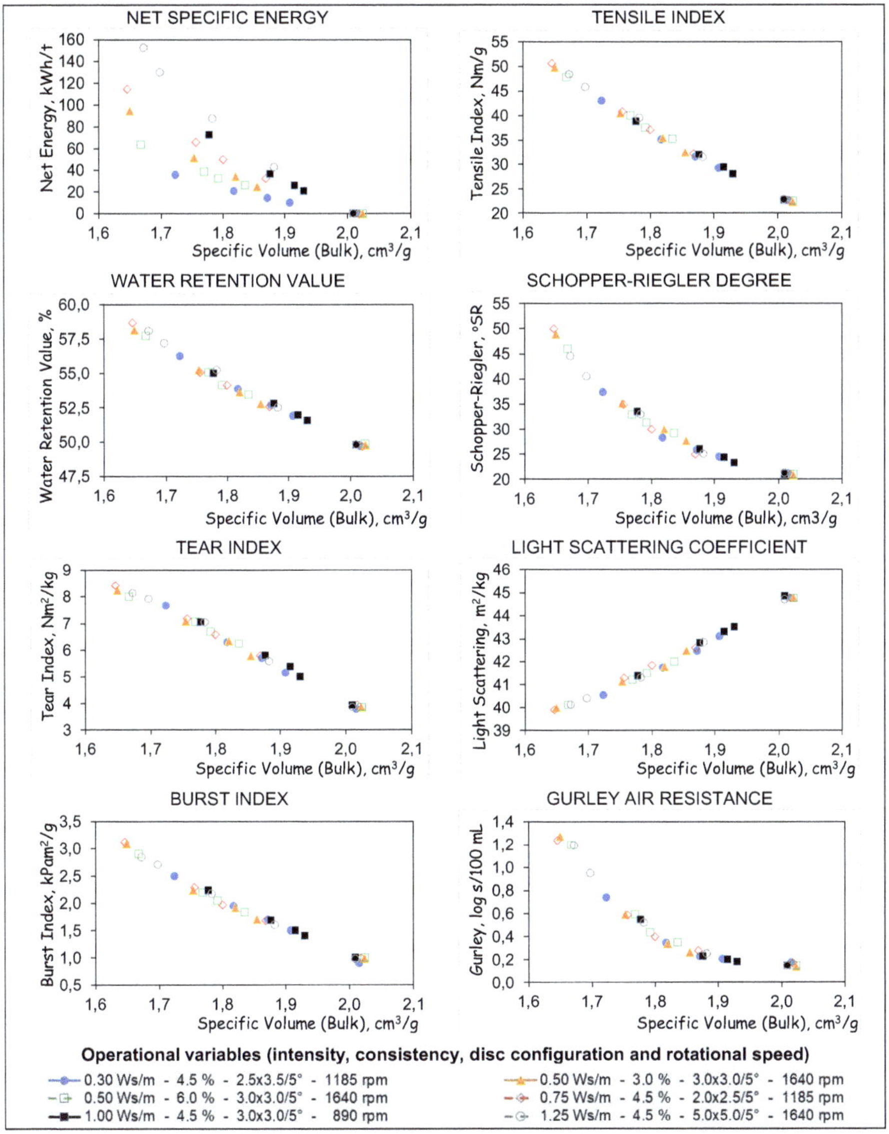

Fig. 10.27 Pulp properties evolution during the refining for same commercial eucalypt kraft pulp under several operating refining conditions. Properties at the same paper compaction (specific volume or bulk)

The cell wall hydration of any pulp is conditioned by physical and chemical parameters, namely:

- Acidic groups amount present in the fibers cell wall and their dissociation constants. The increase in the amount of acid groups results in an increase in the swelling capacity [44, 429, 433, 440, 450, 632, 701–708, 712, 715, 722–734, 736, 750, 1338].
- Fiber cell wall elasticity modulus (flexibility) [38, 443, 450, 603, 663, 703, 704, 706, 711, 713, 721, 735, 740, 771, 1183, 1187, 1361].
- Type and quantity of counter-ions (ions that maintain electrical neutrality with the suspension) in the fiber cell wall [43, 73, 159, 300, 433, 450, 632, 701–704, 706–708, 721, 733, 736, 745–748]
- Ionic concentration of the liquid surrounding the suspended fibers [124, 456, 450, 463, 616, 702, 705–707, 712, 721, 736, 750, 1338].
- pH of the liquid surrounding the suspended fibers [450, 701–703, 705, 707, 736, 752, 753, 765],
- Refining degree (amount of fiber modifications already presented by the fibers) [736].

Refining is traditionally carried out at low pH, which favors fiber fragmentation, generating a greater fines amount and reducing the average fiber length. At pH below 3, the carboxylic and phenolic groups present in pulp are not dissociated, which justifies why there is practically no adsorption of metallic cations from the suspension, not favoring fiber hydration. When the pH increases, the dissociation of the carboxylic groups present in the carbohydrates and residual lignin begins and the adsorption capacity increases until reaching a maximum level in the pH range 8 to 9, when the carboxylic groups are completely dissociated, being the best condition for the swelling of the fibers [22, 433, 632, 701, 702, 704, 736].

Depending on the chemical environment, the acidic groups of the fibers may or may not be dissociated. If the ionic shape is changed, the swelling of the fibers as well as the distribution of pores in the cell wall will also change [433, 443, 450, 701–703, 708, 733, 745, 749].

As the pH increases, more water is adsorbed into the cell wall, which progressively loses rigidity. Refining takes place faster, reducing energy demand and resulting in higher levels of strength and fiber compaction in the refined pulp [43, 52, 115, 214, 219, 335, 429, 433, 450, 454, 511, 606, 663, 692, 703, 704, 712, 713, 721, 733, 736, 740, 824, 828, 908, 924, 1106, 1118, 1183, 1187, 1361].

In Figs. 10.28, 10.29 and 10.30, the results of the evaluation carried out in a pilot refining plant, of the effect of the suspension pH on the refining results from commercial sample of bleached eucalyptus kraft pulp are presented.

The results obtained behaved as predicted in the literature, that is, with lower refining energy consumption for the development of consolidation (mechanical strengths) and compaction of the paper structure when refining was carried out at alkaline pH [22, 702, 704, 736, 737, 1342, 1382].

In the evolution of properties with refining energy (Fig. 10.28), the tear strength (Elmendorf) and elongation are noteworthy. The two properties presented higher

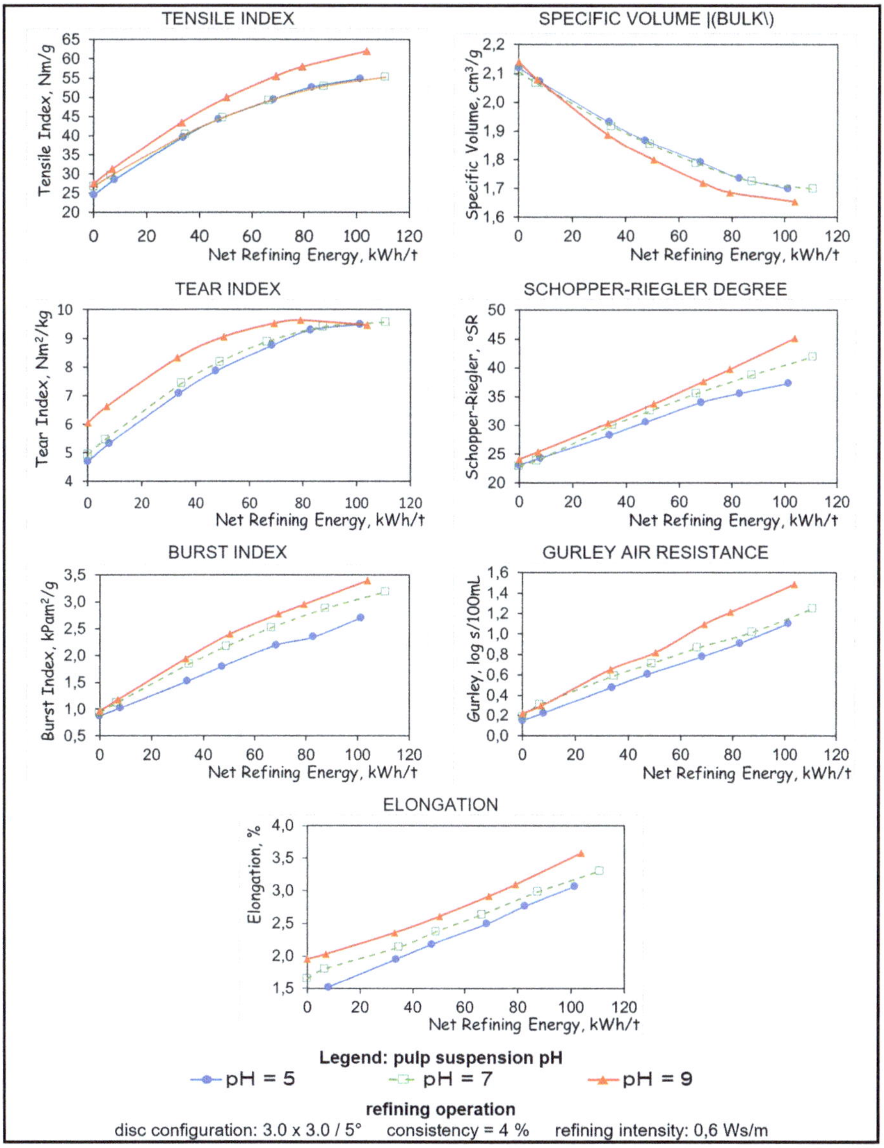

Fig. 10.28 Pulp properties evolution – Effect of the pulp suspension pH. (Adapted from Ref. [1382])

initial values in the refining carried out at alkaline pH. Among the effects of fiber hydration (fiber swelling), we have the increase in the fibers bonding capacity obtained by the flexibilization of the fibers that occurs by the reduction of the cell wall intrinsic resistance.

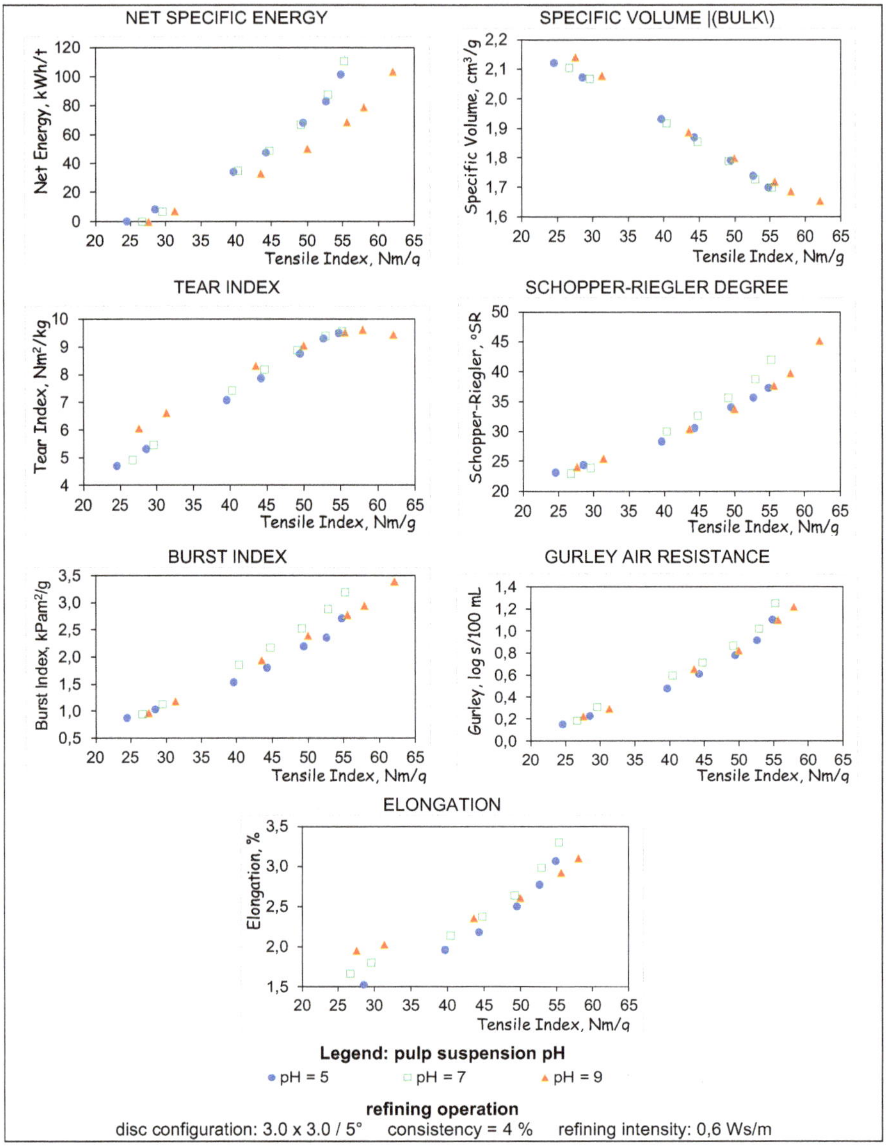

Fig. 10.29 Pulp properties evolution – Effect of the pulp suspension pH. Comparison at the same paper consolidation (tensile index). (Adapted from Ref. [1382])

The tear strength (Elmendorf) and paper elongation (Fig. 10.28) evolutions show greater release of fibrillar fines in refining at alkaline pH, already in the early stages of refining. From 60 to 70 kWh/t of applied net energy, the effect of the cell wall intrinsic strength reduction is observed, which occurred regardless of the suspension pH level, indicating that this is the equilibrium point between the effects from

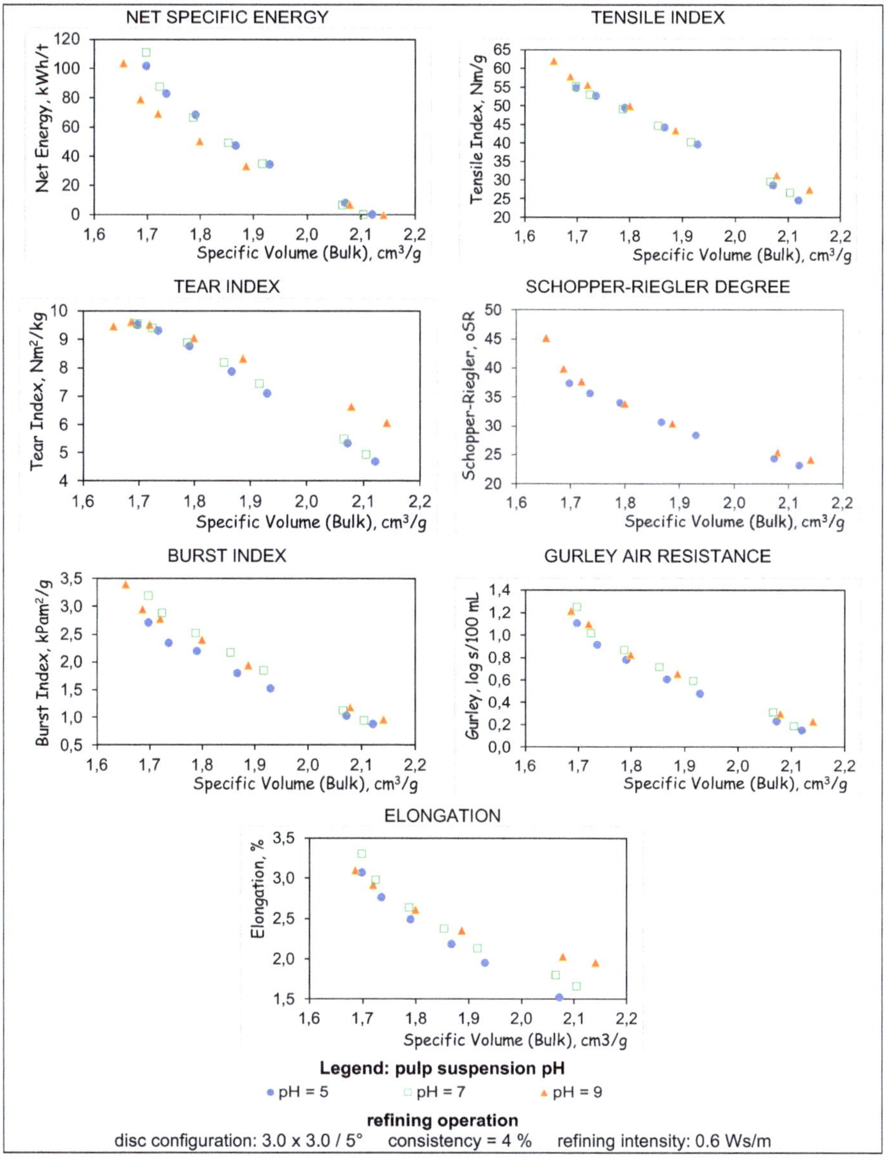

Fig. 10.30 Pulp properties evolution – Effect of the pulp suspension pH. Comparison at the same paper compaction (specific volume/bulk). (Adapted from Ref. [1382])

cell wall intrinsic strength reduction the fibers bonding capacity increase in the tear strength (Elmendorf) measurement.

During the refining, the fibers cell walls begin to break down due to the reduction in intrinsic resistance and fines start to be released from the outer cell wall layers, as indicated by the porosity evolution curves (Gurley air resistance) and Schopper-Riegler degrees.

At higher pH, these actions are pronounced at early refining levels probably because the effects of alkalinity are similar to that of energy application. With higher energy applications the effect of mechanical treatment is the predominant.

Thus, the results obtained in this evaluation confirmed the results from literature. In low consistency refining, which is the most common way of refining the bleached eucalyptus kraft pulp, operating the refiner at alkaline pH favors both the treatment. The practical result is a lower refining energy demand for the development of the desired pulp properties representing a major economic impact for the paper manufacturer.

10.12 Retention Time in the Refiner

The retention time inside the refiner is an important operating variable as it relates to the numbers of impacts on fibers flocs. Higher the retention time greater is the probability to flocs being impacted by intersecting bars [217, 219, 354, 409, 692, 765, 817, 868, 895, 1026, 1075, 1085, 1115, 1130, 1133, 1155, 1170, 1187, 1229, 1291, 1307, 1336, 1384, 1395–1397].

Retention time is controlled by several factors, such as flow rate and suspension consistency, disc configuration, and even disc wear [67, 1085, 1096, 1133, 1117, 1291, 1397]. However, there are controversies as to the effectiveness of their practical results [67, 1133].

As evidenced in the evaluation of the effects of refining consistency (Chap. 10, Sect. 10.9), for the same operating condition (refining intensity, disc rotation speed, disc configuration, and applied energy), the increase on retention time, due to pulp consistency increasing, allowed the reduction of the refining energy consumption to meet the target refined pulp quality.

10.13 Refining Efficiency

The refining action is related to how effectively and efficiently the fibers are impacted to improve their machine behavior and the quality of the produced. Paper. It is highly dependent on the compressive and shear forces applied to the fiber flocs, especially on refiner no-load power and the intensity at which the refining is carried out. In the same intensity (SEL) and net refining energy, all refiners are comparable [15, 178, 1067, 1097, 1105, 1114, 1217] if the pulp type and refining consistency are also be specified.

Refiner energy efficiency (η) is defined as the ratio between net power (P_{net}) and total power (P_{tot}) of the refiner (Eq. 10.3). It represents the fraction of the total power used in the mechanical action required to modify the fibers characteristics [452, 511, 1078, 1315, 1317].

$$\eta = \frac{P_{net}}{P_{tot}} \tag{10.4}$$

where:

η refiner efficiency
P_{net} refiner net power
P_{tot} refiner total power

Due to its lower no-load power, double-disc refiners are more energy-efficient than single-disc refiners. For the same specification (pulp type, production capacity and net specific energy), single-disc refiners support only half the load of a double-disc one [1097, 1187].

In industrial facilities, the refining efficiency is generally higher than 75%, while in experimental disc refiners it is relatively lower, usually in the range of 35% to 50% [1019, 1028, 1187, 1315].

Evidently, a low energy demand is interesting, but a constant refining behavior is probably more important since variations in refining energy demand are difficult to handle in the process [42].

Chapter 11
Refining Process Control

Refining is the only operation in the entire pulp preparation that modifies the fibers morphology giving the desired final product properties.

Process control aims to reduce refined pulp variation preventing operating problems in the paper machine, as well as in the quality of the finished product. With proper process control, it is possible to operate at higher speeds in paper machines through faster drainage, increased wet web strength, and lower steam consumption in paper drying [984, 1354] and, in addition, ensuring paper quality.

Process control in beaters, such as the Hollander one, was done by beating time and applied load, usually measured as weight. The beating process was considered a skill or an art being heavily dependent on the operator's experience. An experienced operator could determine the beating degree and in many cases, measure pulp moisture just by compressing the pulp manually [829]. As beaters were replaced by refiners and batch processing gave way to continuous treatment, the operators' ability to "feel" the mass disappeared giving way to increasingly sophisticated and difficult-to-perform control systems only with the use of experience and practical concepts.

Currently, with paper machines running at 2000 m/min, it is of paramount importance that refining be a fully known and controlled process, so as not to adversely affect machine's productivity or exceed planned productions costs [402, 829, 984, 1350].

The careful selection and sizing of equipment is of fundamental importance. Pumps, pipelines, and valves must be dimensioned and positioned in such a way as to allow the minimum variation in the fiber suspension in terms of consistency, pulp velocity, pressure, refiner load, and final physical properties. There is a hierarchical importance and consistency is the master key [88, 398, 1114, 1354, 1390].

The unrefined pulp characteristics vary according to oscillations in wood characteristics and pulping conditions. The papermaker usually adjusts the operating conditions of the paper machine to the performance limitations presented by the pulp taking advantage of the availability of lower priced pulps [53, 984].

© The Author(s), under exclusive license to Springer Nature Switzerland AG 2024

V. Manfredi, *Eucalyptus Kraft Pulp Refining*,
https://doi.org/10.1007/978-3-031-47285-5_11

Adequate pulp properties characterization is necessary to maintain the quality of the finished product. Normally, anatomical characteristics of the fibers, pulp suspension drainage behavior, and structural or resistance paper properties are monitored.

Tensile strength is one of the parameters used by papermakers to define the appropriate level of refining, however, as there is a certain time interval between the stock preparation step, and obtaining the quality results, control tends to be ineffective [208, 872, 984, 1301–1303].

In this way, refining has been controlled by the applied energy and, indirectly, from the refining effects, the most common being the monitoring of drainage resistance and variations in vacuum levels in the paper machine forming area [298, 1111, 1302, 1343, 1354]. Another measure, widely used by operators working in a Fourdinier machine, is the position of the "water mirror" on the forming screen.

As paper quality is significantly influenced by refining, manual control of this process is considered inadequate to ensure uniform quality in the finished product. Automated controls not only reduce the paper variability but also the refining energy consumption [385, 984, 1302, 1305, 1354].

Another important fact is that the pulp preparation operators need to be trained and have autonomy to make decisions quickly and accurately. Therefore, team training is a crucial factor for the success of the process.

11.1 Drainage Resistance of the Refined Pulp

The "refining degree" is the most traditional form of process control. It characterizes the changes in the pulp drainage rate and is widely used to represent the fibers modification during beating or refining. Each pulp has its specific "refining degree" assuring the desired quality standard in the refined pulp.

Stock preparation includes sieving, cleaning, deaeration, refining, and consistency control. Each one of them influences the pulp performance in the process; however, refining is the most important. Simplistically papermaking is a process of forming a fiber web, followed by consolidation and dewatering. All steps in the manufacturing process can affect drainage and ultimately paper machine productivity [1161].

Paper machine drainage is not a feature of the final paper. However, it is a very complex and major concern for papermakers, so this behavior has been used for process monitoring [262, 1161, 1287, 1298, 1300].

One of the first devices developed to measure pulp drainage rate was designed by Paul Klemm, in 1903, and built by Schopper. It is a device made up of a glass tube about 5 cm in diameter and 30 cm in length. Loaded with 200 ml of water and two grams of dried pulp, the suspension was stirred by rotational movements and drained through one end of the tube measuring the volume after draining [262].

The wait to obtain the results, as well as the difficulties in reading the drained volume, led to the development of a new equipment. In 1910, E. W. L. Skark presented an equipment based on measuring the time needed to drain a fixed volume of water. In this case, the presence of lamellar fines started to have a decisive influence. This was followed by an equipment designed by Riegler and built by Schopper ("The Schopper-Riegler Freeness Tester") and, in 1926, came the equipment developed by the Forest Products Laboratory of Canada, "The Canadian Standard Freeness Tester – CSF" [262].

The Schopper-Riegler and the CSF testers (Fig. 11.1) became accepted worldwide, becoming classic methods for "drainage rate" evaluation.

These two methods present correlated results with the S-R degree being directly proportional to the drainage resistance, while the "freeness" is inversely proportional, hence the nomenclature "freedom"[456, 866, 877, 984, 1125, 1232, 1298, 1239].

Refining reduces suspension freeness. In contrast, refining increases the Schopper Riegler degree, which means lower draining flow. The measurement ranges are between 0 and 100 °SR [456].

The two devices are similar in design. In the Schopper-Riegler, a wire mesh is used instead of a calibrated screen plate. It should be noted that the samples in this method (°SR) have lower concentrations and an inverse scale is used to measure the drainage of the suspension.

Drainage resistance is closely related to several refining effects especially the lamellar and fibrillar fines formation, the reduction in fiber length, and the increase in flexibility due to fiber collapsing. In this way, the drainage resistance can represent the properties of the fibers being able to select the most appropriate level of refining energy required for each type of the paper [82, 108, 179, 281, 470, 721, 774, 866, 877, 984, 996, 1125, 1161, 1228].

The system requires prior knowledge of the relationship between drainage values and the desired properties in the refined pulp, which are obtained through experimental beating or refining curves [877, 984, 1104, 1354].

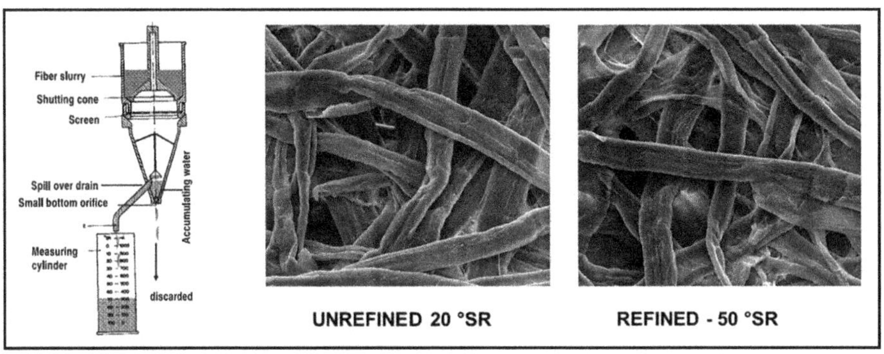

Fig. 11.1 Determination of "freeness" (Schopper-Riegler Method) [1212, 1280]

Although they are easy to determine, the "drainage degree" represents an empirical analysis that, according to some authors, can produce discrepant results in relation to the refined pulp quality. The basic reason for this criticism is that the resistance to drainage, in a non-turbulent regime as determined by traditional methods, is strongly dependent on the type of fines, especially on its external surface, that are not directly related to the development of physical-mechanical resistance properties of pulps [42, 52, 218, 262, 270, 298, 599, 640, 710, 803, 804, 872, 884, 1045, 1097, 1181, 1202, 1256, 1301, 1313].

By this way, refining will be controlled by one of its secondary effects with little or no influence on the paper strength properties. For good refining control, the use of parameters more related to the effects of the process becomes necessary [52, 168, 203, 218, 262, 270, 298, 546, 621, 640, 710, 803, 804, 872, 884, 1075, 1097, 1181, 1202, 1222, 1256, 1301, 1313].

The level of pulp drainage is not a paper goal itself. When the manufacturer produces paper, he wants to achieve good machine running and target product quality, not drainage values [52, 1220].

In the paper machine, pulp drainage is also affected by other factors such as pulp type, paper machine setup, wet end chemistry, type of form and drainage elements, additives, pH, white water circuit, temperature, and consistency of the suspension, among others [39, 82, 262, 266, 398, 470, 640, 664, 702, 1161, 1298].

11.2 Controlling by Vacuum in the Paper Machine Formation Zone

It is also a form of control based on drainage. The fines content in the fibers suspension influences the "drainage rate" through the forming sheet onto the paper machine screen making it difficult to drain and reducing paper porosity while the vacuum in the suction rollers is increased. In practice, this behavior is considered an effect of refining, and many mills even manage to control the refining operation based on the vacuum values in the suction rolls [984].

The control concept uses an actual on-machine measurement (vacuum table) in combination with various types of hardware to develop the final refining power setpoint for the refiner controller. However, there is a delay between refining the pulp and the table roll, as refined pulp is normally mixed in a box before going to the paper machine [219, 829, 1104, 1354].

As with the drain-based control, the suction roll vacuum control also considers the characteristics of the refined pulp and can be used in parallel. If the drain is used as the primary command for refining control, the vacuum control can generate a subroutine to adjust the drain value as a control. However, this type of control is sensitive to other parameters such as weight variation, filler content, and retention in the paper and machine speed [219, 829, 984, 1104].

11.3 Control by the Refiner Power

A low consistency refiner is controlled by adjusting the disc spacing. Forces are applied to the fiber flocs in suspension developing structural changes in the fibers cell wall. Energy consumption is an important indicator for process control [1317].

The automatic refiner power control is one of the simplest systems for controlling the process. The objective is to keep the power applied to the refiner at a constant level. The system controls the distance between the discs, has a low installation cost and allows for quick responses keeping the refiner full protecting the discs against excessive wear. The system does not require dedicated operation (operator direct attention) allowing automatic restart [219, 664, 829, 877, 946, 984, 1104, 1354].

A critical issue to be considered in this type of control is that applied power (kW, Hp, or Amperes) does not consider changes in the flow or consistency of the suspension, which require specific parallel controls [829, 877, 984, 1015]. Incorporating these variables, the control can be done by the specific net refining energy (kWh/t).

11.4 Control by the Refiner Specific Net Energy

The refining specific net energy is the most important parameter to be controlled in the refining process and one of the most used control systems [219, 829, 1096, 1183]. Process control by specific energy corresponds to an improvement of automatic control by applied power.

Three distinct signals are used (applied power, pulp flow rate, and mass consistency) to calculate the specific net refining energy (in kWh/t or HPd/t) which is compared to the target value. If necessary, adjustments in the applied power are performed to avoid variation in the refined pulp properties [219, 298, 829, 984, 1111, 1183, 1390].

The advantages of automatic control by refining specific energy are the automatically responses to flow and consistency variations minimizing delay times and protect discs against excessive wear. The use of flow and consistency inputs and the data processing method greatly increases the accuracy of the control [877, 984, 1015, 1111, 1114, 1354].

11.5 Control by the Discs Gap (Distance Between Discs)

From a practical point of view, control by distance or spacing of the discs is equivalent to control by applied power or specific refining energy. The spacing of the discs is related to the applied power. Greater quantity of fibers between the discs and the more stable the distance between them more uniform is the treatment and lower the refining energy consumption is [765, 883, 984, 1109].

Chapter 12
Refining Strategies

There are several ways to assess the suitability of different refining strategies. The final evaluation, however, must be made after the paper machine by comparing the refining energy consumption, raw material costs, drying energy consumption, production capacity, number of breaks, and quality of the final product [470, 393, 871, 1184].

In paper mills the basic parameter used is the cost-effectiveness of the investments required to optimize the refining process and ensures the quality control in the refined pulp [662, 640, 1302].

12.1 Serial Refining vs Parallel Refining

The selection of a refining system considers the type of pulp that will be used and the desired refined pulp quality. In configuration with two or more refiners in operation, it is possible to switch the installation design, that is, to position the refiners in series, in parallel or in a combination of these two options.

In some cases, especially in those where a high energy demand is required, the paper manufacturer chooses to work with refiners in series, applying the total energy in successive stages. Series refiners are preferred because they are easier to control mass flow, using just a simple valve and flow meter. The number of refiners should be selected to allow the lowest possible no-load power [401, 877, 946, 1096, 1384].

If the incoming suspension flow is divided into two or more refiners, we have the parallel refining, where each refiner operates in a single pass and the refined pulps are mixed. This configuration gives higher production capacity but requires a more complex control system for pulp flow and consistency. The main disadvantage of the parallel configuration, or with combined refiners, is the control of the flow through the refiners because it is necessary to allow the same refining-specific energy despite flow variations [1096, 1111, 1384].

© The Author(s), under exclusive license to Springer Nature Switzerland AG 2024

V. Manfredi, *Eucalyptus Kraft Pulp Refining*,
https://doi.org/10.1007/978-3-031-47285-5_12

A single refiner is used when the refiner can supply the specific energy needed to develop the required pulp properties for all target production. Parallel refiners are used when the total mass flow exceeds the capacity of a single refiner [1015]. It is worth remembering that a single refiner is easier to be operated and that series refining cannot be replaced by a single refined pulp recirculation [1374].

A comparative evaluation for series and parallel refining, as illustrated by Fig. 12.1, was carried out in a pilot plant equipped with a 12-inch disc refiner and discs with a 3.0 × 3.0/5° configuration operating at 1185 rpm.

Refining was carried out at 1.0 Ws/m intensity and 4.5% consistency using commercial bleached eucalyptus kraft pulp.

The refining curves were performed as described in Appendix III, with variation of the specific refining energy obtained, in the case of the simulation of single pass refining, by the variation of the suspension flow through the refiner. In the series refining simulation, this variation was obtained through successive passes through the refiner at a flow rate of 10 m^3/h. The results obtained are illustrated by Figs. 12.2, 12.3, and 12.4 in the sequence.

The results indicate that the specific consumption of net refining energy is the same for the two strategies (Fig. 12.2) and there are no differences in the refined pulp quality, neither when the comparison is made at the same paper consolidation (Fig. 12.3) nor at the same paper structure compaction (Fig. 12.4).

12.2 Using Pulp Mixtures

Due to the better formation and good surface properties of papers produced from short fiber pulps, the objective has been to produce papers with the maximum possible content of this pulp. Obviously, this maximum content as possible depends on its capacity to develop all paper properties, especially the mechanical ones.

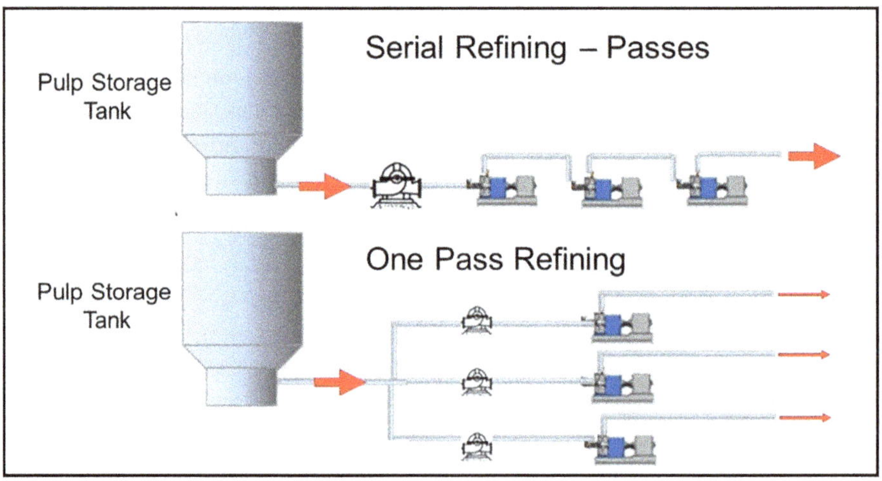

Fig. 12.1 Layout for single pass refining and serial ones

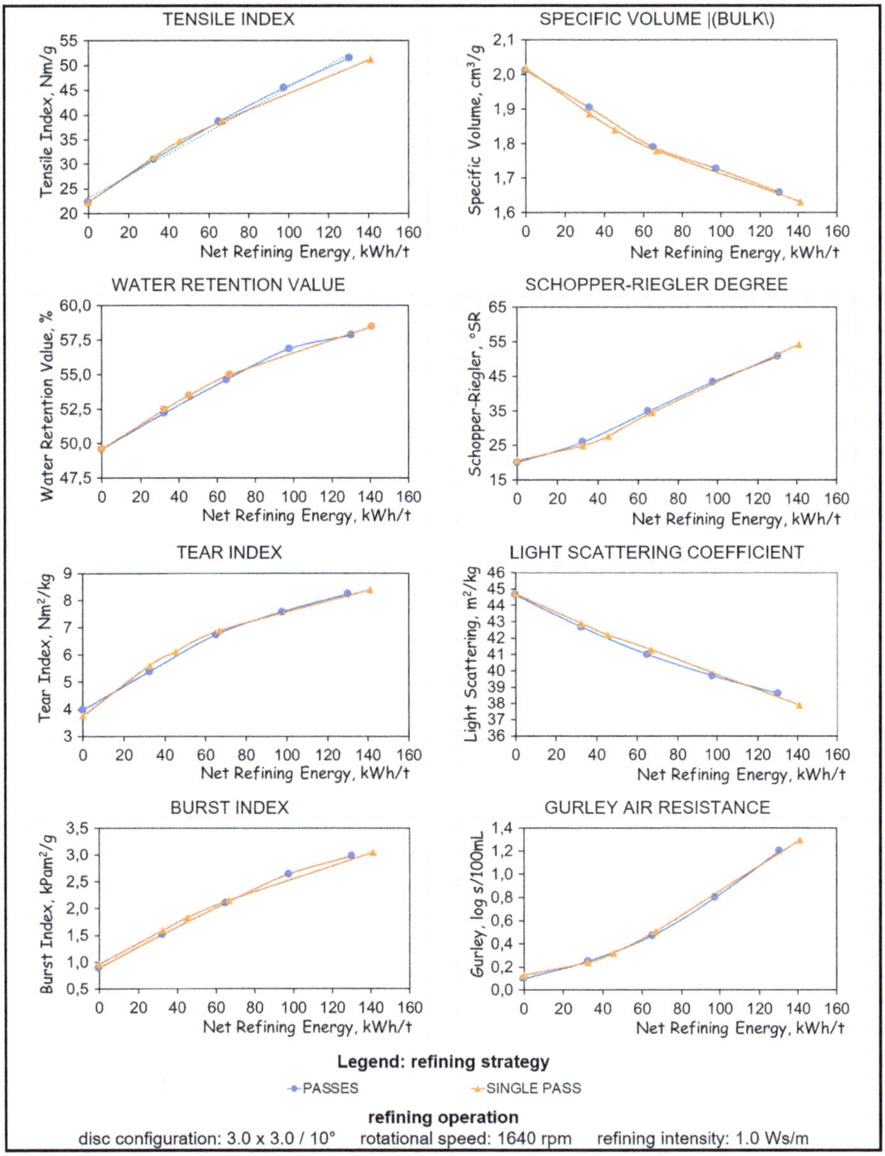

Fig. 12.2 Pulp properties evolution with net refining energy – Single pass vs multiple passes

Pulp mixtures are widely used to obtain paper characteristics, which would not be possible if a single pulp were used. The desired combination of manufacturing costs and paper properties can be achieved by changing the proportions of different fiber types [26, 27, 90, 119, 120, 159, 209, 235, 278, 359, 380, 390, 397, 486, 595, 597, 644, 767, 1137, 1142, 1284, 1311, 1372, 1411].

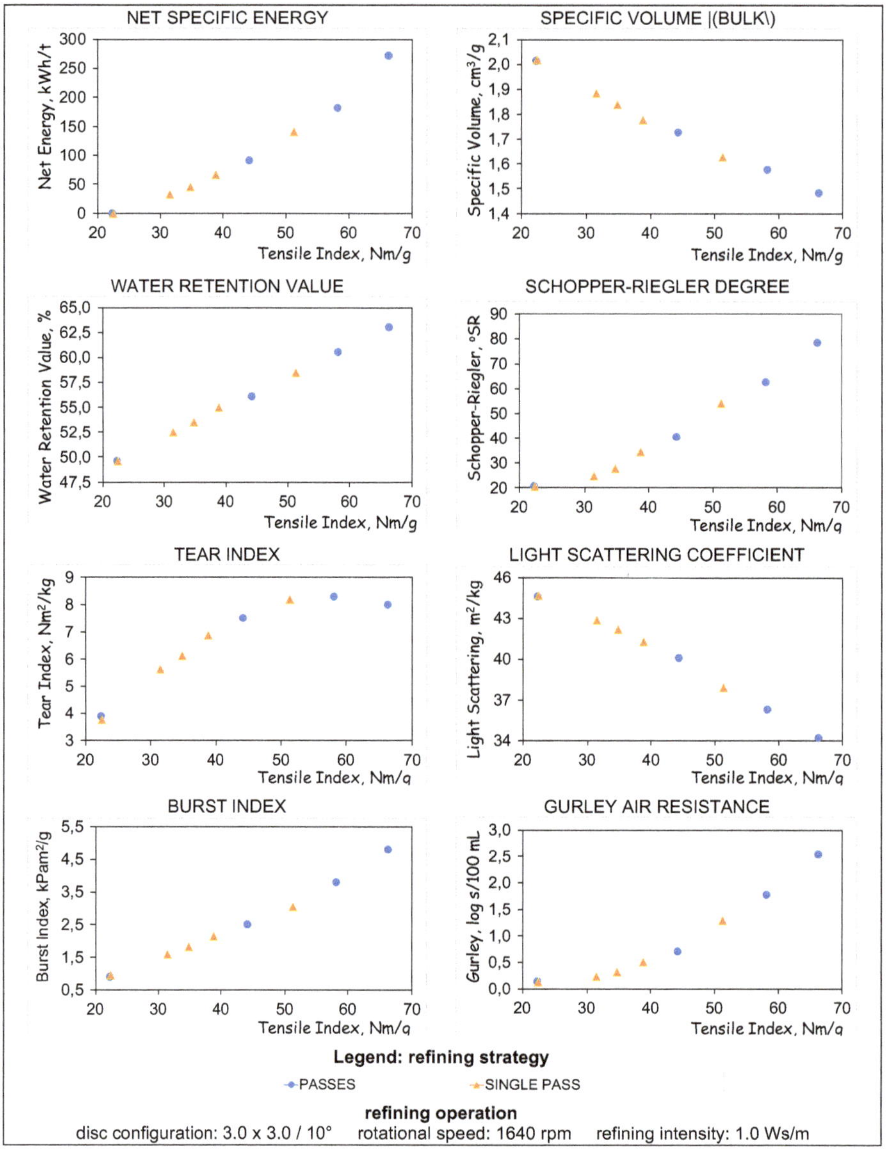

Fig. 12.3 Pulp properties relationship for pulp mixture: single pass vs multiple passes. Comparison at the same tensile index

Eucalyptus pulps have a non-negligible contribution to tear strength, but in papers where high tear strength is desirable, it is common to develop this property through blends of eucalyptus pulps with long fiber pulps [982].

Normally, long fiber pulp is used to reduce drainage resistance and provides wet resistance, ensuring machine performance and mechanical strength of the paper,

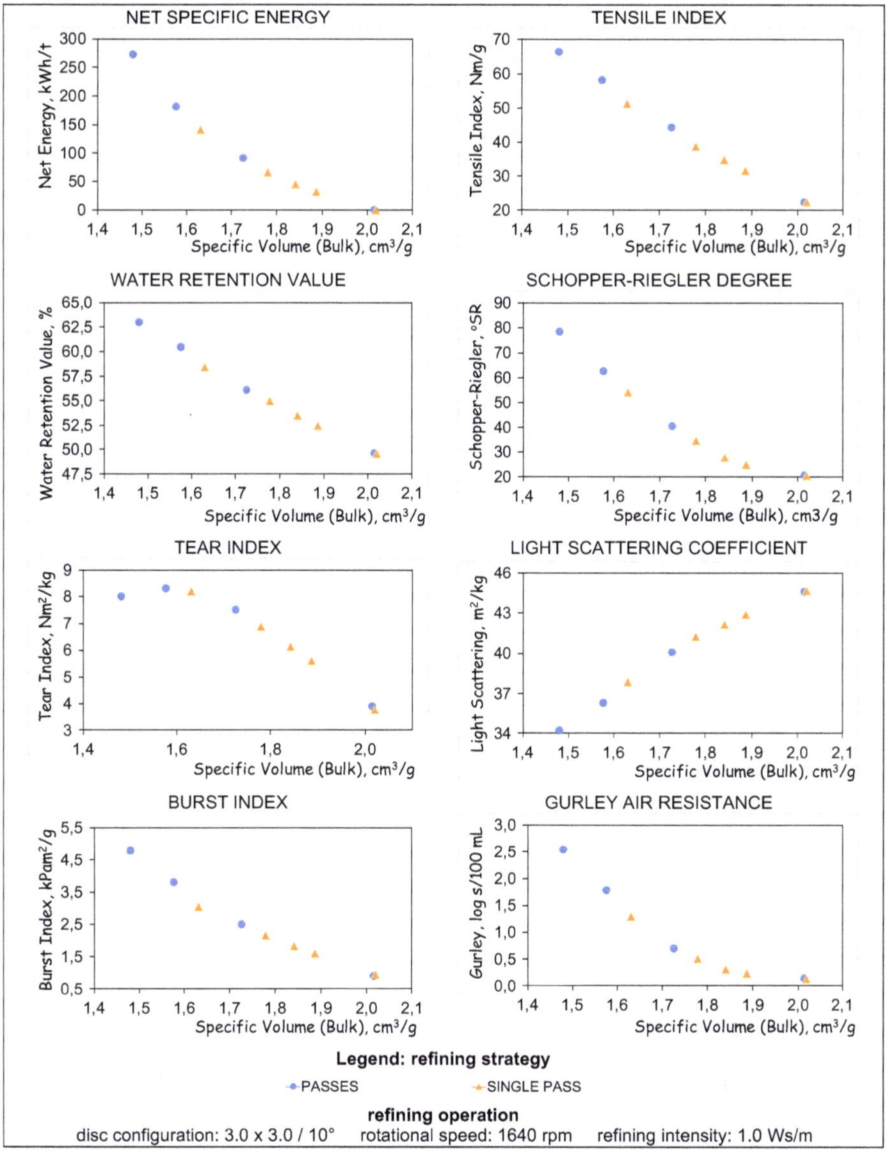

Fig. 12.4 Pulp properties relationship for pulp mixture: single pass vs multiple passes. Comparison at the same specific volume

while short fiber fibers are used to improve printing characteristics, optical properties, paper formation and, in addition, to allow production costs reduction [58, 169, 165, 249, 270, 298, 393, 401, 402, 470, 486, 591, 592, 597, 600, 617, 870, 871, 875, 877, 989, 1007, 1137, 1309, 1411].

The presence of long fibers contributes to the increase in strength while, in the refining of the fiber mixture, it increases the pulp flocs size, accelerates the entrapment of fibers between the bars, and generates, for the same specific refining energy, greater distance between discs. It is believed that, in the refining of fiber blends, the short fibers are protected by the thicker fibers which preferentially receive the refining action [592, 868, 1091, 1118, 1142].

The most used methods when using pulp mixtures are the "mixed refining," when the pulps are mixed before being refined in the same refiner, and the "separate refining," in which the pulps are treated separately for further mixing. These two methods produce different results [209, 262, 402, 470, 742, 766, 869–872, 875, 877, 1104, 1111, 1118, 1136–1143, 1411].

A third way to run the process is a hybrid of the two previous ones, which we call as "sequential refining." The basic process consists of refining the long fiber pulp, which requires a greater amount of energy and different operating conditions, separately and composing the pulp mixture by adding the short fiber pulp to complement the pulp blend [871, 1352].

An illustrative design for these three refining routes for pulps mixtures is presented in Fig. 12.5.

For some blends, refining separately allows greater strength with less energy consumption, but for other blends, refining the mixed pulps can lead to better results [87, 110, 209, 262, 402, 742, 766, 869–872, 875, 877, 1104, 1111, 1113, 1118, 1137–1143].

The final behavior will depend on the type of fibers involved in blends. Thus, the requirements for different pulps must be established considering the final products and the production capacity of the installation, in addition to the flocculation characteristics of the pulps, which influence the hydrodynamic behavior of the fiber suspension [156, 166, 209, 216, 217, 615, 644, 656, 742, 766, 835, 836, 838, 868–878, 1112, 1118].

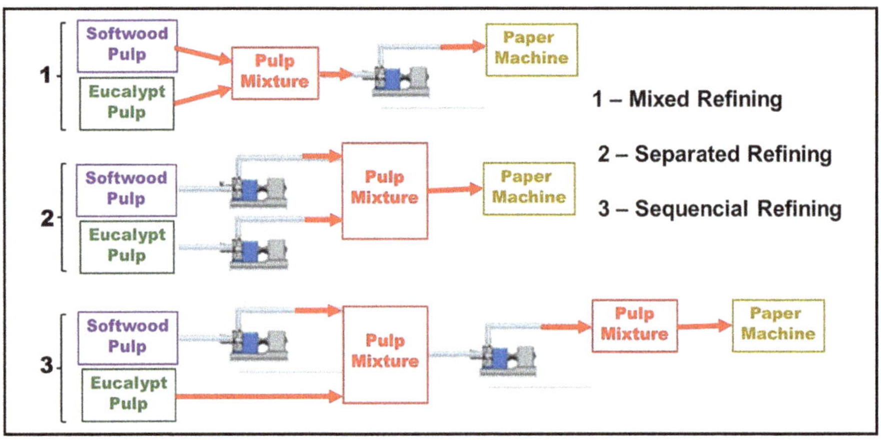

Fig. 12.5 Scheme of the three most common strategies to refining pulp mixtures

The most used strategy is the separated refining, with subsequent mixing of the refined pulps. It is believed that this alternative allows maximizing both the development of paper properties and the refining energy consumption, as it would allow the use of more favorable operating conditions for each type of pulp. However, the specialized literature also reports positive results for the two other forms of [110, 209, 262, 402, 470, 742, 766, 869–872, 875, 877, 1034, 1104, 1111, 1113, 1118, 1137–1143, 1230].

Therefore, the refining of blends of long and short fibers is still a controversial subject.

Short and long fibers pulps have different flocculation tendencies. Therefore, under the same operating conditions, they will show different rheological behaviors. Thus, from a hydrodynamic point of view, they should be treated differently to obtain the maximum potential for each fiber [42, 265, 295, 320, 389, 393, 394, 402, 470, 1085, 1161, 1251].

However, when mixed, two different behaviors can be observed regarding the flocculation tendency [644].

- *Synergism*, which occurs when fibers interact resulting in a property greater than the same property, for the weighted sum based on the fractions of each pulp.
- *Antagonism*, defined for the case where the property of the mixture is lower than the sum of the weighted contributions of the individual components.

The main cause of these differences, as reported by the literature, is related to the flocculation tendency of the mixed pulps. The refining behavior of any pulp is related to its flocculation which is determined by the anatomical and chemical surface fibers characteristics. In this case, mixture refining tends to be advantageous when the addition of long fibers improves the flocculation capacity of the short fibers [156, 166, 209, 216, 217, 615, 644, 656, 742, 766, 835, 836, 838, 868–873, 875, 877, 878, 1112, 1118].

The characteristics related to the fibers collapsibility and their specific surface are, basically, the determining factors of the final paper properties. Thus, results obtained with a given mixture can be greatly modified if the proportion of one of its components is changed or the component itself is replaced. The results obtained cannot be freely generalized, and a case-by-case evaluation is recommended, especially in situations where one or more components of the mixture are replaced.

The evaluation of the refining of mixtures presented in the sequence allowed, in addition to selecting the best refining strategy for the mixture of used pulps, to identify the factor that most contributed to this definition, contributing to clarify the most cited reasons why some authors recommend refining in a mixture, while others recommend refining separately.

Samples of bleached eucalyptus and tropical pine industrial kraft pulps were used, whose main characteristics are shown in Table 12.1. There are, as expected, significant differences in fiber characteristics. In addition to a greater length, long fiber pulp has a greater diameter and cell wall thickness, generating greater coarseness and lower wall fraction than eucalyptus pulp fibers.

Table 12.1 Characterization of the used pulps [1230]

Pulp characteristics		Pulps	
		Eucalyptus	Softwood
Pulp brightness, % ISO		90.6	88.7
Intrinsic viscosity (dm^3/kg)		752	737
Pentosans contents (%)		15.4	8.5
Sedimentation volume – 12 hours (mL)		1370	1860
Kajaani FS 200	Fiber length (mm)	0.72	2.73
	Fiber coarseness (mg/100 m)	9.0	29.8
	Fiber population ($x10^6$)	18.8	2.2
	Fines content in pulp (%)	6.7	20.4
Optical microscopy	Cell wall width (μm)	7.43	11.16
	Lumen diameter (μm)	1.92	15.51
	Cell wall fraction (%)	44.3	29.5

The experiment was carried out in a pilot refining plant equipped with a single 12-inch diameter disc refiner, following the procedure described in Appendix III. All refining curves were performed at 4.5% consistency with the refiner operating at 1185 rpm.

Anatomical evaluation of pulps was done by using optical microscopy or using the Kajaani Fiber Analyzer, according to supplier standards.

In the separate refining, as in the pulp mixture refining, the variation of the specific refining energy was obtained by the variation of the suspension flow through the refiner. In the separate refining of the long fiber pulp, the 5.0 mm × 5.0 mm/5° disc was selected, and in the others, including all the mixtures evaluated, the 3.0 mm × 3.0 mm/5° disc configuration was selected.

In the sequential refining, the tropical pine pulp was first treated as a single pass operation under constant net energy inputs. The refined suspension was collected in the auxiliary tank where eucalyptus pulp was added to meet the target mixture proportion. The consistency was adjusted, and pulp suspension was transferred to the pulper. A new refining curve was made using this pulp blend, in a single pass. The total energy applied was estimated adding the energy used for the softwood pulp to the energy used by the single pass refining taking account of the pulp proportions in the final mixture.

There are several routes for results evaluation. Table 12.2 presents the net specific refining energy demands to meet the target paper structure consolidation as indicated by the tensile index measured in the refined pulps.

In the refining of pure pulps, eucalyptus pulp, when refined under the lowest intensity (0.5 Ws/m), required less specific refining energy to reach the target paper quality, confirming previous results (Chap. 10, Sect. 10.6). Among the mixtures, the refining of the pulp mixture required the lower net specific refining energy to reach the target value of tensile strength than the separated refining (Table 12.2).

The effect of the pulps proportion (mixture composition) was less evident than that of the refining intensity, with the mixture containing 75% eucalyptus

Table 12.2 Net specific refining energy for a target tensile index in the final pulp mixture [1230]

Refining	Pulp in mixture, %		Refining intensity, Ws/m				Net specific energy for target tensile index				
	SW	HW	SW	HW	First pass	Mixture	35 Nm/g	40 Nm/g	45 Nm/g	50 Nm/g	55 Nm/g
Pure pulps	100	–	1.25	–	–	–	36.5	48,6	61,1	74,0	87,3
	–	100	–	0.75	–	–	64.3	117,6	186,2	–	–
				0.50	–	–	34.3	47,8	59,9	10,6	87,8
Pulps in mixtures	25	75	–	–	–	0.50	24.6	33,9	43,5	53,3	63,4
						0.75	31.0	43,2	56,1	69,7	83,9
						1.25	41.8	66,4	96,4	131,6	172,2
	50	50	–	–	–	0.50	23.7	31.7	39.8	48.1	56.6
						0.75	25.8	36.4	48.2	61.1	75.2
						1.25	35.4	53.0	73.7	97.4	124.1
	75	25	–	–	–	0.50	23.9	31.4	39.1	46.9	55.0
						0.75	25.2	36.5	49.5	64.2	80.5
						1.25	29.8	45.1	63.4	84.8	109.3
Mixtures after refining	25	75	1.25	0.50	–	–	45.3	60.7	75.9	90.9	105.7
				0.75	–	–	94.9	154.8	228.6	–	–
	50	50	1.25	0.50	–	–	43.1	59.8	78.0	97.7	119.0
				0.75	–	–	54.8	73.4	92.9	113.1	134.1
	75	25	1.5	0.50	–	–	44.3	59.3	75.2	92.0	109.7
				0.75	–	–	53.2	70.3	87.9	106.1	124.7
Sequential refining	25	75	–	–	1.25 Ws/m in 5 m^3/h	0.50	36.5	48.6	61.7	75.7	90.7
						0.75	43.8	58.4	71.9	84.3	95.7
					1.25 Ws/m in 3 m^3/h	0.50	41.8	53.6	66.5	80.5	95.5
						0.75	43.8	56.6	72.1	90.3	111.3
	50	50	–	–	1.25 Ws/m in 5 m^3/h	0.50	39.1	49.3	60.7	73.3	87.0
						0.75	43.8	57.1	71.0	85.6	101.0
					1.25 Ws/m in 3 m^3/h	0.50	50.0	58.7	68.8	80.4	93.4
						0.75	61.6	67.3	76.1	88.2	103.5

demanding slightly higher energy (2 to 12%) than the mixtures with 50 or 25% of this pulp, in which the consumption of net specific refining energy was similar (Table 12.2).

This result differs from those usually presented in the literature, which indicate a reduction in refining energy consumption when separate refining is adopted [209, 271, 402, 870–872, 875, 877, 1113, 1137, 1352]. Furthermore, separate refining is cited as the most suitable in cases where the hardwood pulp exceeds 50% of the blend [877].

The sequential refining showed intermediate net specific energy consumption. They were lower when the pulp mixtures were refined, in a second pass, under the lowest intensity 0.3 Ws/m. The lower energy consumption of this strategy, when compared to mixing the pulps after refining, is in agreement with information in the literature [169, 389, 470, 486, 598, 871, 875, 877, 989, 1284, 1352].

For the same type of study, the literature [216, 1139] reports that refining pulp blends is advantageous when the addition of long fibers improves the flocculation tendency of the short fiber. This was the trend observed in this study as evidenced by the sedimentation volumes of the mixtures in relation to pure pulps (Fig. 12.6).

The long fiber pulp forms the largest floc sedimentation, while the short fiber pulp forms the smallest one.

In the mixtures, synergy occurred and, probably, the eucalyptus fibers occupied empty spaces inside the longer fiber's flocs forming new flocs with greater internal

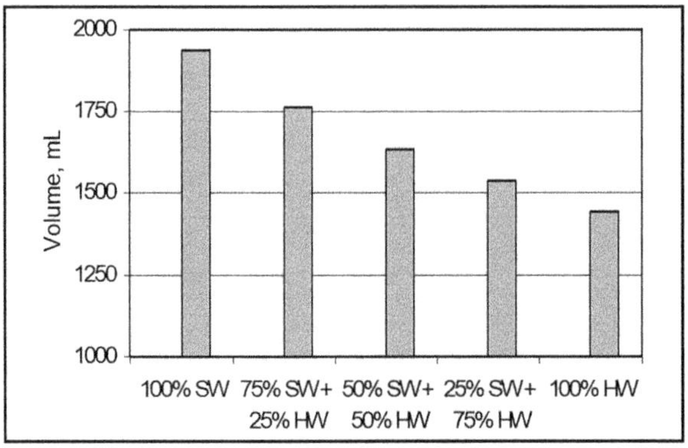

Fig. 12.6 Mixtures and pure pulps sedimentation volumes [1230]

surface for contacts between the fibers, which are easier to be treated in the refining [209, 216, 217, 402, 470, 654, 804, 836, 838, 868, 870–872, 875, 877, 1104, 1112, 1113, 1137].

Another way of observing the same trend is by the wall fraction of the long fibers, which is much lower than the traditional long fibers used in the studies of mixtures, mostly carried out in the Northern Hemisphere. Such studies, as a rule, conclude that separate refining is the most suitable strategy for mixing long fibers and short fibers. These fibers, pine from the Northern Hemisphere, are more rigid than the tropical pine used in this evaluation.

Thus, an interesting way to assess whether two different pulps, when used in a mixture, should be refined in a mixture or separately is to check if there is a synergy flocculation between them. A simple way to do this is to run a sedimentation test; it is suggesting a 12-hour test using pulp mixture at same composition to be used in the paper machine.

In beakers, with the pure pulps and with the mixtures to be evaluated, in suspension in a concentration of 2 g/L, the sedimentation volume after 12 hours resting is measured. If the sedimentation level of the mixtures is intermediate to the pure pulps, there is a synergy flocculation, and the best strategy is to refine the pulps in mixture.

In the presented evaluation, mixture refining was the form of treatment with the lowest refining energy consumption to obtain the target value for the tensile strength (paper consolidation). It is the simplest and most economical way to treat both pulps as indicated by the mixture's sedimentation volume in Fig. 12.6.

The interactions within the paper properties of the mixtures considered, that is, the quality of the final product, are illustrated in Fig. 12.7 in which the evolution of the pulp mixtures properties during refining is plotted as a function of the mixture's composition and the refining intensity at the same paper structure consolidation, the same criterion used to select the blending strategy with the lowest refining energy requirement.

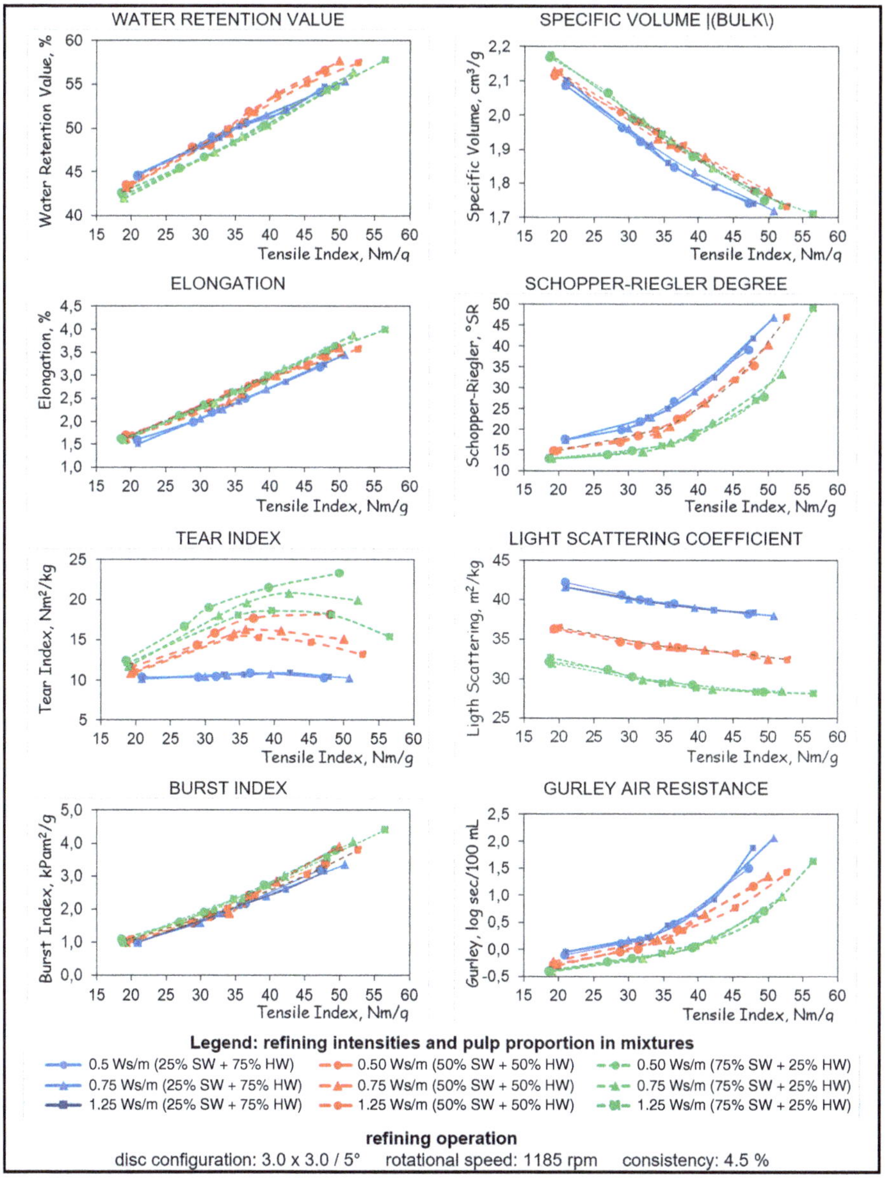

Fig. 12.7 Pulp properties relationship for pulp mixtures as a function of the pulp proportion in mixtures. (Adapted from [1230])

The effect of the proportion of each type of fibers in mixture on paper properties is evident. The most accentuated differences are observed for porosity, light scattering, Schopper-Riegler degrees, and Elmendorf tear strength.

The increase in the short fiber content in the mixture fibers in mixtures resulted in grater values for Schopper-Riegler degrees, light scattering, and porosity. For tear strength, the situation was reversed; mixtures with higher contents of long fibers, which have thicker cell walls, and, therefore, with greater intrinsic strength gave the highest tear strength results.

12.3 Pulp Fractionation

Fractionation is the mechanical process of separating the original pulp into fractions with different physical characteristics such as length (the most common form), width, wall thickness, or coarseness. It provides papermakers with the possibility of selectively processing virgin or recycled fiber separating short fibers stock (fine fraction) and long fibers stock (coarse fraction). As a result, each fiber type, or fraction, can be selectively treated to meet production and product quality requirements [45, 325, 391, 579, 615, 772, 837, 850–858, 1268, 1324].

The two basic types of equipment used in industrial fractionation are screens and hydro cyclones. Generally, screens fractionate according to fiber size and hydro cyclones according to fiber density [851]. Although fractionation has been known for many years, its use has been limited to mechanical pulps and recycled fibers [996–998].

The pulp fractionation before refining aims to separate the fibers reducing the variability in relation to the original pulp to apply the most adequate refining for each fraction obtained, under specific operating conditions to maximize the results of each fraction and then mixed for sending to the paper machine [997]. Driven by the need to improve strength, save energy, and develop new types of paper, the association of fibers fractionation and refining of long fibers has become a flexible tool for optimizing both the chemical and mechanical properties of the fibers [391, 837, 850, 851, 854–858, 996, 1268].

Coniferous wood has growing rings due to different growing rates throughout the year generating spring and late wood. Late wood fibers tend to be longer and have much thicker cell walls. Thus, the fractionation of this pulp generates fractions with different flocculation trends and refining load capacity. The fraction that forms smaller pulp flocs, the springwood fraction, does not tolerate high refining intensities as is the case of the original pulp or latewood fiber fraction [772, 1159]. The long fibers fraction is processed through high or low consistency refining to develop the desired strength. After refining, the long fiber portion is recombined with the short fiber fraction. As a result, only a portion of the entire stock stream is refined, so a smaller refiner is needed. Meanwhile, energy consumption is reduced, that is, the operating cost decreases [619, 855].

Another example, now involving short fiber pulps, is fractionation for vessel elements separation in order to reduce problems with vessel picking or speckles during off-set printing. The vessels, as they have lesser thick walls than the fibers, will be concentrated in the fine fraction of the pulp that may be diverted from the production line and used for conversion into nanocellulose or sent for refining treatment under specific conditions only to destroy vessel elements [336, 851].

In terms of machine performance (runability), reducing the fines content improves drainage in paper formation, consistency after wet presses, and steam consumption in drying. For this reason, fractionation can be considered for managing the pulp fines content giving to operators one of the simplest and most convenient methods to control machine behavior as the properties of the finished product [115].

When a paper mill has two or more machines, the management of fines, via white water, can be even more effective distributing fines in the right dosages in one or another machine, according to the quality of the paper to be produced [115, 855].

Compared to whole pulp refining, the association of pulp fractionation followed by a fractional selective refining can be used to improve the paper strength and surface properties allowing the achievement of unique paper properties [391, 400, 772, 837, 996].

The long fiber fraction can be used as a reinforcing component since low intensity refining improves the tensile strength-drainage ratio. The short fiber fraction can be used, after light refining, in the production of board surface layers as it provides a smooth surface [772].

12.3.1 Fractionation by Pressurized Screens

The most efficient and practical way to fractionate fibers by length is by means of pressurized sieves. Its operation involves the preferential dynamic passage of pulp fiber suspensions and other particulate solids through perforated cylinders with the aid of a fast-moving rotor [853, 1220].

Pressurized screens work well for long fibers and OCC. As the fibers of eucalyptus pulps are much smaller than those long fibers, the use of screens has limitations being recommended the use of fractionation by density difference, which is carried out by hydro cyclones.

12.3.2 Fractionation by Hydro Cyclones

Hydro cyclones, also called “centrifugal cleaners,” use centrifugal forces generated by the swirling motion of the pulp suspension to separate solid particles in the suspended fluid. They are low cost, smaller than other separators, simple in design, easy to operate, and have low maintenance costs [851].

It has been known since the early 1960s that hydro cyclones can be used not only to remove pulp contaminants but can also be used to separate fibers and fiber components based on their different flux densities. Fibers with thick walls are preferentially segregated in the “rejected fraction” (underflow), and fibers with thinner walls are segregated as “accepted” fraction (uperflow).

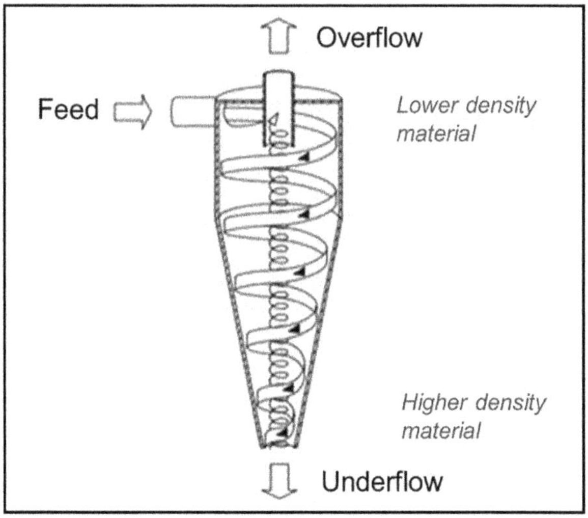

Fig. 12.8 Flow pattern in a hydrocyclone. (Adapted from [1070])

The difference in the fibers apparent density is the main controlling factor [325, 615, 847–851, 1249, 223] (Fig. 12.8).

Only a few studies have been carried out on eucalyptus fractionation using hydro cyclone showing that the different fractions have different chemical and morphological compositions and, therefore, also different papermaking properties.

Hydro cyclones have relatively high vessel separation efficiency, with high treatment capacity and selective tendency to separate larger vessel elements as reject (underflow).

The "accepted fraction" tends to produce denser papers, with greater tensile and tear resistances than the "reject fraction." The fibers in the "accepted fraction" have greater conformability and, therefore, better bonding capacity [391, 400, 615]. In this way, there is the possibility of separating the pulp into two distinct fractions, with different impacts on the paper properties development.

In hydro cyclones, the fibers are subjected to two main forces in the radial direction, generated by the rapid rotation of the suspension. There are two forces involved in the hydro cyclone segregation: a net centrifugal force (FC), causing the fibers to move away from the wall if the density of the fiber is greater than that of water, and a drag force (DF), acting in the opposite direction to the net centrifugal force, determining fibers movement due to the pressure and shear action on the fibers surface [56, 843–849, 850].

$$FC = \left(\rho_p - \rho_l\right)\frac{\mu_t^2}{rp}V_p \tag{12.1}$$

$$FD = \frac{1}{2}\rho_l U_s^2 A_p C_d \tag{12.2}$$

where:
ρ_p *density of particles*
ρ_l density of fluid,
μ_t tangential velocity of fluid,
rp radial position,
V_p volume of particle,
A_p projected area of particle on a plane perpendicular to the direction of motion,
U_s relative radial velocity between particle and fluid,
C_d drag coefficient

12.4 Fiber Swelling

Free swelling is the effect of water adsorption by pulp fibers. In this case, free swelling refers to the treatment only with the alteration of the ionizable charges of the cell wall (fiber swelling), without mechanical action or agitation of the suspension. Free swelling has a significant impact both on water retention by the fibers and on paper mechanical properties as discussed in Chap. 7, Sect. 7.2.2.

There is the possibility of substituting part of the mechanical action in the refiner for the pulp suspension-free swelling through the introduction of chemical charges in the slurry [300, 747, 1171, 1172]. This was demonstrated when assessing the effect of suspension pH on refining results (Chap. 9, Sect. 9.2.10).

Changing the pulp suspension pH, as a pre-treatment of the fibers, reduces the refining energy demand. The fibers would be placed in the sodium form by sodium hydroxide treatment increasing the pH to dissociate the carboxyl groups from the cell wall improving the swelling properties of the fibers [159, 433, 582, 704, 733, 737].

The more efficient refining of pulps in the Na^+ form can be explained as an interaction between a higher osmotic pressure in the cell wall and the mechanical stress applied during refining. This is a "assisted refining by electrostatic repulsion" because when mechanical forces are applied on the fibers, the electrostatic repulsion forces due to ionization contribute to the delamination of the cell wall, increasing its swelling [433].

However, we must consider there is a retention time for ion exchange to occur. Therefore, we should not consider the pH effect if the pulp suspension is sent to the refining facility right after the addition of sodium hydroxide but after adequate retention time. In other words, it is not the pH of the suspension that defines the result but the pH inside the cell wall.

There is also the possibility of applying a similar treatment, but after refining.

If the suspension pH was changed after refining, an increase in fiber hydration (WRV) is also expected [712, 737, 785, 1172], and the refining energy could be reduced proportionally to the hydration obtained with the free swelling treatment after refining.

This refining strategy deserves to be better evaluated. However, the practical inhibitor for its industrial implementation is the pulp suspension pH control since this step would occur exactly where chemicals and various papermaking inputs that require specific pH levels are currently added in the fibers suspension to be sent to the paper machine.

12.5 Pulp Recirculation Trough the Refiner

Due to financial evaluation, the most common pulp preparation installations are those requiring the lowest initial investment and allow the lowest operational cost as possible.

Oversized systems can be installed in situations where future increases in production are projected. However, such systems are typically inefficient as they do not allow satisfactory results in addition to increasing operational and maintenance costs [1390]. Cavitation occurs when the refiner is oversized as the fiber layer thickness between the discs is small, and even absent, in some regions of the discs. This process is common in low consistency refining and occurs because the fluid pressure reaches a level below the liquid vapor pressure. Vapor cavities, or bubbles, are formed in the fluid due to pressure reduction. When the pressure is re-established, these vapor bubbles implode [67, 909, 935, 1051, 1390].

To avoid cavitation, the refiner internal pressure must be increased which is normally done recirculating the pulp suspension. This recirculation increases the outflow maintaining a constant flow through the refiner. Properly dimensioned, mass recirculation also provides operational advantages during system start-up or during sheet breaks in paper machines [67, 714, 841, 891, 1015, 1100, 1104, 1318, 1372, 1374, 1390].

The recirculation flow must be directed to the pump suction and not to the feed tank. In addition, the refining energy (kWh/t) must be allocated considering the incoming flow and not the flow through the refiner [841].

Even if the specific energy is balanced for recirculation, equivalent to that of a single pass, the total energy consumption is higher in relation to refining without recirculation because, in the case of recirculation, the no-load power component is constant at each pass through the refiner [1318, 1387].

In addition to increasing costs, another disadvantage in recirculating refined pulp is the fines content increase. This contributes to make more difficult drainage in paper machine.

In the case of automatic refining control based on pulp drainage rates, for example, this increasing on fines content introduces variation in the amount of refining energy applied as the flow through the refiner [67, 96, 1096].

12.6 High Consistency Refining

The process was developed in the 1960s to improve paper's stretching characteristics. This is particularly important for some paper grades such as sack and knapsacks being the actual technical standard for these stretch paper grades [862, 1315].

The pulp is pumped from storage at low consistency to a dewatering device such as a screw or press where it is dewatered to high consistency (20–35%) and transferred to a leveling conveyor that feeds the refiner [860, 861].

A high consistency refiner differs significantly from refiners used in low consistency refining. It has low no-load power and operates at speeds of around 1500–1800 rpm. The high speed of rotation and the greater length of bar edges increase the number of crossings points between bars reducing the refining intensity [862].

Under high consistency, the pulp is less susceptible to refiner loading. High consistency refining favors fiber hydration because a less drastic action prevails on their surfaces due to the greater number of fibers per refining surface unit resulting in a lower load per fiber. The refining is less intense making the fibers more flexible and generating a small reduction in the drainage capacity due to a lower tendency to form fines [52, 59, 312, 315, 484, 861, 879, 1023, 1026, 1090, 1106, 1359].

Compared to low consistency conditions, the main characteristics of high consistency refining include greater average fiber length, more fiber deformations, greater wet mat stretching, greater tear resistance, and higher specific volume but lower resistance to traction and modulus of elasticity [216, 243, 1271, 1359].

With high consistency refining, vessel elements from short fiber pulps can be destroyed or size reduced [862] reducing problems whit vessel picking and speckles.

Chapter 13
Refining Process Optimization

All theoretical models, or "theories" of refining, are based on the type and extent of the mechanical action exerted on the suspended fibers. Most of these models consider the need to use two process parameters to obtain a satisfactory correlation with the pulp properties development. These two parameters are the net refining energy and the treatment intensity, as defined by the specific edge load.

The energy-based models consider the applied energy (specific refining energy) and the way in which this energy is applied (refining intensity) during the disc bar. As these models consider how energy is transferred by the refining elements, this intensity is classified as a "machine intensity."

In models based on the forces developed inside the refiner, the rheological behavior of the fiber suspension is considered. This behavior depends on the characteristics of the fibers, especially those determining their flocculation tendency. The treatment reference becomes how the fibers receive the energy transmitted by each impact, being important the number and length of impacts as well as the specific energy per impact. The refining intensity, in these cases, is defined as "specific intensity," that is, the amount of energy received by pulp mass unit.

There is also the concept of refining as a fatigue process, which considers the number of stress cycles applied and the specific energy applied in each compression cycles.

Several published studies indicate that two parameters alone are not adequate to characterize refining. The initial models, based on energy and machine intensity, did not consider the contribution of fiber characteristics, whose importance has been continuously reported.

In this book, we detail, based on experimental results, the effects of several refining operating variables. As the specific edge load (SEL) model is the most used by papermakers and researchers, it was used as reference. The results confirmed that two parameters are not enough to adequately characterize the refining process in addition to the specific refining energy (SEC) and the refining intensity (SEL), it is

© The Author(s), under exclusive license to Springer Nature Switzerland AG 2024

V. Manfredi, *Eucalyptus Kraft Pulp Refining*,
https://doi.org/10.1007/978-3-031-47285-5_13

necessary to consider two additional parameters: the refining consistency and the fiber characteristics, that is, the pulp to be refined.

The pulp to be refined adds information about the fiber flocculation tendency, which, associated with the pulp consistency, defines the rheological behavior of the suspension.

In papermaking, pulps typically represent the main component of operating costs, but inputs such as energy, water, steam, and chemicals also have a significant impact on total costs. The mass preparation is the biggest consumer of electric energy in a paper machine, being the refining under low consistency the great villain in this story.

Thus, improvements in refining that allow the reduction of energy demand will result in significant savings, improving the paper machine competitiveness in the markets in which it operates [300, 401, 619].

In stock preparation, refiners usually have a high operating cost generated both by energy consumption and by the used disc configuration and size, sealing water consumption and maintenance costs. Opting for a better refiner size operating under optimized way can result in significant savings, minimizing operating and maintenance costs [401, 883, 886, 983].

There are two types of refining optimization. The quantitative optimization, which corresponds to the reduction of energy consumption to obtain the desired pulp properties and the qualitative optimization, obtained by increasing the bonding capacity between the fibers by straightening the fibers and by the reduction of fines generation, that is, by obtaining the desired balance between the pulp properties. In practice, these optimizations are required simultaneously [505, 1109, 1389].

Methods for reducing energy consumption vary in complexity and investment. Refining must be operated under optimal conditions to ensure maximum operating efficiency. However, many refiners are not properly sized or are not properly maintained, which results in high no-load power reducing their efficiency [678, 710, 1096, 1111, 1317, 1320].

From the four operating variables that must be used for the refining characterization, three (the pulp itself, the refining intensity, and the refining consistency) must be managed so that the objective of reducing energy consumption, which corresponds to the fourth variable, is obtained.

Energy optimization does not mean a reduction in energy consumption, but a better use of the used energy. The energy consumed in refining is composed of empty power plus net power, which is the fourth variable that characterizes the process.

Pulp is the main component of paper manufacturing cost. Thus, it must be carefully selected in such a way that, even in the unrefined stage, it presents properties closer to those desired for the finished product, which allows for a lower application of specific energy in refining. However, in some situations like produce paper whit a specific properties relationship, this is not possible.

In these cases, there is the possibility of combining different pulps in mixtures, or even fractioning the pulp using only the fraction that allows the desired interaction between the paper properties. This is the case, for example, of increasing the specific volume (bulk) without harming the physical strength of the paper.

The pulp mixture or pulp fraction corresponds to a "new pulp" and the specific energy required will be determined by its characteristics, in particular its flocculation tendency (rheological behavior). The pulp consistency and suspension flow rate through the refiner are also important operating variables related to the specific refining energy.

For the same nominal production of the refiner, maintaining the net specific energy and the refining intensity, the pulp consistency is related to the permanence time of the fibers inside the refiner. The installation should be operated at the maximum stable consistency possible to ensure the longest retention time and thus increase the probability of impacting the fibers [1085, 1116, 1372]. However, it is essential to associate consistency with the disc configuration so that its increase does not compromise the pulp flow through the disc channels.

Flow variations through the refiner must be avoided as they also cause variation on the specific refining energy and, consequently, in the quality of the refined pulp. Thus, the flow must be constant and fixed at a value that, associated with consistency, fills the refiner's nominal capacity, ensuring high operating efficiency.

As demonstrated in discussions of refining operating variables effects, intensity reduction (SEL) is beneficial in all cases and for any pulp type. Therefore, the refining intensity should be the lowest as possible.

It is at this point where the main difficulties arise for the papermaker because, considering the operating variables set in which he is used to working, the reduction of intensity means reducing the useful refining power or increasing the rotation speed of the discs [1005, 1114, 1386].

Net refining power reduction, as a way of reducing the refining intensity while maintaining the specific energy applied, leads to production losses because production must be reduced proportionately to keep the same finished product quality. This increases the specific cost of the paper produced because, in addition to lower production, the refiner empty power remains the same.

To maintain production, the alternative is to increase the refiner rotational speed, increasing the operating edges length allowing an increase in the net power applied while maintaining the refining intensity. However, the specific cost also increases due to the higher refiner no-load power.

The exchange of refiners for another one with a larger diameter is an alternative to be considered only in cases of the facility rebuilding aiming at increasing production or, if allowing, a reduction in the number of refiners in operation. In these two cases, there is an increase in refiner no-load power that must be carefully analyzed. What needs to be implemented is the reduction of the installation's no-load power (lower specific no-load power, in kW/ton), using combinations with the lowest no-load power as possible [816, 1085, 1115, 1372].

If switching to larger refiners allows for such a reduction in relation to the sum of refiners at the previous facility, and this reduction amortizes the investment in new equipment, then it is a viable strategy. In the case of installation rebuilding, remember that double disc refiners require less refining energy than single disc ones, which require less energy than conical refiners [883, 946, 983, 1096, 1097, 1317, 1372].

For facilities with two or more refiners, one way to optimize energy consumption is to divide production between them, keeping as many refiners as possible operating at rated capacity and keeping a single refiner for surplus production, even if below the nominal capacity but operating with the target net specific energy. The efficiency of this refiner will be lower than the others, but it will be only one refiner operating outside the rated capacity and with a higher specific consumption of empty power. However, as discussed in Chap. 12, Sect. 12.1, changes in the quality of the refined pulp are not expected.

This refiner can be equipped with a speed variator so that the rotation speed variation allows, considering the production complement in this refiner, to operate with the same useful specific energy and the same refining intensity of the other refiners in the installation. In this way, all refiners will operate in the ideal condition of specific energy and refining intensity, optimizing operating costs and refined pulp quality [872, 1067, 1096, 1104, 1106, 1111, 1372, 1393]. This is a strategy that ensures uniformity in the quality of the refined pulp and is advisable in relation to the practice of recirculating refined pulp.

The recirculation of the already refined suspension is traditionally used in installations with oversized refiners. It ensures that refiners operate at full capacity but spend more refining energy.

Another strategy, which is the most common in paper mills, is to change the discs toothers with longer bar edges length using thinner configuration discs, that is, with narrower bars and channels. Other options are to increase the bar inclination angle or, still, the disc rotational speed [221, 816, 1005, 1068, 1085, 1096, 1372, 1390].

Discs with greater edges bar length will allow the reduction of the refining intensity and the consequent reduction of the net specific energy, as long as the other operating conditions are kept constant.

Thus, to optimize refining energy demand, first optimize the disc configuration and then increase consistency as much as this configuration allows without obstructing the pulp flow through the channels [1015].

The increase in the refiner rotation speed must be carefully evaluated because it increases the no-load power. This increasing cannot be greater than the savings in net refining power provided by the reduction of refining intensity due to higher operating edges length allowed by the higher refiner rotational speed.

When changing the discs, some care must be observed to avoid problems. We cannot compromise the pulps suspension flow through the refiner; hence, we must respect a minimum channels width. For eucalyptus pulp, due to the average size of the flocs, the channels must have a width in order of 2.0 to 2.5 mm. With narrow channels, the refiner no-load power increases going in the opposite direction to the optimization. If the channels are wider, the flocs flow through the channels without receiving adequate treatment.

A strategy to increase the bars edge length, without changing the refiner and keeping the minimum channels width and a proper pulp flow through the refiner, is using curved bar discs.

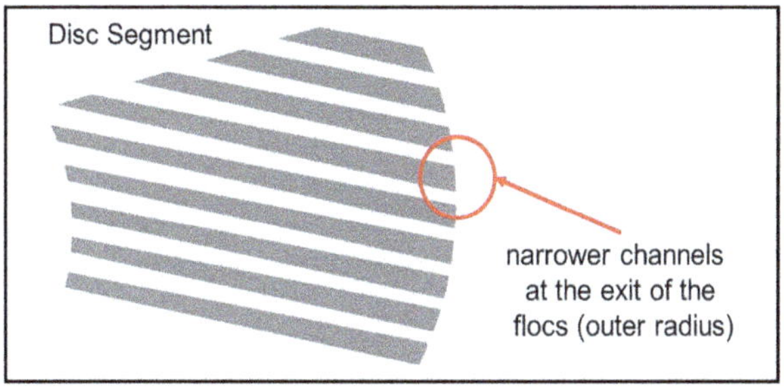

Fig. 13.1 Illustration of the reduction in the width of the channels in the direction from the inner radius to the outer radius of the discs

Comparing discs with the same refining area (external radius minus internal radius), the curved bar discs have greater bar edges length. This is because curved bars with arc-shaped leading edges are longer than a straight bar in the same useful radius.

There is another possibility to increase refining efficiency. During the treatment, the pulp suspension apparent viscosity is reduced [178, 289, 454, 866, 986, 1099, 1125, 1165, 1166, 1252, 1254, 1337]. This means that, in the outside of the discs, there is lower flocs trapping by the bars, as they flow more easily through the channel, than in the disc internal side. Evidence of this lower flocs capture is the more accentuated wear on the bar's surfaces in the external region of the working areas of the discs.

We need to increase the capture of these lower average floc diameter. This can be done by reducing the channel width on the outside of the refiner's working area, allowing better use of refining energy and reducing wear on the bas surfaces that improves the disc time life. This channel width reduction must be gradual from the disc inner radius, as illustrated in Fig. 13.1.

This is a theoretical possibility. In our literature review, we did not identify references about this width channel progressive reduction. However, it makes sense and is a viable alternative for the refining optimization. This is our suggestion.

Legends Used in Equations

A	fiber cross-sectional area, cm^2
AA	average contact angle
AF	crowding factor
AMF	average mechanical force
A_p	projected area of particle on a plane perpendicular to the direction of motion
a_c	radius of curvature of the edge of the bar (sharpness)
B	bar width
BEL	discs edge bar total length
BEL_R	edge bar total length in the rotor disc
BEL_S	edge bar total length in the stator disc
BF	length of impact
BW	mean basis weight, g/m^2
B_s	bar width in stator disc
B_r	bar width on rotor disc
CWS	cell wall surface
C_d	drag coefficient
C_f	pulp consistency, fraction
C_v	volumetric concentration of the pulp suspension
D	channel average deep, m
DD	effective discs diameter
DEF	damage or defect in the fiber cell wall
D_r	channel deep in the rotor disc
D_s	channel deep in the stator disc
d_0	fiber diameter without compression, m
d	fiber diameter, m
E_{net}	specific refining energy, energy/mass unit
E_P	average stripping energy associated with fiber pull-out
E_R	average rupture energy
E_{long}	fiber longitudinal modulus of Young

© The Editor(s) (if applicable) and The Author(s), under exclusive license to Springer Nature Switzerland AG 2024

V. Manfredi, *Eucalyptus Kraft Pulp Refining*,
https://doi.org/10.1007/978-3-031-47285-5

FP	fiber perimeter, cm
FR	normal reaction force
F_c	edge force
F_{max}	maximum shear force per impact
F_n	bar normal force
F_s	bar shear force
f	fraction of the bar that traps fibers
G	groove width
G_r	groove width in rotor disc
G_s	groove width on stator disc
g	acceleration due to gravity, cm/s^2
g_o	distance from the discs at no load power
g_f	distance from the bar edge covered by captured fibers, m
I	energy transferred to fibers by the bar impact
I_e	specific energy (energy per mass unit)
K	channel fraction where fibers are captured
L	length of the refining zone, m
LP	Laplace's pressure
L_n	distance from internal and external radii of the disc working area
$L_{s\gamma}$	extended edge length per second
l	average fiber length
l_B	average bas length
M	fiber mass
N	number of impacts
NF	formation index
N_m	specific number of impacts, number per mass unit
N_P	total number of fibers along the failure line that pulled out of the network
N_R	total number of fibers along the failure line that rupture
n	number of bars/unit arc length m^{-1}
n_c	number of contacts points per fiber
n_r	number of bars/unit arc length of rotor, m^{-1}
n_s	number of bars/unit arc length of stator, m^{-1}
OBL	operational bar length, m/s
P_{net}	liquid (net) refiner power
P_{nol}	no-loaded refiner power
Pt_{ot}	total refiner power
q	pulp flow through the refiner (mass/time unit)
q_m	mass of fibers to be treated
RA	total area of the disc refining zone
RBA	relative bonded area
r	average radius of the refining area
r_1	internal radius of the disc working zone
r_2	external radius of the disc working zone
rp	radial position
rpm	rotational speed, rpm
S	impact relative severity (intensity)

SB shear bond strength per unit area, dynes/cm^2 = 0.1 Pa
SEC specific energy consumption, kWh/t
SEL specific edge load, Ws/m or J/m
SSL specific surface load
S_{eff} cell wall surface available for tensions
s distance on bar movement where force is exerted
T disc gap – distance
TS tensile strength of the sheet (as breaking length)
U work done on fibers
U_s relative radial velocity between particle and fluid
V fiber volume
V_L lumen volume per fiber mass unit (m^3/kg)
V_p particles volume
X_w amount of water absorbed by the cell walls of the fibers (kg of water/kg of fibers)
W_r bar width plus channel – rotor
W_s bar width plus channel – stator
w fiber coarseness
ZS zero-span tensile strength, cm
z_f fraction of bar length occupied by fibers
z_r number of bars in rotor disc
z_s number of bars in stator disc
β disc sectorization angle
γ^* characteristic "cutting" angle
Ø angle of inclination of the disc bars
$Ø_r$ angle of inclination of the rotor disc bars
$Ø_s$ angle of inclination of the stator disc bars
ε ratio of the inter-crossing area versus the global area
ε_x stress on fibers
η refiner energy efficiency
θ angle of conical refiner
μ_E equivalent coefficient of friction
μ_s coefficient of friction of bars
μ_t tangential velocity of fluid
ρ pulp suspension density ($\approx$1000 kg/m^3)
ρ_a paper apparent density
ρ_b apparent density (mass by volume of suspension) (kg/m^3)
ρ_f cell wall density (g/cm^3)
ρ_l density of fluid
ρ_p density of particles
ρ_w water density (kg/m^3)
σ applied stress
σ^{eff} effective stress applied to the cell wall surface
σ_x stress generated in the fiber
τ^f time that a fiber remains in the refiner
ω angular rotational velocity

Appendix I: Traditional Nomenclature for Disc Characterization

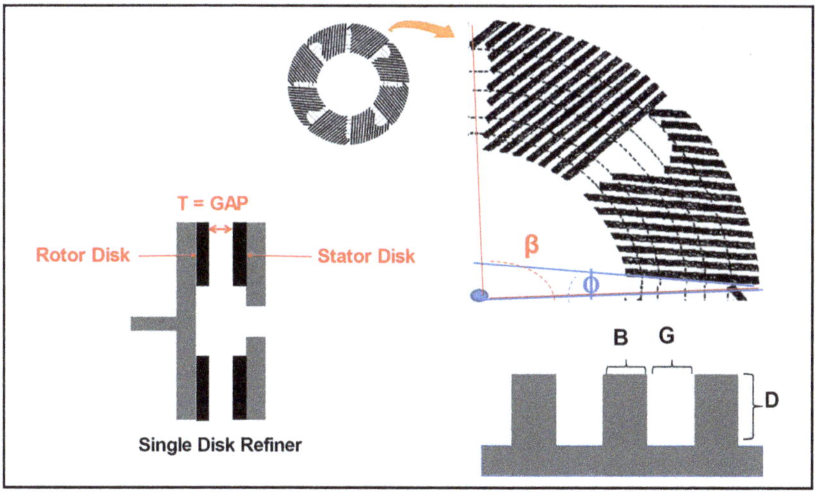

B – width of bars
G – channel width
D – channel depth
Ø – angle of disc bars
β – disc sectorization angle
T – gap clearance (discs distance)

© The Editor(s) (if applicable) and The Author(s), under exclusive license to Springer Nature Switzerland AG 2024
V. Manfredi, *Eucalyptus Kraft Pulp Refining*,
https://doi.org/10.1007/978-3-031-47285-5

Appendix II: Operational Bar Edge Length Example of A Determination Method

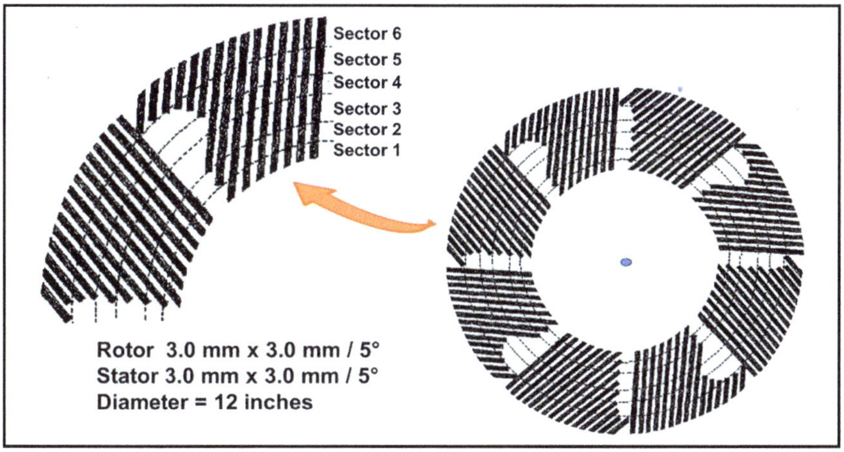

Sector	Number of sectors by disc		Number of bar segments in discs		Average length of bar segments, cm	Bar edge length, cm/rpm
	Rotor	Stator	Rotor disc	Stator disc		
1	4	4	14	14	0.94	2,948
2	4	4	17	17	1.01	4,670
3	4	4	19	19	1.6	9,742
4	4	4	20	20	1.12	7,168
5	4	4	25	25	1.79	13,900
6	4	4	30	30	1,76	25,344
Edge bar length, cm/revolution						63,272 cm/rev
Operational edge length, km/s		890 rpm	$= (63{,}272 \times 10^{-6}) \times 890/60$			9.39 km/s
		1185 rpm	$= (63{,}272 \times 10^{-6}) \times 1185/60$			12.50 km/s
		1640 rpm	$= (63{,}272 \times 10^{-6} \times 1640/60$			17.29 km/s

© The Editor(s) (if applicable) and The Author(s), under exclusive license to Springer Nature Switzerland AG 2024

V. Manfredi, *Eucalyptus Kraft Pulp Refining*,
https://doi.org/10.1007/978-3-031-47285-5

In the mold obtained directly from the discs, five concentric semi-circles were traced, considering six sectors from the inner end of the bars. In each sector, all the attack bars lengths were measured one by one, given the above table.

The sum of the lengths generated in each sector corresponds to the length of the edges of the bars generated by the disc during one rotation.

In this example, disc with configuration 3.0 mm × 3.0 mm/5°, 12 inches in diameter and with the same configuration on the rotor and stator was used. The length of the edges of the bars was determined to be 63,272 cm/revolution.

To calculate the operating length of the bar edges, necessary for the determination of the refining intensity, this value is converted into kilometers and multiplied by the rotation speed (rpm). The values obtained for the three speeds taken into account (890, 1185, and 1640 rpm), which were used in the refining curves presented in this book, are also presented in the table.

Appendix III: Refiner Pilot Plant Configuration and Operating Procedures

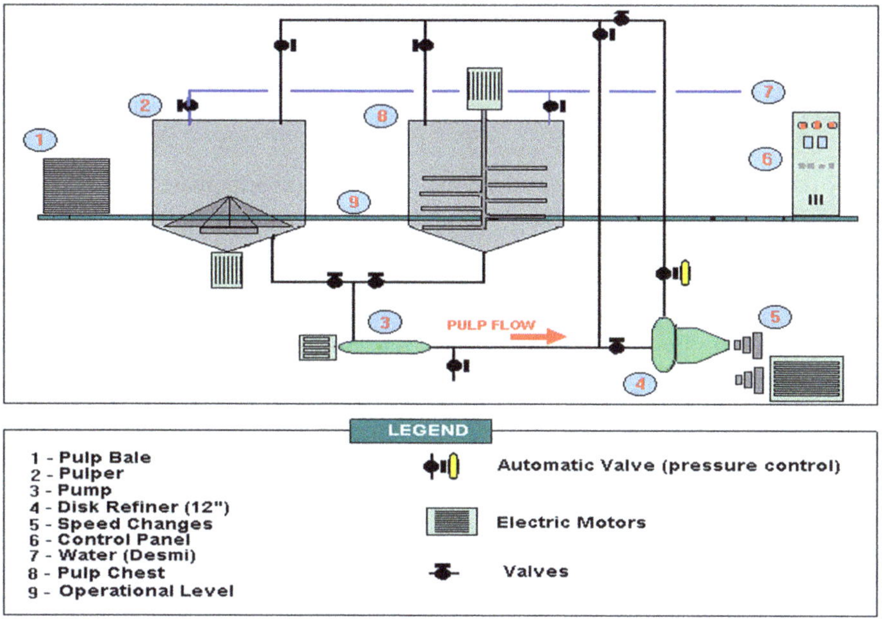

The Refining Facility and How Pulp Was Prepared for Trials

The refining curves were carried out in a pilot plant equipped with a single disc refiner (4), 12 inches in diameter, and equipped with an automatic system to ensure positive pressure inside the refiner. The refiner can be operated at different rotation speeds by means of a pulley system (5) that assures a stable rotational speed during the refining curves.

© The Editor(s) (if applicable) and The Author(s), under exclusive license to Springer Nature Switzerland AG 2024
V. Manfredi, *Eucalyptus Kraft Pulp Refining*,
https://doi.org/10.1007/978-3-031-47285-5

Suspension flow is controlled by a positive displacement pump (3) and can be changed to allow different levels of liquid energy to be applied. The flow rates used were 2.5, 5.0, 7.5, and 10.0 m^3/h, which, keeping the consistency constant, allowed obtaining four different levels of specific energy (kWh/t) applied.

The installation allows refining in a single pass as well as in multiple passes (serial refining) in nominal consistencies ranging from 2.0% to 6.0 %.

The pulp suspension to be refined is prepared by adding the quantity of industrial pulp sheets (1) in the pulper (2) already loaded with the volume of water necessary to meet the objective consistency.

After disaggregation, the suspension was circulated through the refiner (2→3→4→2) during five minutes at a flow rate of 10 m^3/h to homogenize the suspension. During this period, the refiner's no-loaded power was measured, and three samples of the unrefined pulp were collected at the exit of the return pipe to the pulper.

Single-Pass Refining

After the suspension homogenization, the discs were approximated to obtain the total energy required by the desired refining intensity simultaneously with the flow deviation to the pulp chest (8).

At this point, there is an important difference to be observed when comparing industrial refining operation and pilot refining trials.

Pilot refining studies are usually performed at constant specific edge load despite the variation in specific energy. In a mill situation, an increase in specific energy is usually obtained by increasing the power. In other words, at the same time as the specific energy is increased, the specific edge load (intensity) is also increased.

After sampling the refined pulp at the outlet of the feed pipe of the pulp chest (9) the pulp suspension flow was changed to 7.5 m^3/h and the discs were approximated to obtain the total energy required by the desired refining intensity and, after a period of stabilization of the total power, a new sampling was performed at the outlet of the pulp chest supply line (8). The same procedure was repeated for flows of 5 m^3/h and 2.5 m^3/h. After completing the first cycle, the operation was repeated, now in increasing order of flow.

After sampling at 10.0 m^3/h, the cycle was repeated until the pulper suspension was completely consumed. This procedure allows at least three samples of refined pulp per each pulp suspension flow (specific refining energy).

Unrefined and refined pulp samples were treated in the laboratory. The consistency was determined for all samples the consistency was evaluated for the exact determination of the specific refining energy tested.

Subsequently, the samples were prepared for the pulp properties determinations (water retention index and Schopper-Riegler degrees) and used in the formation of laboratory sheets to the paper properties evaluations.

For the analysis of the pulps and the formation and evaluation of the test sheets, standardized methods (Scan Methods) were used.

Serial Refining

The basic procedure was the same as described above. The difference is that the slurry suspension was recirculated through the installation passing through the refiner several times, always at a constant flow rate (5.0 m^3/h).

In the first pass, the suspension was completely transferred from the pulper (2) to the pulp chest (8) with three sampling during this period.

After transferring the entire suspension volume, the pulper and pipes were drained and washed before the experimental sequence.

For the second pass, keeping the same operating conditions, the suspension was transferred from the pulp chest (8) to the pulper (2). After transferring the entire suspension volume, the mass tank and pipes were drained and washed for the experimental sequence. The procedure was repeated as long as suspension was available.

The analysis procedures of the unrefined and refined samples were identical to the samples of refines in a single pass.

References

1. Gatti, T. H. – A história do papel artesanal no Brasil. ABTCP, São Paulo, 2007 127 pgs.
2. Portal São Francisco, A história do papel. https://www.portalsaofrancisco.com.br/historia-geral/historia-do-papel
3. Enciclopedia.com – The spread of papermaking technology into Europe. https://www.encyclopedia.com/science/encyclopedias-almanacs-transcripts-and-maps/spread-papermaking-technology-europe
4. Teixeira, M. B. D.; Oliveira, R. A.; Gatti, T. H. and Suarez, P. A. Z. – O papel: uma breve revisão histórica, descrição da tecnologia industrial de produção e experimentos para obtenção de folhas artesanais. *Revista Virtual de Química, 9* (3): 1364–1380 (2017) http://static.sites.sbq.org.br/vq.sbq.org.br/pdf/v9n3a28.pdf
5. Blechschmidt, J.; Heinemann, S. and Süss, H-U. – Mechanical pulping. *In:* Handbook of Pulp. Volume 2, Section II Ed. Sixta, H. WILEY-VCH Verlag GmbH &Co. KGaA, Weinheim, 2006 pgs. 1069–1148.
6. Bloom, J. M. – Paper before print: the history and impact of paper in the islamic world. Yale University Press, citado por O Estado de São Paulo em "A influência muçulmana na divulgação do papel" (01/27/2002)
7. Whitney, R. P. – The dtory of paper. Tappi Press, Atlanta, 1984, 28 pgs.
8. Scott, W. E., Abbott, J. C. and Trosset, S. – Properties of paper: an introduction. Tappi Press, Atlanta. 191 pgs. (1995).
9. Atchinson, J. E. and McGovern, J. N. – History of paper and the importance of non-wood plant fibers. *In:* Pulp and Paper Manufacture – Volume 1 – Properties of Fibrous Raw Materials and Their Preparation fo Pulping. Ed.Kocureck, M. J. and Stevens, C. F. B. The Technical Committee of the Paper Industry, Atlanta/Montreal, 1983 pg. 154–156.
10. Fritoli, C. L.; Kruger, E.; Carvalho, S. K. de P. – História do papel: panorama evolutivo das técnicas de produção e implicações para sua preservação. *Revista Ibero-americana de Ciência da Informação, 9* (2): 475–502 (2016).
11. The invention of paper: the birth of papermaking. https://paper.gatech.edu/early-papermaking
12. Norman's, J. – Louis-Nicolas Robert invents the papermaking machine. History of Information.com www.historyofinformation.com/detail.php?id=433.
13. Haunreiter, K. J. – 200th Anniversary of the paper machine – the first 100 years. *Tappi Journal, 80* (10): 87–96 (1997).
14. Recicloteca – Papel: história, composição, tipos, produção e reciclagem. *In:* Recicloteca. https://www.recicloteca.org.br/material-reciclavel/papel/
15. Danforth, D. W. – Analyzing stock preparation. *Chemical 26 – Magazine of Paper Processing, 4* (11): 30 (1968)

© The Editor(s) (if applicable) and The Author(s), under exclusive license to Springer Nature Switzerland AG 2024

V. Manfredi, *Eucalyptus Kraft Pulp Refining*,
https://doi.org/10.1007/978-3-031-47285-5

16. Núcleo de Negócios da ABTCP – A história da indústria de celulose e papel no Brasil. ABTCP, São Paulo, 2004 151 pgs.
17. Kanazawa, T. and Fujita, O. K. – The latest refiner and deflaker designed with a twin rotor and stator system. *Jappan Tappi Journal, 52* (4): 62–71 (1998).
18. Clarke, C. R. E. and Wessels, A. M. – Variation and measurements of pulp properties in Eucalyptus wood. *In:* Eucalypt Plantations: Improving Fibre Yield and Quality. CRCTHF/IUFRO, Hobart: 19-24/02/1995 pgs. 93–100.
19. Vaillancourt, L. – Introduction to paper machine terminology. *In:* TECH95 Theory & Practice of Papermaking Course. CPPA, Ottawa: 26/11-01/12/1995 14 pgs.
20. Cotterill, P. P. and Brolin, A. – Improving *Eucalytpus* wood, pulp and paper quality by genetic selection. *In:* Conference on Silviculture and Improvement of Eucalypts. IUFRO, Salvador: 24-29/08/1997 Volume 1 pgs. 1–13.
21. Baker, C. F. – Advances in the practicalities of refining. *In:* Refining and Mechanical Pulping conference. PIRA, Barcelona: 02-03/03/2005 Paper 2.
22. Bugajer, S.; Silva, O. F. and Pires, F. – Influência do pH na refinação de pastas celulósicas. *In:* 17° Congresso Anual da ABCP. São Paulo. 1984 pgs. 31–39.
23. Cotterill, P. and Macrae, S. – Improving eucalyptus pulp and paper quality using genetic selection and good organization. *Tappi Journal, 80* (6):82-89 (1997).
24. Raymond, C. A.; Banham, P. and MacDonald, A. C. – Whitin tree variation and genetic control of basic density, fibre length and coarseness in *Eucalyptus regnans* in Tasmania. *Appita Journal, 51* (4): 299–305 (1998).
25. Schmidt, E. A. – A practical model relating kraft pulping costs to hardwood chemical properties and morphology. *Appita Journal, 58* (3): 218–224 (2005).
26. Johansson, A. – Correlations between fibre properties and paper properties. Master's Thesis in Pulp Technology. Royal Institute of Technology, Stockholm, 2011 49 pgs www.diva-poal.org/smash/get/diva2D:505453/FULLTEXT01
27. Sorieul, M.; Dickson, A.; Hill, S. J. and Pearson, H. – Plant fibre: molecular structure and biomechanical properties, of a complex living material, influencing Its deconstruction towards a biobased composite. *Materials, 9* (618): 1–36; (2016).
28. Rodés, L. – Ensaios históricos – influência da cultura árabe sobre a refinação das polpas celulósicas. *O Papel, 55* (11): 13–15 (1994).
29. Pihlajamäki, P. and Hytonen, H. – Mixed tropical hardwood – a minor and declining source of fibre for paper. Twogether, n° 17: 2–6 (2004).
30. Kellinson, R. C. – Improved wood properties from plantation eucalypt. *In:* Process and Product Quality Conference. TAPPI, Savannah: 17-20/10/1994 pgs. 39–44.
31. Valente, C. A.; de Sousa, A. P. M. de; Furtado, F. P. and Carvalho, A. P. – Improvement program for *Eucalyptus globulus* at Portucel: technological component. *Appita Journal, 45* (6): 403–407 (1992).
32. Lammi, L. and Svedman, M. – Tailoring pulp quality in the pulping process. *In* 2° Seminário de Deslignificação. ABTCP, Ribeirão Preto: 02-03/09/1999 Paper 2 8 pgs.
33. Coradin, V. T. R. and Camargos, A. A. – A estrutura anatômica da madeira e princípios para sua identificação. IBAMA, Laboratório de Produtos Florestais. Brasília, 2002 28 pgs.
34. Hubbe. M. A – Flocculation and redispersion of cellulosic fiber suspensions: a review of effects of hydrodiynamic shear and polyeledtolytes. *BioResources*, 2 (2): 296–331 (2002).
35. Chevalier-Billosta, V.; Joseleau, J.-P.; Cochaux, A. and Ruel, K. – Tying together the ultrastructural modifications of wood fibre induced by pulping processes with the mechanical properties of paper. *Cellulose, 14*:141 –152 (2007).
36. Almgren, K. M. – Wood-fibre composites: Stress transfer and hygroexpansion. Doctoral Thesis No. 9 KTH Fibre and Polymer Technology School of Chemical Sciences and Engineering Royal Institute of Technology, Stockholm, 2010 51 pgs www.diva-portal.org/smash/get/diva2:309582/FULLTEXT01

37. Fahlén J, Salmén L. – Cross-sectional structure of the secondary wall of wood fibers as affected by processing. *In:* 11th International Symposium on Wood and Pulping Chemistry. Nice, 2001 pg. 585.
38. Åkerholm, M. – Ultrastructural aspects of pulp fibers as studied by dynamic FT-IR spectroscopy. Doctoral Thesis. Royal Institute of Technology. Stockholm, 2003 71 pgs http://www.diva-portal.org/smash/get/diva2:9438/FULLTEXT01.pdf.
39. Lindström, T. – Fundamentals of papermaking. *In:* 9th Fundamental Research Symposium. PIRA. Cambridge, 1989 pg. 311.
40. Clarke, C. R. E.; Shaw, M. J. P.; Wessels, A. M. and Jones, W. R. – Effect of differences in climate on growth, wood, and pulp properties of nine eucalypt species at two sites. *Tappi Journal, 82* (7): 89–99 (1999).
41. Arjas, A. – Printability and runnability. *World Pulp and Paper Technology*: 157–158. (1994).
42. Annergren, G. – Fundamentals of pulp fiber quality and paper properties. *In:* Pulping Conference. TAPPI, Orlando. 1999 pgs: 29–39
43. Antolovich, E. – Method of treating bleached pulp on a washer with calcium ions to remove sodium ions. U.S. Patent no 5,273,625 (1993) https://patents.justia.com/patent/5273625.
44. Ampulski, R. S. – Influence of fiber surface charge on tensile strength. *In:* Papermakers Conference. TAPPI, Denver. 1985 pg. 9.
45. Paavilainen, L. – Influence of fibre morphology and processing on the softwood sulphate pulp fibre and paper properties. Doctoral Thesis, Helsinki University of Technology, Helsinki 1993. 155 pgs. www.https://www.elibrary.ru/item.asp?id=6878557
46. Beghello, L. – The tendency of fibers to build flocs. Abo Akademi University. 1998 58 pgs. http://web.abo.fi/fak/tkf/pap/newpaf/luciano-thesis.pdf
47. Lundqvist, S.O.; Grahn, T.; Hedenberg, O.; Hansen, P. – Visions and tools for forest and mill integration. Part 2: Some new approaches under development at STFI for prediction of paper properties. Eurofiber Seminar, STFI, Oral presentation. (2003). http://www.stfi.se/upload/3439/02_d14-3_stfi-lundqvist2_Part%202x.pdf
48. Lyytikainen, K.; Saukkonen, E.; Kajanto, I. and Kayhko, J. – The effect of hemicellulose extraction on fibre charge properties and retention behavior of kraft pulp fibres, *Bioresources*, 6 (1): 219–231 (2011).
49. Magaton, A. S.; Colodette, J. L.; Gouvea, A. F. G.; Gomide, J. L.; Muguet, M. C. S. and Pedrazzin, C. – Eucalyptus wood quality and its impact on kraft pulp production and use. *Tappi Journal, 8* (8): 32–39 (2009).
50. Xiang, Y.; Bousfield, D. W.; Hassler, J.; Coleman, P. and Osgood, A. – Measurement of local variation of ink tack dynamics. *Journal of Pulp and Paper Science, 25* (9): 326–330 (1999).
51. Rossi Filho, S. – Impressões offset: soluções práticas. SENAI, Porto Alegre, 1997 127 pgs.
52. Foelkel, C. E. B. and Dalmolin, I. – Improving eucalyptus pulp refining through the control of pulp consistency and stock pH: comparisons at given bulk and given tensile strength. *In:* Papermakers Conference, TAPPI, Atlanta: 01-04/03/1999 Volume 2 pgs: 751–754
53. Paavilainen, L. – European prospects for using nonwood fibers. *Pulp and Paper International, 40* (6): 61 (1998).
54. Grossmann, U. and Valbusa, L. A. – Formação da folha com o duoformer D e prensas com sapata do tipo flexonip, na fabricação de papéis linearboard, testliner, miolo e cartão. *In:* XXVII Congresso Anual de Celulose e Papel. ABTCP, São Paulo: 07-11/11/1994 pgs. 543–561.
55. Waech, T. G. – Lint reduction by high temperature calendering. *Pulp and Paper Canada 93* (10): 44–47 (1992).
56. Paavilainen, L. – Quality – competitiveness of Asian short-fibre raw materials in different paper grades. *Papperi Ja Puu, 82* (3): 156 (1998).
57. Foelkel, C. – Madeiras para uso celulósico-papeleiro: formação, ultraestrutura, química e topoquímica. *Eucalyptus Newsletter* (84) 645 pgs (2020) http://www.eucalyptus.com.br/news/pt_dez2020.pdf

58. Kibblewhite, R. P.; Bawden, A. D. and Hughes, M. C. – Hardwood market kraft fibre and pulp qualities. *Appita, 44* (5): 325–332 (1991).
59. Chen, T.; Xie, Y.; Wei, Q.; Wang, X.; Hagman, O.; Olov Karlsson, O. and Liu, J. – Effect of refining on physical properties and paper strength of *Pinus massoniana* and China Fir cellulose fibers. *BioResources 11* (3): 7839–7848 (2016).
60. Ferahi, M.; Kortschot, M. T. and Dodson, C. T. J. – Effect of anisotropy on the fracture behaviour of newsprint. *Journal of Pulp and Paper Science, 22* (11): 430–446 (1996).
61. Navaee-Ardeh, S. – A new model for maximizing the bending stiffness of a symmetric three-ply paper or board. *Pulp and Paper Canada, 108* (4): 45–47 (2007).
62. Joutsimo, O. P. and Asikainen, S. – Effect of fiber wall pore structure on pulp sheet density of softwood kraft pulp fibers. *BioResources, 8* (2):2719–2737 (2013).
63. Setasith, S, – Effect of compressive and abrasive refining on structural changes in fiber and paper. Master's Thesis. Aalto University. Espoo, 2014 84 pgs https://aaltodoc.aalto.fi/bitstream/handle/123456789/13458/master_Setasith_Suchart_2014.pdf?sequence=1&isAllowed=y
64. Ghosh, A. K. – Enhancement of paper strength of container materials using wet end additives in acidic system. *Appita Journal, 47* (3): 227–230, 237 (1994).
65. Perez, S. R.;Treier, G. and Rutland, D. – Design to customer requirements: the Xerox paper program. *In:* Process and Product Quality Conference. TAPPI, Savannah, 1994 pgs. 83–98.
66. Morley, M. J. – Wet-end management. *World Pulp and Paper Technology:* 115–116, 118 (1992).
67. Baker, C. – Refining and improved paper machine runnability. *In:* African Pulp and Paper Week. TAPPSA (2004) http://www.tappsa.co.za/archive2/APPW_2004/Title_2004/Refining_and_improved_paper_machine_runnability.html
68. Sigl, R. – Low intensity refining of hardwood and deinked pulps with a new generation of filling. *Twogether – Paper Technology Journal* (8): 7–11 (1999).
69. Ora, M. and Maloney. T. C. – The effect of moisture and structure on wet web strength and its variation – a pilot scale approach using dry and rewetted mill made papers. *In:* 15th Fundamental Research Symposyum, Cambridge, 2013 pgs. 71–100.
70. Uesaka, T. – Principal factors controlling web breaks in pressrooms – quantitative evaluation. *Appita Journal, 58* (6): 426–432 (2005).
71. Ko, Y. C. and Park, J. M. – Engineering cellulose fibers for high-value added Products for pulp & paper industry. *Journal of Korea TAPPI,* 47 (6): 22–40 (2015)
72. Omholt, I. – The effects of curl and microcompressions on the combination of sheet properties. *In:* International Paper Physics Conference. TAPPI, San Diego, 1999 pg 499–515.
73. Lindqvist, H. – Improvement of wet and dry web properties in papermaking by controlling water and fiber quality. Academic Dissertation, Åbo Akademi University. Åbo., 2013 85 pgs
74. Miller, P. R. – Fibre and sheet properties resulting from refining – stock consistency variaton. *Appita Jorunal, 42* (2): 125–130 (1998).
75. Bektals, I.; Tutus, A. and Eroglu H. – A study of the suitability of calabrian pine (Pinus brutia Ten.) for pulp and paper manufacture. Turkish Journal of Agriculture and Forestry 23 (3): 589–597 (1999).
76. Pikulik, I. I. – Wet pressing – operating aspects. *In:* TECH95 Theory and Practice of Papermaking Course. CPPA, Ottawa, 1995 Section C3 29 pgs.
77. Tuderman, M. – Offline paper testing. *World Pulp and Paper Technology*: 183–184 (1991).
78. Tiikkaja, E. – Fibre properties and paper machine runnability. *In:* Engineering/Process and Product Quality Conference & Trade Fair. TAPPI, Anaheim, 1999 Volume 3 pgs. 1241–1242.
79. Riki, J. T. B., Sotannde, O. A. and Oluwadare, A. O. – Anatomical and chemical properties of wood and their practical implictions in pulp and paper production – a review. *Journal of Research in Forestry, Wildlife & Environment, 11* (3): 358–368 (2019).
80. Tiikkaja, E.; Kauppinen, M. and Glorigiano, P. – Fibre dimensions, their effect on paper properties and required measuring accuracy. *In:* XXXI Congresso Anual de Celulose e Papel. ABTCP, São Paulo, 1998 pgs. 397–402.

81. Kramer, J. D. – Pulping eucalyptus – a review. *In:* XXXI Congresso Anual de Celulose e Papel. ABTCP, São Paulo, 1998 pgs.: 615–632.
82. Demler, C. L. and Pitz, M. – Comparison of conventionally, ECF and TCF bleached hardwood pulp refining response. *In:* III International Refining Conference and Exhibition. PIRA/IPST, Atlanta, 1995 Paper 9 8 pgs.
83. Seth, R. S. – The importance of fibre coarseness for pulp properties. *In:* 77th Annual Technical Meeting. CPPA, Montreal: 1991 pg. 251.
84. Malan, F. S.; Male, J. R. and Venter, J. S. M. – Relationship between the properties of eucalyptus wood and some chemical, pulp and paper properties. *Paper Southern Africa, 14* (1): 6–16 (1994).
85. Barrichelo, L. E. G. – Estudo das características físicas, anatômicas e químicas da madeira de *Pinus caribeae* Mor. Var. hondurensis Barr. Golf para a produção de celulose kraft. Tese de Livre Docência – ESALQ/USP (1979).
86. Kärenlampi, P. – The effect of pulp fiber properties on the tearing work of paper. *Tappi Journal, 79* (4): 211–216 (1996).
87. Silva Jr., F. G. da; Valle, C. F. do e Muner, J. C. G. – Programa de qualidade da madeira da Votorantim Celulose e Papel – VCP. *O Papel, 57* (1): 35–43 (1996).
88. literature review. *In:* IV International Refining Conference. PIRA, Fiuggi, 1997 Paper 1 pgs. 5–43.
89. Kärenlampi, P.; Retulainen, E. and Kolehmainen, H. – Properties of kraft pulps from different forest stands – theory and experiment. *Nordic Pulp and Paper Research Journal, 9* (4):214–218 (1994).
90. Retulainen, E.; Moss, P. and Nieminen, K. – Effect of fines on the properties of fibre networks. *In:* 10th Fundamental Research Symposium Proceedings. Mechanical Engineering Publications, London, 1993 Volume 2 pgs. 727–769.
91. Seth, R. S. – The importance of fibre straightness for pulp strength. *Pulp and Paper Canada, 107* (1): 34–42 (2006).
92. Crotogino, R. H. – Machine calendering – recent advances in theory and practice. *In:* TECH95 Theory & Practice of Papermaking Course. CPPA, Ottawa, 1995 Section F1 13 pgs.
93. Luce, J. – Paper structure. *In:* Wet-End Operations Short Course. TAPPI, Cincinnati, 1994 Paper 2 pgs. 9–23.
94. Fonseca, S. M. da; Oliveira, R. C. de e Silveira, P. N. – Industrial tree selection: procedures, risks, costs and benefits. *In:* Eucalypt Plantations: Improving Fibre Yield and Quality Conference. CRCTHF/IUFRO, Hobart, 1995 pgs. 15–19.
95. Raymond, C. A. – Genetic control of wood and fibre traits in eucalypts. *In:* Eucalypt Plantations: Improving Fibre Yield and Quality Conference. CRTHF/IUFRO, Hobart, 1995 pgs. 49–52.
96. Piirainen, R. and Paavilainen, L. – Fiber length measurement in the pulp and paper industry. *In:* International Process and Materials Quality Evaluations Conference. TAPPI, Atlanta, 1986 pgs. 67–73.
97. Bawden, A. D. and Kibblewhite, R. P. – Effects of multiple drying treatments on kraft fibre walls. *Journal of Pulp and Paper Science, 23* (7): 340–346 (1997).
98. Sjöström, K. and Teder, A. – Changes within the modified kraft process aiming to improve TCF Bleachability. *Journal of Pulp and Paper Science, 25* (5): 176–182 (1999).
99. Hatton, J. V. and Cook, J. – Kraft pulps from second-growth Douglas Fir: relationships between wood, fiber, pulp, and handsheet properties. *Tappi Journal, 75* (1): 137–144 (1992).
100. Hatton, J. V. – Kraft pulping of second-growth Jack pine. *Tappi Journal, 76* (5): 105–113 (1993).
101. Hatton, J. V. and Gee, W. Y. – Kraft pulping of second- growth lodgepole pine. *Tappi Journal, 77* (6): 91–102 (1994).
102. Philipp, P. and D'Almeida, M. L. O. – Celulose e papel – tecnologia de fabricação de papel – Volume 2. SENAI/IPT, São Paulo, 1988 402 pgs.

103. Pinheiro, A. L. – Considerações sobre a taxonomia, filogenia, ecologia, genética, melhoramento florestal e a fertilização mineral e seus reflexos na anatomia e qualidade da nadeira. SIF, Viçosa, 1999 144 pgs.
104. Purdy, T. – New techniques for surface assessment. *In:* Papermaking. World Pulp and Paper Technology, 1994/1995: 181–183.
105. Ferreira, P. J. and Figueiredo, M. M. – Efeito do cozimento e da refinação nas dimensões transversais das fibras de *E. globulus*. *O Papel, 62* (01: 73–80 (2000).
106. Alber, W. and Erhard, K. – Differentiation of commercial paper pulps in terms of their behavior in paper manufacturing. *Papier, 50* (10A): 24–30 (1996).
107. Miranda, I.; Almeida M. H. and Pereira, H. – Variation of fibre biometry in different provenances of eucalyptus globulus Labill. *Appita Journal,* 54 (3): 272 (2001).
108. Green, S. I. – Pulp fibre drag coefficient. *Appita Journal, 59* (2): 120–126 (2006).
109. Pere, J.; Sika-Aho, M. and Vikar, L; – Biomechanical pulping with enzymes: Response of coarse mechanical pulp to enzymatic modification and secondary refining. *Tappi Journal, 83* (5): 8 pgs (2000).
110. Ramires, H.; Oliveira, R. C.; Rubini, B.; Demuner, B. J.; Pavan, P. – Aumento da proporção de fibra de eucalipto na formulação de papeis tissue e avaliação de suas propriedades. *In:* 46th International Pulp and Paper Congress, ABTCP, São Paulo, 2013 9 pgs.
111. Fiserová, M.; Gigac J.; Stankovská e Opálená, E. – Influence of bleached softwood and hardwood kraft pulps on tissue paper properties. *Cellulose Chemistry and Technology., 53* (5–6): 469–477 (2019)
112. Mithrush, T. L. – An experimental study of fluid flow in a low consistency refiner. Master's Thesis. The University Of British Columbia, Vancouver, 2013 102 pgs www.open.library.ubc.ca/soa/cIRcle/collections/ubctheses/24/items/1.0165714
113. Ratnieks, E. and Foelkel, C. E. B. – Uma discussão teórioco-prática sobre polpas de eucalipto para a fabricação de papel "tissue". *In:* XXIX Congresso Anual de Celulose e Papel. ABTCP, São Paulo, 1996 pgs. 717–734.
114. Gigac, J. and Fišerová, M. – Influence of pulp refining on tissue paper properties. *Tappi Journal, 7* (8): 27–32 (2008).
115. Foelkel, C. – The *Eucalyptus* fibers and the kraft pulp quality requirements for paper manufacturing. *Eucalyptus Online Book & Newsletter.*- Chapter 3 (Feb/Mar) 2007 – 42 pgs www.eucalyptus.com.br/capitulos/ENG03_fibers.pdf
116. Stankovská, M.; Fišerová, M.; Gigac, J. and Opálená, O. – Blending impact of hardwood pulps with softwood pulp on tissue paper properties. *Wood Research, 65* (3): 447–458 (2020).
117. Gonçalves, D. – The eucalyptus fiber for tissue papers. *In:* 7th Brazilian Symposium on the Chemistry of Lignin and Other Wood Components. UFMG, Belo Horizonte, 2001 p. 317–323.
118. Ratnieks, E. – The effect of refining consistency on *eucalyptus* pulp strength. *In:* XXVI Congresso Anual, ABTCP, São Paulo (1993). 12 pgs.
119. Watson, P. – High quality NBSK fibre for premium tissue paper grades. *Tissue World Magazine*, May/June, (2014).
120. Sacon, V.; Menocelli, S. and Ratnieks, E. – Misturas de polpas brasileiras com o eucalipto. *In:* XXVII Congresso Anual de Celulose e Papel, ABTCP, São Paulo, 1994 pgs. 463–473.
121. Zhang, C.; Chen, M.; Keten, S.; Coasne, B.; Derome, D. and Carmeliet, J. – Hygromechanical mechanisms of wood cell wall revealed by molecular modeling and mixture rule analysis. *Materials Science, 7* (37): eab8919 (2021).
122. Xu, L.; Filonenko, Y; Li, M. and Parker, I. – Measurement of wall thickness of fully collapsed fibres by confocal microscopy and image analysis. *Appita Journal, 50* (6): 501–504 (1997).
123. Fellers, C. and Carlsoon, L. A. – Handbook of physical and mechanical testing of paper and paperboard, Volume 1, p. 323, Marcel Dekker Inc., New York, 1993.
124. Botková, M.; Šutý, Š.; Jablonský, M.; Kučerková, L. and Vrška, M. – Monitoring of kraft pulps swelling in water. *Cellulose Chemistry and Technology, 47* (1–2): 95–102 (2013).

125. Guo, X.; Dong, J.; Liu, H.; Duan, C.; Yang, R. and Qi, K. – Effect of combined refining plates with different bar angles on paper properties during mixed pulp refining. *Journal of Korean Wood Science and Technology, 48* (5): 581–590 (2020).
126. Page, D. H. – A theory for tensile strength of paper. *Tappi, 52* (4): 674–681 (1969).
127. Koubaa, A e Koran, Z. – Measure of the internal bond strength of paper/board. *Tappi Journal, 78* (3): 103–111 (1995).
128. Niskanen, K.; Kajanto, I. and Pakarinen, P. – Paper structure. *In:* Papermaking Science and Technology. Volume 16 – Paper physics, Chapter 1. Ed. Gullichsen, J. e Paulapuro, H. Finnish Paper Engineers' Association/Paperi ja Puu Oy, Helsinki, 2000.
129. Sjöberg, J. C.; Häggquist, M.; Wikström, M.; Lindström, T. and Höglund, H. – Effects of pressurized high consistency refining on sheet density. *Nordic Pulp and Paper Research Journal, 23* (1): 39–45 (2008).
130. Wathén, R.; Rosti, J.; Alava, J. M. and Salminen, L. – Fiber strength and zero-span strength statistics – some considerations. *Nordic Pulp and Paper Research Journal, 21* (2): 193–201 (2006).
131. Suurnäkki, A.; Oksanen, T.; Grönqvist, S.; Orlandi, M.; Canevali, C.; Pere, J. and Viikari, L. – Target modification of fibre surface properties. *In:* 59th Annual Conference. APPITA, 2005 pgs. 315–316.
132. Pulkkinen, I.; Alopaeus, V.; Fiskari, J. and Joutsimo, O. – O uso de dados sobre espessura da parede da fibra para predição das propriedades de folhas de laboratório de polpa de eucalipto. *O Papel, 69* (19): 71–85 (2008).
133. Waterhouse, J. – Formation measurements and paper quality. *World Pulp and Paper Technology,* (9): 107 (1998).
134. Komarov, V. I.; Kazakov, Y. V. – Svyaz' fundamental'nykh svoitstv (po Klarku) nerazmolotoi sul'fatnoi nebelenoi tsellyulozy s kharakteristikami deformativnosti i prochenotsi. *Izv. Vyssh. Uchebn. Zaved. Lesn. Zh.* (2/3): 112–116 (1993).
135. Xu, J. – A physical model for three-D paper structure, drainage, formation and retention. *In:* Papermakers Conference. TAPPI, Philadelphia, 1996 pg. 5.
136. Kiviranta, A. J. – Fiber and forming-related mechanisms affecting formation. *In:* Papermakers Conference. TAPPI, Philadelphia, 1996 pgs. 239–245.
137. Condon, M. – Mechanical Aspects of Forming and formation, a current review. *In:* Papermakers Conference. TAPPI, Philadelphia, 1996 pgs. 253–273.
138. Waterhouse, J. F. – On-line formation measurements and paper quality. *In:* Papermakers Conference. TAPPI, Philadelphia, 1996 pgs. 275–288.
139. Laleg, M. and Nguyen, N. – The specific perimeter of fibre flocs – a criterion of paper formation. *Journal of Pulp and Paper Science, 21* (10): 356–361 (1995).
140. Norman, B. and Wahren, D. – Mass distribution and sheet properties of paper. *In:* The Fundamental Properties of Paper Related to Its Uses. Ed. Bolam., F. BPPIF, London, 1976 pgs. 7–73.
141. Jokinen, O. and Ebeling, K. – Flocculation tendency of papermaking fibres. *Paperi ja Puu, 67* (5): 317, 318, 320, 321, 323–325 (1985).
142. Jang, H. F. – Measurement of fibril angle in wood fibres with polarization confocal microscopy. *Journal of Pulp and Paper Science, 24* (7): 224–230 (1998).
143. Chirat, C. and Lachenal, D. – Brushing up on bleaching techniques. *Pulp & Paper International, 41* (10): 41–43 (1999).
144. Johnson, J. A.; Bennett, K. A. and Montrey, H. M. – Failure phenomena. *In:* Handbook of Physical and Mechanical Testing of Paper and Paperboard. Ed. Mark, R. E., Marcel Dekker Inc, New York, 1983 pgs. 145–253.
145. Farnood, R. R.; Loewen, S. R. and Dodson, C. T. J. – Estimation of intra-floc forces. *Appita, 47* (5): 391–396 (1994).
146. Lyne, M. B. and Hazell, R. – Formation testing as a means of monitoring strength uniformity. *In:* The Fundamental Properties of Paper Related to Its Uses Symposium. BPBIF, Cambridge, 1976 Volume 1 pgs. 75–103.

147. Hashemi, S. J.; Tawfik, R. P.; Berniė, J-P. and Douglas, W. J. M. – Partitioning formation into its components for analysis of paper strength properties. *In:* Engineering/Process and Product Quality Conference & Trade Fair. TAPPI, Anaheim, 1999 Volume 2 pgs. 665–683.
148. Kajanto, I. M.; Komppa, A. and Ritala, R. K. – How formation should be measured and characterized. *Nordic Pulp and Paper Research Journal, 1* (3): 219–118 (1986).
149. Norman, R. J. – Dependence of sheet properties on formation and forming variables. *In:* Consolidation of Paper Web Symposium, BP&BMA, London, 1966 pgs; 269–298.
150. Dodson, C. T. J. – A survey of paper mechanics in fundamental terms. *In:* The Fundamental Properties of Paper Related to Its Uses Symposium, BP&BIF, London, 1976 Volume 1 pgs. 202–226.
151. Algar, W. H. – Effect of structure on mechanical properties of paper. *In:* Consolidation of the Paper Web Symposium, BP&BMA, London, 1965 Volume 2 pgs. 814–851.
152. Batten Jr. G. L. and Nissan, A. H. – Unified theory of the mechanical properties of paper and other H-bond-dominated solids. Part 1. *Tappi Journal, 70* (9): 119–123 (1987).
153. Paavilainen, L. – Importance of cross-dimensional fibre properties and coarseness for the characteristics of softwood sulphate pulp. *Paperi ja Puu, 75* (5): 35–43 (1993)
154. Korteoja, M. J.; Salminen, L. I.; Niskanen, K. J. and Alava, M. – Statistical variation of paper strength. *Journal of Pulp and Paper Science, 24* (1): 1–6 (1998).
155. Wong, L.; Kortschot, M. T. and Dodson, C. T. J. – Effect of formation on local strain fields and fracture of paper. *Journal of Pulp and Paper Science, 22* (6): 213–219 (1996).
156. Yan, H.; Norman, B.; Lindström, T. and Ankefors, M. – Fibre length effect on fibre suspension flocculation and sheet formation. *Nordic Pulp and Paper Research Journal, 21* (1): 30–35 (2006).
157. Yan, H. and Söderberg, D. – Two-dimensional wavelet flocculation analysis of fibre suspensions and paper sheets. *Nordic Pulp and Paper Research Journal, 21* (1): 13–18 (2006).
158. Huber, P.; Roux, J-C.; Mauret, E. and Carré, B. – Characterizing the floc size distribution of flowing suspensions: from fibre flocculation to sheet formation. *Appita Journal, 59* (1): 37–43 (2006).
159. Lindholm, C.-A. – Determining optimum combinations of mechanical pulp fractions. *Paperi ja Puu, 65* (4): 243–245, 247–250 (1983).
160. Vuorinen, T.; Buchert, J.; Teleman, A.; Tenkanen, M. and Fagerström, P. – Selective hydrolysis of HexA groups and its application in ECF and TCF bleaching of kraft pulps. *In:* International Pulp Bleaching Conference. TAPPI, Atlanta, 1996, pgs. 43–51.
161. Ramezani, O. and Nazhad, M. M. – The effect of refining on paper formation. TAPPSA Technical Articles http://www.tappsa.co.za/html/The_effect_of_refining_on_paper_formation.html
162. Kajanto, I. – The effect of formation on print quality with wood free offset papers. *Nordic Pulp and Paper Research Journal, 4* (1): 8 (1989).
163. Helmer, R. J. N.; Covey, G. H.; Raverty, W. D.; Vanderhoek, N, Sardjono, A. A. and Chau, T. – Laboratory simulation of the effects of refining on paper formation. *Appita Journal, 59* (4): 291–296 (2006).
164. Ramezani, O. and Nazhad, M. M. – The effect of coarseness on paper formation. *In:* African Pulp and Paper Week. TAPPSA, (2004) http://www.tappsa.co.za/archive2/APPW_2004/Title2004/The_effect_of_coarseness_on_paper_formation.
165. Gurnagul, N., Page, D. H. and Seth, R. S. – Dry sheet properties of Canadian hardwood kraft pulps. *Journal of Pulp and Paper Science, 16* (1): 36–41 (1990).
166. Kerekes, R. J. and Schell, C. J. – Effects of fiber length and coarseness on pulp flocculation. *Tappi Journal, 78* (2): 133–139 (1995).
167. Higgins, H. G. – Speed reduction in disc refining: A strategy for saving energy. *In:* TAPPI Papermakers Conference. TAPPI, Portland, 1983 pgs. 141–142.
168. Fahey, H. – Testing and control. *In:* Pulp and Paper Manufacture, Volume 2, Mechanical Pulping. Third Edition, Chapter 21 R. A. Leask, Ed. Joint-Textbook Committee, Technical Section. CPPA/TAPPI. pgs. 252–271.

169. Walmsley, M. R. W.; Weeds, Z. and Atkins, M. – Pulp consistency and rotor back-flushing effects on pressure screen reject thickening. *Journal of Pulp and Paper Science, 34* (1): 59–68 (2008).
170. Tamolang, F. N. and Wangaard, F. F. – Relationships between hardwood fiber characteristics and pulp-sheet properties. *Tappi Journal, 44* (3): 201–206 (1961).
171. MacLeod, J. M. – Kraft pulping from Canadian wood species. *Pulp and Paper Canada, 87* (1): 76–81 (1986).
172. MacLeod, J. M. – Comparing pulp strengths. *Pulp and Paper Canada, 81* (12): 128–132 (1980)
173. Blechschmidt, J. – Papierqualitat: begriffe und anforderungen. *Wochenblatt Papierfabrik. 122* (3): 73–75 (1994).
174. Chatterjee, A.; Dodson, C. T. J. and Whiting, P. – Formation and distribution of fiber length and flexibility. *Nordic Pulp and Paper Reserch Journal, 9* (2): 120–124 (1994).
175. Cresson, T. and Luner, P. – Simulation of paper formation. *Espra Research Reports,89* (III): 104–105 (1988).
176. Görres, J; Luner, P e Grant, P. – Simulation of paper structure using an image analysis system. *Espra Research Reports, 80* (XI): 84 (1984).
177. Freeman, W. L. – Chemical related factors influencing the balance of formation, drainage, and retention. *In:* Papermakers Conference. TAPPI, Philadelphia, 1996 pgs. 247–252.
178. Lundin, T. – Tailoring pulp fibre properties in low consistency refining. Academic dissertation, Åbo Akademi University, Abo, 2008 259 pgs www.researchgate.net/publication/261286869_Tailoring_pulp_fibre_properties_in_low_consistency_refing_diss/link/543cfc4e0cf2c432f7423263
179. Pikulik, I. I.; McDonald, J. D.; Mentele, C. J. and Lange, D. V. – The refining, forming, and pressing on fine paper quality. *Tappi Journal, 81* (6): 122–130 (1998).
180. Bither, T. W. and Waterhouse, J. – Strength development through refining and wet pressing. *Tappi Journal, 75* (11): 201–208 (1992).
181. Foelkel, C. E. B.; Mora, E. and Menochelli, S. – Densidade básica: sua verdadeira utilidade como índice de qualidade da madeira de eucalipto para a produção de celulose. *In:* VI Congresso Florestal Brasileiro. SBS/SBEF, Campos do Jordão: 1990 pgs. 719–728.
182. Lynd, L. R.; Weimer, P. J.; van Zyl, W. H. and Pretorius, I. S. – Microbial cellulose utilization: fundamentals and biotechnology. *Microbiology and Molecular Biology Reviews, 66,* (3): 506–577 (2002) https://doi.org/10.1128/MMBR.66.3.506-577.2002
183. Görres, J., Amiri, R.; Grondin, M. and Wood, J. R. – Fibre collapse and sheet structure. *In:* 10th Fundamental Research Symposium, PIRA, Oxford, 1993 pgs. 385–310.
184. Kibblewhite, R. P. and McKenzie, C. J. – Kraft fibre property variation among 29 trees of 15 year old *Eucalyptus fastigata* and comparison with E. nitens. *Appita Journal, 52* (3): 218–225 (1999).
185. Pulkkinen, I.; Fiskari, J. and Aittamaa, J. – The effect of refining on transverse dimension distributions of eucalypt pulp speies. *O Papel, 67* (5): 58–72 (2007).
186. Hubbe, M. A.; Venditti, R. A.; Barbour, R. L. and Zhang, M. – Changes to unbleached kraft fibers due to drying and recycling. *Progress in Paper Recycling, 12* (3):11–20 (2003).
187. Yan, N. and Aspler, J. – Surface texture controlling speckle-type print defects in a hard printing nip. *Journal of Pulp and Paper Science, 29* (11): 357–362 (2003).
188. Fiadeiro, P.T.; Pereira, M.J.T. Jesus, M.E.P. and Silvy, J.J. – The surface measurement of fibre orientation anisotropy and misalignment angle by laser diffraction. *Journal of Pulp and Paper Science, 28* (10): 341–346 (2002).
189. Nanri, Y. and Uesaka, T. – Dimensional stability of mechanical pulps – drying shrinkage and hydroexpansivity. *Tappi Journal, 76* (6): 62–66 (1993).
190. Pulkkinen, I. – From eucalypt fiber distributions to technical properties of paper. Doctoral Thesis. Aalto University School of Science and Technology, Espoo, 2010 74 pgs https://research.aalto.fi/en/publications/from-eucalypt-fiber-distributions-to-technical-properties-of-pape

191. Courchene, C. E.; Peter, G. F. and Litvay, J. – Cellulose microfibril angle as a determinant of paper strength and hygroexpansivity in *Pinus taeda L. Wood and Fiber Science, 38* (1): 112–120 (2006).
192. Hagerty, G. A.; Walkinshaw, J. W. and Foley, P. M. – Design of a modern air permeability instrument comparable to Gurley-type instruments. *Tappi Journal, 76* (2): 97–100 (1993).
193. Dodson, C. T. J. and Sampson, W. W. – The effect of paper formation and grammage on its pore size distribution. *Journal of Pulp and Paper Science, 22* (5): 165–169 (1996).
194. DuPlooy, A. B. J. – The relationship between wood and pulp properties of *E. grandis* (Hill ex-maiden) grown in South Africa. *Appita, 33* (4): 257–264 (1980).
195. Law, K. N. and Koran, Z. – Effect of press drying on paper properties. *Appita, 34* (5): 38–390 (1981).
196. Silva, R. P. and Oliveira, R. C. de – O efeito da refinação na recuperação das propriedades físico-mecânicas de papéis reciclados de pinus. *O Papel, 63* (8):87–99 (2003).
197. Settlemeyer, L. A. – A new laboratory method for testing paper plister in web-offset lithography. *Tappi Journal, 78* (2): 187–191 (1995).
198. Pikulik, I. I. – Pressing – theoretical aspects. *In:* Theory & Practice of Papermaking Course. CPPA, Ottawa, 1995 Section C1 18 pgs.
199. Castanho, C. G. and Oliveira, R. C. de – Recuperação e avaliação do rejeito fibroso industrial da polpação kraft de eucalipto para produção de papel. *In:* III Seminário de Tecnologia Papeleira. ABTCP, São Paulo: 2000 pgs. 96–106.
200. Beuther, P. D.; Veith, M. W. and Zwick, K. J. – Characterization of absorbent flow rate in towel and tissue. *Journal of Engineered Fibers and Fabrics, 5* (2) (2010)
201. Chen, Y.; Wan, J.; Zhang, X.; Ma, Y. and Wang, Y. – Effect of beating on recycled properties of unbleached eucalyptus cellulose fiber. *Carbohydrate Polymers, 87*: 730–736 (2012)
202. Dean, G. H. – Objectives for wood fibre quality and uniformity. *In:* Eucalypt Plantations: Improving Fibre Yield and Quality Conference. CRCTHF/IUFRO, Hobart: 19-24/02/1995 pgs. 5–9.
203. Forgacs, O. L. – The Characterization of Mechanical Pulps. *Pulp and Paper Magazine Canada, 64* (Convention Issue): 89–115 (1963).
204. Kärenlampi, P.; Rantanen, R.; Hämäläinen, T. and Suur-Hamari, H. – Opacitt, smoothness and toughness of mechanical printing papers: the effect of softwood kraft pulp properties. *In:* Papermakers Conference. TAPPI, Philadelphia, 1996 pgs. 153–160.
205. Seth, R. S.; Jang, H. F.; Chan, B. K. and Wu, C. B. – Transverse dimensions of wood pulp fibres and their implications for end use. *In:* Fundamentals of papermaking materials, Ed. Baker, C. B., Pira International, Leatherhead, 1997 Volume 1 pgs. 473–503.
206. Park, J. Y.; Melani, L.; Lee, H. and Kim, H. J. – Effect of pulp fibers on the surface softness component of hygiene paper. *Holzforschung, 74* (5): 497–504 (2019)
207. Kibblewhite, R. P., Evans, R. and Riddell, M. J. C. – Handsheet property prediction from kraft-fibre and wood-tracheid properties in eleven radiata pine clones. *Appita Journal, 50* (2): 131–138 (1997).
208. Retulainen, E. – Fibre properties as control variables in papermaking? Part 1: fibre properties of key importance in the network. *Paperi ja Puu, 78* (4): 187–194 (1996).
209. Kibblewhite, R. P. – Reinforcement and optical properties of separate and co-refined softwood and eucalypt market kraft pulps. *Appita Journal, 47* (2): 149–153, 158 (1994).
210. Rundlöf, M.; Hoglund, H.; Htun, M. and Wagberg, L. – Effect of fines quality on paper properties – new aspects. *In:* International Mechanical Pulping Conference. EUCEPA, TAPPI, PTF/SPCI/SPIY/CPPA, Ottawa, 1995 pgs. 109–118.
211. Middleton, S. R. and Scallan, A. M. – Filler retention and the specific surface area of fibres. *Nordic Pulp and Paper Research Journal, 9* (3): 156–160 (1994).
212. Gulya, T. and Warner, P. – Advances in paper machine clothing. World Pulp and Paper Technology, 1998/1999: 71.

213. Ragnar, M. – The importance of the structural composition of pulp for the selectivity of ozone and chlorine dioxide bleaching. Nordic *Pulp and Paper Research Journal, 16* (1): 72-xx (2001).
214. Santos, A.; Anjos, O. M. and Simões, R. M. S. – Influence of kraft cooking conditions on the pulp quality of *Eucalyptus globulus. Appita Journal, 61* (2): 148–155 (2008).
215. Stratton, R. A. – Characterization of fibre-fibre bond strength from out-of-plane paper mechanical properties. *Journal of Pulp and Paper Science, 19* (1): 7–12 (1993).
216. Gurnagul, N. and Seth, R. S. – Wet-web strength of hardwood kraft pulps. *In:* 83rd Annual Meeting Technical Section. CPPA, Montreal, 1997 Book B pgs. B137–145.
217. Demuner, B. J.; Manfredi, V. and Claudio-da-Silva Jr., E. – O refino da celulose de eucalipto – uma análise fundamental. *O Papel, 52* (8): 44 – 54 (1990).
218. McKenzie, A. W. – Interpretation of pulp evaluation results. *Appita Journal, 38* (4): 284–290 (1985).
219. Stevens, W. V. – Refining. *In:* Pulp and Paper Manufacture – Volume 6. Ed. Hagemeyer, R.W., Manson, D.W. and Kocurek, M.J. The Joint Textbook Committee of the Paper Industry, Atlanta/Montreal, 1992: 187–219 (1992).
220. Joutsimo, O. and Robertsén, L. – The effect of mechanical treatment on softwood kraft pulp fibers.- pulp and fiber properties. *Maderas, Ciencia y Tecnologia, 18* (3): 17 pgs (2016).
221. Koskenhely, K. – Refining of chemical pulp fibres. *In:* Papermaking Science and Technology. Volume 8 – Paper making part 1 – stock preparation and wet end, Chapter 4. Ed. Paulapuro, H. Finnish Paper Engineers' Association/Paperi ja Puu Oy, Helsinki: 2007.
222. Sharkawya, K. E.; Haavistob, S.; Koskenhely, K. and Paulapuroa, H. – Effect of fiber flocculation and filling design on refiner loadability and refining characteristics. *BioResources, 3* (2): 403–424 (2008).
223. Johnston, R. E.; Li, M. and Waschl, R. – Eucalypt fibre size fractions: modeling and measuring their effect on sheet properties. *Appita Journal, 50* (4): 307–312, 318 (1997).
224. Fergus, B. J.; Procter, A. R.; Scott, J. A. N. and Goring, D. A. I. – The distribution of lignin in spruce wood as determined by ultraviolet microscopy. *Wood Science Technology* (3): 117–138 (1969).
225. Kärenlampi, P. – Tensile strength of paper – A simulation study. *Journal of Pulp and Paper Science, 21* (6): 209 (1995)
226. Mohlin, U-B. – Market SBK and refining response. *In:* Refining & Mechanical Pulping Conference. PIRA, Barcelona, 2005 Paper 8.
227. Johnston, R. E. – Further developments of the Shalhorn and Karnis model. *Appita Journal, 50* (5): 400 (1997).
228. Blomstedt, M.; Kontturi, E. and Vuorinen, T. – Optimising CMC sorption in order to improve tensile stiffness of hardwood pulp sheets. *Nordic Pulp and Paper Research Journal, 22* (3):336–342 (2007).
229. Campbell, W. B. – The cellulose-water relationship in papermaking. Dept. of Interior, Forest Service Bulletin, 84. Ottawa, F.A. Ackland. (1933).
230. Komarov, V. I. – Analiz zavisimosti napryazheniedeformatsiya pri ispytaniyii na ractyazhenie tsellyuloznodumazhnykh materialov. *Izvestija vyssich ucebnych zavedenij/Lesnoj zurnal,* (2/3): 123–131 (1993).
231. Klungness, J. H.; Sykes, M. S.; Tan, F.; Abubakr, S. R. and Eisenwasser, J. – Effect of FL on paper properties. *In:* Papermarkers Conference. TAPPI, Chicago, 1995 pgs. 533–538.
232. Page, D. H.; El-Hosseiny, F.; Winker, K. and Lancaster, A. P. S. – Elastic modulus of single wood pulp fibers. *Tappi Journal, 60* (4): 114–117 (1977).
233. Page, D. H.; Seth, R. S. and De Grace, J. H. – The elastic modulus of paper. *Tappi Journal, 62* (9): 99–102 (1979).
234. Kerekes, R. J. – Rheology of fibre suspensions in papermaking: an overview of recent research. *Nordic Pulp and Paper Research Journal, 21* (5): 598–612 (2006).
235. Corson, S. R. – Fiber and fines fractions influence strength in TMP. *Pulp & Paper Canada, 81* (5): 108–112 (1980).

236. Davidson, R. W. – The weak link in paper strength. *Tappi, 55* (4): 567–575 (1972).
237. Tydeman, P. A. and Hiron, A. M. – What we are doing. *BPBI Research Association* (35): 9–21 (1964).
238. Moffatt, J. M.; Beath, L. R e Mihelich, W. J. – Major factors governing newsprint strength. *In:* The Fundamental Properties of Paper Related to Its Uses Symposium, BPBIF, London: 1976 Volume 1 pgs. 104–136.
239. Setterholm, V. C. and W. A. Chilson. – Drying restraint: its effect on the tensile properties of 15 different pulps. *Tappi 48* (11):634–640 (1965).
240. Tanaka, A.; Hiltunen, E.; Kettunen, H. and Niskanen, K. – Inter-fiber bonding effects, starch or filler. *Nordic Pulp and Paper Research Journal, 16* (4): 306–312 (2001).
241. Wikipedia – Módulo de Young. https://pt.wikipedia.org/wiki/M%C3%BModulo_de_Young
242. Horn, R. A. – Morphology of pulp fiber from hardwoods and influence on paper strength. Forest Products Laboratory, Research Paper 312, 12 pgs (1978) https://www.fpl.fs.fed.us/documnts/fplrp/fplrp312.pdf
243. Sjöberg, J. C. and Höglund, H. – Refining systems for sack paper pulp – Part I: HC refining under pressurized conditions and subsequent LC refining. *Nordic Pulp and Paper Research Journal, 20* (3):320–328 (2005).
244. Ryu, J-H.; Kim, C-H.; Lee, J-H.; Lee, J-S.; Lee, C-H. and Park, J-H. – Study of refining effect of mixed pulps using refiner plates with different bar patterns. *Applied Sciences, 12*: 11445 (2022).
245. Görres, J.; Amiri, R.; Wood, J. R. and Karnis, A. – The shear bond strength of mechanical pulp fibres. *Journal of Pulp and Paper Science, 22* (5): J161–164 (1995).
246. Niskanen, K.; Alava, M. J.; Seppälä, E. T. and Åströn, J. – Fracture energy in fibre and bond failure. *Journal of Pulp and Paper Science, 25* (5): 167–169 1999.
247. Seth, R. S. (2005). Understanding sheet extensibility. *Pulp and Paper Canada, 106* (2), 33–40 (2004).
248. Kärenlampi, P.; Niskanen, K. J. and Alava, M. – Fracture toughness of paper: role of fiber properties and fiber bonding. *In:* International Paper Physics Conference. CPPA/TAPPI, Niagara-on-the-lake, 1995. pgs. 39–46.
249. Seth, R. S. – Optimizing reinforcement pulps by fracture toughness. *Tappi Journal*, 79 (1): 170–178 (1996).
250. Seth, R. S. and Page, D. H. – Fracture resistance: a failure criterion of paper. *Tappi Journal, 58* (9): 112–117 (1975)
251. Seth, R. S. – Measurement of in-plane fracture toughness of paper. *In:* Papermakers Conference. TAPPI, Philadelphia, 1996 pgs. 395–402.
252. Niskanen, K.; Kärenlampi, P. and Alava, M. – Stochastic analysis of paper toyghness. *Journal of Pulp and Paper Science*, *22* (10): 392–397 (1996).
253. Peng, Y.-X. – Fracture toughness of paper. *Appita Journal, 48* (5): 343–346 (1993).
254. Page, D. H. and Seth, R. S. – The problem of pressroom runnability. *Tappi Journal, 65* (8): 92–95 (1982).
255. Wågberg, L.; Forsberg, S.; Johansson, A. and Juntti, P. – Engineering of fibre surface properties by application of the polyelectrolyte multilayer concept. Part I: modification of paper strength. *Journal of Pulp and Paper Science, 28* (7): 222–8 (2002).
256. Yan, N. and Kortschot, M. T. – Modelling of out-of-plane tear energy absorption of paper. *Appita, 49* (3): 177–180 (1996).
257. Kettunen, H. and Niscanen, K. – On the in-plane tear test. *Tappi Journal, 83* (4): 83 (2000).
258. Wanigaratne, D.M.S.; Batchelor, W.J. and Parker, I. H. – Comparison of fracture toughness of paper with tensile properties. *Appita Journal, 53* (5): 369–374, 385 (2002).
259. Eklund, D. and Lindström, T. – Paper chemistry – an introduction. DT Paper Science Publications, Grankulla, Finland, 1991 305 pgs.
260. Czibula, C.; Ganser, C.; Seidlhofer, T.; Teichert, C. and Hirn, U. – Transverse viscoelastic properties of pulp fibers investigated with an atomic force microscopy method. *Journal of Mater Sciince, 54*:11448–11461 (2019).

261. Lindström, T.; Wågberg, L. and Larsson, T. – On the nature of joint strength in paper – a review of dry and wet strength resins used in paper manufacturing. *In:* 13th Fundamental Research Symposium, Cambridge. 2005 pgs. 457–562.
262. Clark, J. d'A. – Mill beating and Refining, *In:* Pulp Technology and Treatment of Paper. Miller Freeman, San Francisco, 1978 516 pgs.
263. Seth, R. S.; Page, D. H.; Barbe, M. I. C. and Jordan, B. D. – The mechanism of the strength and extensibility of wet webs. *Svensk Papperstiding, 87* (6): 36–43 (1984).
264. Kulachenko, A. and Uesaka T. – Direct simulations of fiber network deformation and failure. *Mechanics of Material, 51*:1–14 (2012).
265. Miettinen, P. P. J.; Keetoja, J. A. and Klingenberg, D. J. – Simulated strength of web fibre *networks. Journal of Pulp and Paper Science, 33* (4): 198–205 (2007).
266. Seth, R. S. – Fiber quality factors in papermaking – II: the importance of fibre coarseness. *In:* Materials Interactions Relevant to the Pulp, Paper and Wood Industries. Research Society Symposium, San Francisco, 1990 Volume 197 pgs. 143–161.
267. Page, D. H. – A quantitative theory of the strength of wet webs. *Journal of Pulp and Paper Science, 19* (4): J175–176 (1993).
268. Lazar, D. R.; Foelkel, C. E. B. and Seiter, D. F. – Effective TCF bleaching of eucalyptus to market brightness. *In:* International Pulp Bleaching Conference. TAPPI/CPPA/SPCI/EUCEPA, Vancouver:, 1994 pgs. 69–76.
269. Liu, H.; Roux, J-C.; Dong, J.; Duan, C. and Qi, K. – Physical meaning of cutting edge length and limited applications of specific edge load in low consistency pulp refining. *Nordic Pulp & Paper Research Journal. 37* (2): 250–263 (2022).
270. McKenzie, A. W. – The tear-tensile relationship in softwood pulps. *Appita, 42* (3): 215 (1989).
271. Zhang, G.; Niskanen, K.; Laine, J. E. and Paulapuro, H. – Effect of relaxation during on the final sheet properties. *Appita Journal, 54* (2): 126–131 (2003).
272. Cowan, W. F. – Comparing wet zero span tensile testing with conventional laboratory pulp evaluation. *In:* Process and Product Quality Conference. TAPPI, Savannah, 1994 pgs. 47–50.
273. Swinehart, D. and Broek, D. – Tenacity and Fracture Toughness of Paper and Board. *Journal of Pulp and Paper Science, 21* (11): 389–397 (1995).
274. Cowan, W.F. – Explaining handsheet tensile and tear in terms of fiber-quality numbers. *Tappi Journal, 78* (1) 101–106 (1995).
275. van den Akker, J. A.; Wink, W. A. and Van Eperen, R. H. – Tearing strength of paper. Part III: tearing resistance by the in-plane mode of tear. *Tappi Journal, 50* (9): 466–470 (1967)
276. Mai, Y.-W. – On the tear resistance of paper. *Appita Journal, 36* (6): 461 (1983).
277. Kazi, S. M. and Kortschot, M. T. – The fracture behavior of TMP/kraft pulps. *Tappi Journal, 79* (5): 197 (1996).
278. Kärenlampi, P.; Cichoracki, T.; Alava, M.; Pylkkö, J. and Paulapuro, H. – A comparison of two test methods for estimating the fracture energy of paper. *Tappi Journal, 81* (3): 154–160 (1998).
279. Ibrahem, A. A.; Yousef, M. A. and El-Meadawy, S. A. – Effect of beating on fibre crystallinity and physical properties of paper sheets. *Journal of Islamic Academy of Sciences, 2* (4): 295–298 (1989).
280. Kärenlampi, P.; Suur-Hamari, H. T.; Alava, M. J. and Niskanen, K. J. – The effect of pulp fiber properties on the in-plane tearing work of paper. *Tappi Journal, 79* (5): 203–210. (1996).
281. Silva Jr., F. G. da e Braga, E. P. – Potencialidade da seleção cde *E. urophylla* em função da qualidade da madeira destinada à produção de celulose. *In:* XXX Congresso anual de Celulose e Papel. ABTCP, São Paulo, 1997 pg. 281–292.
282. Howard, R. C.; Poole, R. and Page, D. H. – Factors analysis applied to the results of a laboratory beating investigation. *Jornal of Pulp and Paper Science, 20* (5): 137–144 (1994).
283. Seth, R. S. and Page, D. H. – Fibre properties and tearing resistance. *Tappi Journal, 71* (2): 103–107 (1988).

284. Retulainen, E. and Ebeling, K. – Effects of paper structure on the total elongation behavior of fibre-to-fibre bonds. *In:* 8th Fundamental Research Symposium Proceedings. Mechanical Engineering Publications, Oxford, 1995 pgs. 229–263.
285. van den Akker, J. A. – Some theoretical considerations on the mechanical properties of fibrous structures. *In:* Formation and Structure of Paper – Volume I. British Paper and Board Markers Association, Oxford, 1961.
286. Button, A. – Fiber-fiber bond strength: a study of linear elastic model structure. Doctoral Thesis. Institute of Paper Chemistry, Atlanta, 1979 346 pgs https://scholar.google.com.br/scholar_url?url, https://smartech.gatech.edu/bitstream/handle/1853/5746/button_af.pdf
287. Ebeling, K. I. – Distribution of energy consumption during the straining of paper. *In:* Fundamental Properties of Paper Related With Its Use – Volume I. British Paper and Board Markers Association, Oxford, 1961 pgs. 304–335.
288. Willians, M. D.; Beadle, C. L.; Turnbull, C. R. A.; Dean, G.H. and French, J. – Papermaking potential of plantation eucalypt. *In:* Eucalypt Plantations: Improving Fibre Yield and Quality Conference. CRCTHF/IUFRO, Hobart, 1995 pgs. 73–78.
289. Bourmaud, A.; Morvan, C.; Bouali, A.; Placet, V.; Perre, P. and Baley, C. – Relationships between micro-fibrillar angle, mechanical properties and biochemical composition of flax fibers. *In:* Composites Week @ Leuven and Texcomp Conference, Leuven. 2013 pgs. 2–4.
290. Fardim, P. and Duran, N. – Resistência ao rasgo e composição química da polpa kraft: Um estudo de análise de componentes principais. *In: XXXII* Congresso Anual. ABTCP, São Paulo, 1999 9 pgs.
291. Peng, Y. X. and Seth, R. S. – Fracture toughness of paper and the effects of fiber length on it. *China Pulp and Paper, 13* (4): 10–14 (1994).
292. Broderick, G.; Cacchione, E. and Hérox, Y. – The importance of distribution statistics in the characterization of chip quality. *Tappi Journal, 81* (2): 131–142 (1998).
293. Ferreira, M. and Santos, P. E. T. dos – Melhoramento genético florestal dos Eucalyptus no Brasil – breve histórico e perspectivas. *In:* Silviculture and Improvement of Eucalypts Conference. IUFRO, Salvador, 1997 Volume 1 pgs. 15–34.
294. Giertz, H. W. – Fundamentos da fabricação de papel (papermaking fundamentals). ABTCP. Itu, 1989 136 pgs.
295. Kibblewhite, R. P. – The qualities of radiata pine papermaking fibres. *Appita, 35* (4): 289–298 (1982).
296. Tamolang, F. N.; Wangaard, F. F. and Kellogg, R. M. – Hardwood fibre strength and pulp-sheet properties. *Tappi Journal, 51* (1): 19 (1968).
297. Wangaard, F. F. and Willians, D. L. – Fibre length and fibre strength in relation to tearing resistance of hardwood pulp. *Tappi Journal, 53* (11): 2153 (1970).
298. Paavilainen, L. – Importance of coarseness and fiber length in papermaking. *In:* Process Engineering Handbook. Ed. Process Engineering Committee of the Engineering Division. Tappi Press, Atlanta, 1992, 2nd Edition Appendix: 98–108.
299. Seth, R. S. – Beating and refining response of some reinforcement pulps. *Tappi Journal, 82* (3): 147–155 (1999).
300. Melander, E. – The effect of charged groups on the beatability of pulp fibres. Bachelor Thesis, KTH, Stockholm, 2011 31 pgs https://www.diva-portal.org/smash/get/diva2:425918/FULLTEXT01.pdf
301. Jonhston, R. E.; Li, M. and Waschl, R. – Eucalypt fibre size fractions: modeling and measuring their effect on sheet properties. *Appita Journal, 50* (4):307–311 (1997).
302. Area, M. C.; Benitez, J and Felissia, F. E. – Componentes da resistência à tração de polpas kraft de *Eucalyptus grandis*. *O Papel,71* (8): 48–62 (2010)
303. Praire, B. C. – Measurement of forces in a low consistency refiner. Master's Thesis, Department of Mechanical Engineering. University of Victoria, Victoria, 2005 151 pgs
304. Valeri, S.; Aguiar, I. B. de; Banzatto, D. A. and Alvarenga, S. F. – Variação da densidade básica da madeira de E. grandis Hill ex-Maiden com a altura do caule e aplicação de fósforo e calcário dolomítico. *In:* VI Congresso Florestal Brasileiro. SBS/SBEF, Campos do Jordão, 1990 Volume 3: 746–756.

305. Sabourin, J. J. and Ayer, J. – Rapid characterization of fiber coarseness and pulp properties from optical width measurement. *In:* 48th Annual General Conference. APPITA Melbourne, 1994 Paper 4B12 pgs. 591–600.
306. Batchelor, W. J.; Kure, K. A. and Oullet, D. – Refining and the development of fiber properties. *In:* Papermakers Conference. TAPPI, Philadelphia, 1996 pgs. 207–214.
307. Kärenlampi, P. – Effect of distributions of fibre properties on tensile strength of paper: a closed-form theory. *Journal of Pulp and Paper Science, 21* (4): 138–143 (1995).
308. Paavilainen, L. – Effect of sulphate cooking parameters on the papermaking potential of pulp fibres. *Paperi ja Puu, 71* (4): 356–363 (1989).
309. Behera, N. C. and Basu, S. – Studies on the improvement of pulp strength by fibre fractionation and blending technique. *Appita, 34* (6): 485–491 (1981).
310. Paavilainen, L. – Conformability, flexibility and collapsibility of sulphate pulp fibres. *Paperi ja Puu, 75* (9–10): 680 (1993).
311. van den Akker, J. A., Lathrop, L. A., Voelker, M. H. and Dearth, L. H. – Importance of fiber strength to sheet strength. *Tappi, 41* (8): 416 (1958)
312. Gurnagul, N.; Ju, S.; Shallhorn, P. and Miles, K. – Optimizing high-consistency refining conditions for good sack paper quality. Appita *Journal, 59* (6): 476–480 (2006).
313. Brännvall, E. and Lindström, M. E. – A study on the difference in tensile strength between industrially and laboratory-cooked pulp. *Nordic Pulp and Paper Research Journal, 21* (2): 222–226 (2006).
314. Fiserová, M.; Gigac, J. and Balbercak, J. – Relationship between fibre characteristics and tensile strength of hardwood and softwood kraft pulps. *Cellulose Chemistry and Technology., 44* (7–8), 249–253 (2010).
315. Gao, W. H.; Chen, K. F.; Yang, R. D.; Li, J.; Yang, F.; Rao, G. H. and Tao, H. – Effects of beating on tobacco stalk mechanical pulp. *Cellulose Chemistry and Technology, 46* (3–4): 277–282 (2012).
316. Byrd, V. L. and Fahey, D. J. –How to reduce vessel element picking in printing papers containing Oak. *Paper Trade Journal, 153* (47): 54–59 (1969).
317. Kartusch, B. – Einfluss des kambiumalters auf verteilung unf grosse der gefassflachen im buchenholz. *Holzforsch. Holzverwert, 46* (4): 58, 66–9 (1994).
318. Singh, S. V. and Rai, A. K. – Suface property inter-relationship in wood-free paper. *IPPTA, 5* (4): 9–15 (1993).
319. Higgins, H. G. – Pulp and paper. *In:* Eucalyptus for wood production. Ed. Hillis, W. E. and Brown, A. G., CSIRO/Academic Press. London Capitulo 13 pg 290–316.
320. Shallhorn, P. M. and Heintze, H. U. – Hardwood vessel picking in the offset printing of uncoated fine papers. *Pulp & Paper Canada, 98* (10): 21–24 (1997)
321. Frazão, F. J. L. – Características da madeira e da polpa kraft não branqueada de *Eucalyptus deglupta Blume* introduzido na região de Manaus-AM. *In:* Congresso Anual de Celulose e Papel. ABTCP, São Paulo, 1986 pgs. 79–87.
322. Colley, J. – Factors affecting the vessel picking tendency of hardwood pulps. *Appita, 28* (6): 392–398 (1975).
323. Queiroz, S. C. S. and Gomide, J. L. – Efeito das características anatômicas e químicas na densidade básica da madeira de clones híbridos de *Eucalyptus grandis* x *Eucalyptus urophylla. O Papel, 63* (6): 79–84 (2003).
324. Josefsson, A. – Ultrasonic refining of chemical pulp fibres. Master's Thesis. Chalmers University of Technology, Göteborg, 2010 48 pgs
325. Corson, S.R. – Process impacts on mechanical pulp fibre and sheet dimensions. *Pulp and Paper Canada, 103* (2): 20–27 (2002).
326. Gopichand, K.; Maheshwari, S. – Vessel picking of printing papers – a problem in need of greater attention. *IPPTA, 2* (1): 19–24 (1990).
327. Chinga, G. and Syverud, K. – On structual properties affecting the picking tendency of newsprints. *Nordic Pulp and Paper Research Journal, 22* (4): 447–451 (2007).

328. Blomstedt, M.; Panula-Ontto, S.; Kontturi, E. and Vuorinen, T. – Um método para reduzir o arrancamento de valos de folhas de polpa de eucalipto mediante modificação com carboximetilcelulose. *O Papel, 69* (2): 35–44 (2008).
329. Alves, E. F.; Oliveira, R. C. de; Silva, L. H. M. da e Colodette, J. L. – Interação de fibras e elementos de vasos de polpa Kraft de eucalipto com tintas de impressão offset. *In:* XXXIX Congresso Anual. ABTCP. São Paulo, 2006 18 pgs.
330. Karlsson, H. – New technique for measurement of fibre properties including vessel cells and mix of fibre species. *Appita Journal, 61* (3): 192–196 (2008).
331. Parham, R. A. – Wood structures – hardwood. *In:* Pulp and Paper manufacture – Volume I – Properties of Fibrous Raw Materials and their Preparation for Pulping. E.: Kocurek, M. J. and Stevens, C. F. B. The Joint Textbook Committee of the Paper Industry, Atlanta/Montreal, 1983, 28–34.
332. Tateishi, M.; Seino, T.; Ona, T.; Oshima, J.; Adachi, K.; Yokota, S. and Yoshizawa, N. – Rapid assessment of vessel anatomical features by pyrolysis-gas chromatography. *In:* Eucalyptus in a Changing World Conference. IUFRO, Aveiro, 2004 2 pgs.
333. Moore, G. K. and Jopson, R. N. – The onward march of eucalyptus. *Paper 360°, September*): 14–15 (2008).
334. Colley, J. – The influence of vessel elements on the picking tendency of eucalypt pulps. *Paper Technology, 14*: 293–296 (1973).
335. Carvalho, H. G. de; Oliveira, R. C. de; Gomide, J. L. and Colodette, J. L. – Efeito da idade de corte da madeira e de variáveis de refino nas propriedades da celulose kraft branqueada de eucalipto. *In:* XXXI Congresso Anual de Celulose e Papel. ABTCP, São Paulo, 1998 pg. 367–381.
336. Richardson, J. D.; Wong, K. K. Y. and Clarck, T. A. – Modification of mechanical pulp using carbohydrate-degrading enzymes. *Journal of Pulp and Paper Science, 24* (4): 125–128 (1998).
337. Lobben, H. T. – The tensile stiffness of paper. Part 1: A model based on activation. *Norsk Skogindustri, 29* (12): 311–315 (1975).
338. Ohsawa, J. – Vessel picking in printing papers. *In:* Tropical Wood Symposium. Singapore, 1988 pg. 220 [358].
339. Ohsawa, J.; Ohtake, T.; Komatsu, Y.; Yoneda, Y. (1982): Prevention of vessel picking trouble in tropical hardwood pulps, I Effective vessel separation methods. *Japan Tappi, 57* (10): 57 (10) 975–984 (1982).
340. Kerekes, R. J. – Pulp flocculation in decaying turbulence: a literature review. *Journal of Pulp and Paper Science, 9* (3): 86–91 (1983)
341. Kibblewhite, R. P. – Qualities of kraft and thermomechanical radiata pine papermaking fibres. *In:* 8th Fundamental Research symposium. BP&BMA, Oxford, 1985 Volume 1 pgs. 93–131.
342. Shaw, M. J.; Clarke, C. R.; Pallett, R. N. and Morris, A. R. – Differentiating timber to optimize the pulping process. *Appita Journal, 51* (6): 456–460 (1998).
343. Scott, W. E. – Principles of wet end chemistry. Tappi Press, Atlanta, 1996 185 pgs.
344. Sandercock, C. F.; Sands, R.; Ridoutt, B. G.; Wilson, L. F. and Hudson, I. – Factors determining wood microstructure in eucalyptus. *In:* Eucalypt Plantations: Improving Fibre Yield and Quality Conference. CRCTHF/IUFRO, Hobart, 1995 pgs. 12—135.
345. Trugilho, P. F.; Bianchi, M. L.; Rosado, S. C. da S. and Lima, J. T. – Qualidade da madeira de clones de espécies e híbridos naturais de *Eucalyptus*. *Scientia Forestalis, 73*: 55–62 (2007)
346. Zani Filho, J.; Stape, J. L.; Ribeiro, F. de A. and Balloni, E. A. – Programa de melhoramento genético de *E. urophylla* S. T. Blake, através de seleção precoce. *In:* VI Congresso Florestal Brasileiro. SBS/SBEF, Campos do Jordão: 22-27/09/1990 Volume 3: 464–471.
347. Mokfiensk, A.; Gomide, J.L.; Colodette, J. L. and Oliveira, R.C. de – Importância da densidade e do teor de carboidratos totais na madeira de eucalipto no desempenho da linha de fibras. *In* Colóquio Internacional sobre Celulose Kraft de Eucalipto, UFV, Viçosa, 2003 pgs. 15–38.
348. Silva, R. P. and Oliveira, R. C. de – A reciclagem de papéis: uma aboradagem técnica. *Folha Florestal* (93): 9–11 (1999).

349. Barrichelo, L. E. G. – Densidade básica e características das fibras de madeira de *E. grandis*. *In:* III Congresso Latino-Americano de Celulose e Papel. ABTCP, São Paulo, 1983 pgs 113–125.
350. Faust, T. D.; Clark III, A.; Courchene, C. E.; Shiver, B. D. and Beli, M. L. – Effect of intensive forest management practices on wood properties and pulp yield of young, fast growing Southern pine. *In:* International Environment Conference. TAPPI, Nashiville, 1999 Volume 2 pgs. 501–512.
351. Amidon, T. E. – Effect of the wood properties of hardwoods on kraft paper properities. *Tappi Journal, 64* (3): 123–126 (1981).
352. Greaves, B. L. and Borralho, M. G. – The influence of basic density and pulp yield on the cost of eucalypt kraft pulping: a theoretical model for tree breeding. *Appita, 49* (2): 90–95 (1996).
353. Giertz, H. W. – The effect of beating on individual fibres. *In:* Fundamentals of Papermaking Symposium. PIRA Cambridge, 1961 pgs. 389–402.
354. Arjas, A.; Huuskonen, J. and Rythi, N – Principles of the evaluation of the performance of a beating-machine and of the beating result. Part II. *Paperi ja Puu,* 52 (4): 268y, 379 (A1970) and *52* (6): 379 (1970)
355. Maloney, T. C. and Paulapuro, H. – The formation of pores in the cell wall. *Journal of Pulp and Paper Science, 25* (12): 430–436 (1999).
356. Campinhos Jr., E. and Claudio-da-Silva Jr., E. – Silvicultura e melhoramento genético Fforestal. *In:* VI Congresso Florestal Brasileiro. SBS/SBEF, Campos do Jordão, 1990 Volume I: 83–94.
357. Hillis, W. E. – Wood quality and utilization. *In:* Eucalyptus for Wood Production. Ed. Hillis, W. E. and Brown, A. G., CSIRO/Academic Press, London. Chapter 12 pg 259–289 (1984).
358. Taylor, F. W. – Variation in the anatomical properties of South African grown Eucalyptus grandis. *Appita Journal, 27* (3): 171–184 (1973).
359. Carneiro, C. J. C.; Amaral Santos, C. A. S. do e Manfredi, V. – Caracterização da variabilidade longitudinal da árvore visando a produção de celulose. *In:* XXX Congresso Anual de Celulose e Papel da ABTCP, São Paulo, 1997 pgs. 271–280.
360. Flores, D. M. M.; Frizzo, S. M. B. and Foelkel, C. E. B. – Estudo da variabilidade de um povoamento clonal entre árvores de *Eucalyptus saligna* Smith. *In:* XXXI Congresso Anual de Celulose e Papel. ABTCP, São Paulo, 1998 pgs. 383–395.
361. Turner, C. H.; Balodis, V. and Dean, G. H. – Variability in pulping quality of *E. globulus* from Tasmanian provenances. *Appita Journal, 36* (5): 371–376 (1983).
362. Wilkins, A. P. – Influence of silvicultural treatment on growth and wood density of *Eucalyptus grandis* grown on a previous pasture site. *Australian Forestry,* 53: 168–172 (1990).
363. Manfredi, V. and Barrichelo, L. E. G. – Variação do rendimento em celulose sulfato ao longo do tronco do eucalipto. *In:* XVIII Congresso Anual de Celulose e Papel da ABTCP, São Paulo, 1985 pgs. 5–29.
364. Pereira, H.; Almeida, M. H.; Tomé, M. and Pereira, J. S. – *Eucalyptus globulus* plantations: genetic, silvicultural and environmental control of fibre yield and quality. *In:* Eucalypty Plantations: Improved Fiber Yield and Quality Conference. CRCTHF/IUFRO, Hobart, 1995 pgs. 47–48.
365. Cort, J. B.; Sabourin, M. J. and Musselman, R. L. – Optimized low intensity post refining: lower energy and superior pulp quality. *In:* III International Refining Conference and Exhibition. PIRA/IPST, Atlanta, 1995 Paper 19 10 pgs.
366. Arrieta, I. – El aprovechamiento de la madera y la importancia de la calidad de astilas en la fabricación de la celulosa. *El Papel* (77): 46–49 (1999).
367. Valente, C. A. and Furtado, F. P. – O melhoramento do *Eucalyptus globulus*. Uma abordagem ecológica. *Pasta e Papel* (5): 33–37 (1992).
368. Silveira, V.; Rosado, S. C. S.; Trugilho, P. F. and Carvalho, D. de – Interação entre clones de *Eucalyptus* e ambientes definidos pela qualidade de sítio e espaçamentos. *In:* Silviculture and Improvement of Eucalypts Conference. IUFRO, Salvador, 1997 Volume 2 pgs. 245–252.

369. Greaves, B. L.; Borralho, N. M. G. and Raymond, C. A. – Breending objective for plantation eucalypts grown for production of kraft pulp. *Forest Science* (43): 465 (1997).
370. Yllner, S. and Enström, B. – Studies of the adsorption of xylan on cellulose fibers during the sulphate cook – Part 1. *Svensk Papperistiding, 59* (6): 229–232 (1956).
371. Hans, A. S. and Burley, J. – Wood quality in *Eucalyptus grandis* (Hill) Maiden in a fertilizer trial at Siamambo, Zambia. *East African Agriculture and Forest Journal, 38* (2): 157 (1972).
372. Bhat, K. M.; Bhat, K. V. and Dhamodaran, T. K. – Wood density and fiber length of *Eucalyptus grandis* grown in Kerala, India. *Wood Fiber Science, 22* (1): 54 (1990).
373. Bamber, R. K.; Floyd, A. G. and Humphreys, F. R. – Wood properties of flooded gum. *Australian Forestry,* (33): 3 (1969)
374. Riddout, B. G. and Sands, R. – Whitin tree variation in cambial anatomy and wylem cell differentiation in *Eucalyptus globulus. Trees:Structure and Function* (8):18 (1993).
375. Flores, D. M. M.; Cardoso, G. V.; Foelkel, C. E. B. and Frizzo, S. M. B. – Amostragem de árvores para estudos tecnológicos da madeira para produção de celulose: tamanho da amostra, número mínimo de repetições e variabilidade das popriedades para um clone de *Eucalyptus saligna* Smith. *O Papel, 61* (3): 44–49, 52, 55 (2000).
376. French, J.; Conn, A.B.; Batchelor, W.J. and Parker, I.H. – The effect of fibre fibril angle on some handsheet mechanical properties. *Appita Journal, 53* (3): 210–226 (2000).
377. Reme, P. A. – Some aspects of wood characteristics and the pulping process in mechanical pulp fibres. Doctoral Thesis. Norwegian University of Science and Technology, Trondheim, 2000 121 pgs (2000) https://ntnuopen.ntnu.no/ntnu-xmlui/handle/11250/248066
378. Dickson, A. R.; Corson, S. R. and Dooley, N. J. – Fibre collapse and decollapste determined by cross-sectional geometry. *Journal of Pulp and Paper Science, 32* (4): 1–5 (2006).
379. Miranda, I.; Gominho, J.; Lourenço, A. and Pereira, H. – Heartwood, extractives and pulp yield of three *Eucalyptus globulus* clones grown in two sites. *Appita Journal, 60* (6): 485–488, 500 (2007).
380. Retulainen, E.; Nieminen, K. and Nurminen, I. – Enhancing strength properties of kraft and CTMP fibre networks. *Appita Journal, 46* (1): 33–38 (1993).
381. Ferreira, C.; Fantini Júnior, M.; Oliveira, R.C. de; Colodette, J.L. and Gomide, J.L. – Critérios de seleção de clones para maximizar rendimento e qualidade da celulose. *In:* Colóquio Internacional sobre Celulose Kraft de Eucalipto, UFV, Viçosa, 2003 pgs. 39–58.
382. Xu, E. C. and Saborin, M. J. – Evaluation of APMP and BCTMP for market pulps from South American eucalyptus. *Tappi Journal, 82* (12): 75–82 (1999).
383. Vale, A. T do; Moura, V. P. G.; Martins, I. S. and Rezende, D. C. A. de – Densidade básica média, em função da profundidade de penetração do pino do "Pilodyn"e da classe diamétrica, e variação axial da densidade básica em *Eucalyptus grandis* Hill ex Maiden. *Revista Árvore, 19* (1): 80 (1995).
384. Pryor, L. O. – Biology of eucalyptys. Edward Arnold, London, 1976. 82 pgs.
385. Ilvessalo-Pfäffli, M. –S. – Fiber Atlas – Identification of Papermaking Fibres. Spring Series in Wood Science, Series Ed. Timell, T. E. Spring-Verlag, Berlin, 1995. 400 pgs.
386. Hillis, W. E. – Properties of Eucalyptus woods of importance to the pulp and paper industry. *Appita Journal, 26* (2): 113–122 (1972).
387. Wang, X.; Maloney, T. C. and Paulapuro, H. – Internal fibrillation in never-dried and once-dried chemical pulps. *Appita Journal, 56* (6): 455–459 (2003).
388. Brindley, C. L. and Klimbewhite, R. P. – Refining effects on eucalyptus and mixed hardwood and softwood market pulps and blends. *In:* 48th Annual General Conference. APPITA, Melbourne, 1994 Paper 3B41 pgs. 493–500.
389. Beck, M. V. – The importance of wet end equipment and its influence on retention. *In:* Retention of Fines and Fillers During Papermaking. Gess, J. M. Ed. Tappi Press, Atlanta, 1998, Chapter 7 pgs. 129–158.
390. Garlet, V. N.; Andreotti, V. and Sacon, V. – Misturas de fibras curtas e longas: influência da fibra de eucalipto nas propriedades. *In:* XXIV Congresso Anual de Celulose e Papel. ABTCP, São Paulo, 1991. pgs. 37–43.

391. Demuner, B. J. – Opportunities for market pulp differentiation via fractionation. *In:* 5th International Paper and Board Industry Conference. PIRA, Vienna, 1999 pg 4–18.
392. Gigac, J.; Kuña, V. and Schwrtz, J. – Effects of fibers and fillers on the optical and mechanical characteristics of paper. *Tappi Journal, 78* (2): 162–167 (1995).
393. Baker, C. F. – Good practice for refining the types of fiber found in modern paper furnishes. *Tappi Journal, 78* (2): 147–157 (1995).
394. Koran, Z. – Different process can be used to enhance hardwood pulping quality. *Pulp and Paper Canada, 90* (2): 20–21 (1989).
395. Levlin, J. E. – Different hardwood pulps in papermaking. *Paper, 9* (206): 218 (1986).
396. Demuner, B. J.; Viana Doria, E. L.; Claudio-da-Silva Jr., E. and Manfredi, V. – The influence of eucalypt fiber characteristics on pulp and paper properties. *In:* International Paper Physics Conference. TAPPI, Kona, 1991 Book 1 pgs.185–196.
397. Rudie, A. W. – Wood and how it relates with wood products. *Tappi Journal,* 81 (5): 223–228 (1998).
398. Sepke, P.W., Metzer, F.P. and Selder, H. – Refining of acacia. *In:* Scientific and Technical Advances in Refining & Mechanical Pulping – Course Notes. PIRA, Stockholm, 2003 Paper 7.
399. Busker, L. H. and Cronin, D. C. – The relative importance of wet press variables in water removal. *Pulp and Paper Canada, 85* (6): 138–147 (1984).
400. El-Sharkawy, K.; Koskenhely, K. and Paulapuro, H. – The fractionation and refining of eucalyptus kraft pulps. *Nordic Pulp and Paper Research Journal, 23* (2): 172–180 (2008).
401. Kimmo, H.; Markku, P. and Håkan, S. – New trends and technology in refining. *IPPTA Journal, 24* (1): 109–113 (2012).
402. Baker, C. F. – Refining review – changes in refining practice with new sources of fibre. *World Pulp and Paper Technology:* 95–97, 99 (1992).
403. Lavezzo, E. F.; Dueñas, R. S.; Saucedo, J. T.; Prado, M. C. J. and Becerra, M. C. B. – Evaluacion de eucalyptus procedentes de palantaciones técnicas para la produccion de pulpa e papelera. *El Papel,* (71):40–42 (1998).
404. Silva Jr, F. da; Gomes, I. M. de B.; Barrichelo, L. E. G. and Oda, S. – Avaliação da qualidade da madeira de híbridos de *Eucalyptus grandis x Eucalyptus camaldulensis* visandoa produção de celulose. *In:* VII Congresso Florestal Brasileiro. SBS/SBEF, Curitiba, 1993 Volume 2 pgs.581–585.
405. Sidaway, S. – The availability and use of eucalyptus pulps. *Tappi Journal, 71* (12): 47–51 (1988).
406. Foelkel, C – Papermaking properties of *Eucalyptus* trees, woods, and pulp fibers. *Eucalyptus Online Book & Newsletter.*- Chapter 14 (Jul) 2009 110 pgs www.eucalyptus.com.br/eucaliptos/ENG14.pdf
407. Colodette, J. L.; Gomes, C. M.; Rabelo, M. S.; Eiras, K. M. M.; Gomes, A. de F. and Oliveira, K. D. – Eucalyptus Kraft pulp bleaching: state-of-art and new developments. *Tappi Journal, 7* (2): 18A-18M (2008).
408. Morud, B. – Papermaking properties of hardwood pulps. *Norsk Skogindustri, 26* (2): 30–32 (1972).
409. Lundqvist, S-O. – Efficient wood and fiber characterization – a key factor in research and operation. *Annals of Forest Science, 59*: 491–501 (2002).)
410. Demuner, B. J., Ratnieks, E. and Robinson, D. – Ultra low intensity refining of eucalyptus pulps. *In:* Refining and Mechanical Pulping Conference. PIRA, Barcelona, 2005 Paper 7.
411. Shapiro, S. – Formation and fiber orientation sensors for web quality optimization. *In:* Papermakers Conference. TAPPI, Philadelphia: 24-27/03/1996 pg. 17.
412. Yuan, L.; Wan, J.; Ma, Y.; Wang, Y.; Huang, M. and Chen, Y. – The content of different hydrogen bond models and crystal structure of eucalypty fibers during beating. *BioResources 8* (1), 717–734 (2013).
413. Salmén, L. – Micromechanical understanding of the cell-wall structure (Micromécanique de la structure de la paroi cellulaire) *Comptes Rendus Biologies, 327* (9–10): 873–880 (2014).

414. Koran, Z. – Fiber characteristics. *In:* TECH95 Theory & Practice of Papermaking Course. CPPA, Ottawa, 1995 Section A1 8 pgs.
415. Sands, P. J. – Speculations on modeling eucalypt pulp-wood quality. *In:* Eucalypt Plantations: Improving Fibre Yield and Quality Conference. CRCTHF/IUFRO, Hobart, 1995 pgs. 136–139.
416. Lindström, N. and Fardim, P. – Chemistry and surface chemistry of vessels in eucalyptus kraft pulps. *O Papel, 73* (9): 65–72 (2012).
417. Foelkel, C. – Vessel elements and eucalyptus pulps. 54 pgs (2007) http://www.eucalyptus.com.br/capitulos/ENG04_vessels.pdf
418. Wilson, L.; Hudson, I. and VanBeveren, K. – Vessel distribution at two percentage heights from pith to bark in a seven year old *Eucalyptus globulus* tree. *Appita Journal, 50* (6): 495–500 (1997).
419. Dadswell, H. E. – The anatomy of eucalypt woods. CSIRO Division of Applied Chemistry Technological Paper (66), 1972, 3 pgs.
420. Willians, M. D. – Chemimechanical pulps from plantation eucalypts. *Appita, 46* (2): 137–142 (1994).
421. Malan, F. S. – Genetic variation in some growth and wood properties among 18 fullsib familes of South African grown *Eucalyptus grandis*: a preliminary investigation. *South African Forestry Journal* (146): 38 (1988).
422. Marcoccia, B. S. – The theoretical background to Lo-Solids™ pulping. *In:* 82nd Annual Technical Meeting. CPPA, Montreal, 1996 pgs. 265–274.
423. Core, H. A.; Cote, W. A. and Day, A. C. – Wood structure and identification. Syracuse University Press, 2nd Edition, 1979.
424. Rakkolainen, M.; Kontturi, E.; Isogai, E.; Enomae, T.; Blomstedt, M.; Vuorinen, T. – Carboxymethyl cellulose treatment as a method to inhibit vessel picking tendency in printing of eucalyptus pulp sheets. *Industrial and. Engineering Chemistry Research, 48* (4): 1887–1892 (2009).
425. Chen, F. and Evans, R. – A robust approach for vessel identification and quantification in eucalypt pulpwoods. *Appita Journal, 58* (6): 442–447 (2005).
426. Ververis, C.; Georghiou, K.; Christodoulakis, N.; Santas, P. and Santas, R. – Fiber dimensions, lignin and cellulose content of various plant materials and their suitability for paper production. *Industrial Crops and Products, 19*: 245–254 (2004).
427. Fengel, D. and Wegener, G. – Wood chemistry, ultrastructure, reactions. De Gruyter, Berlin, 1983
428. Pydimalla, M. and Reddy, K. – Effect of pulping, bleaching and refining process on fibers for paper making – A review. *International Journal of Engineering Research & Technology, 9* (12): 310–316 (2020).
429. Stark, H. – Pulp properties of TCF pulps. *In:* Emerging Pulping & Bleaching Technologies Workshop. TAPPI, Durhan, 1995 Section Bleaching III.
430. Stone, J. E. and Scallan, A. M. – A structural model for the cell wall of water swollen sood pulp fibres based on their accessibility to macromolecules. *Cellulose Chemistry Technology* (2): 343–358 (1968).
431. Smith, T. J.; Wearne, R. H. and Wallis, F. A. – Factors influencing the amount of chlorinated phenols formed during bleaching of eucalypty kraft pulps. *Holzforchung* (48): 125–132 (1994).
432. Ruel, K.; Billosta, V. C.; Guillemin, F.; Sierra, J. B. and Joseleau, J-P. – The wood cell wall at the ultrastructural scale – formation and topochemical organization. Maderas. *Ciencia y tecnología. 8*(2):107–116 (2006).
433. Bäckström, M – The effect of environment on refining efficiency of kraft pulps. Doctoral Thesis in Fibre and Polymer Science. KTH Royal Institute of Technology, Stockholm, 2020 53 pgs www.diva-portal.org/smash/getdiva2:1456955/FULLTEXT01
434. Fahlén, J. – The cell wall ultrastructure of wood fibres – effects of the chemical pulp fibre line. KTH Royal Institute of Technology, Stockholm, 2005 70 pgs. https://www.diva-portal.org/smash/get/diva2:7109/FULLTEXT01.pdf

435. Gratzl, J. S. – Delignification and bleaching principles. *In:* Emerging Pulping & Bleaching Technologies Workshop. TAPPI, Durhan, 1995 – section Pulping and Delignification II.
436. Taher, M. R. B. – Tailored low consistency refining for targeted fiber properties. Master's Thesis, Abo Akademi, Gadolinia, 2020 50 pgs. www.doria.fi/bitstream/handle/10024/177080/taher_md.pdf?sequence=2&isAllowed=y
437. Josephson, W.; Jansson, U.; Sezgi, U. S. and Fagerströn, K. – Low consistency refining of a non-conventionally cooked pulp. *In:* Papermakers Conference. TAPPI, Atlanta, 1999 Book 2 pgs. 729–739.
438. Borralho, N. M. G.; Cotterill, P. P. and Kanowski, P. J. – Breeding objectives for pulp production of *Eucalyptus globulus* under different industrial cost structures. *Canadian Journal of Forest Research, 23*: 649 (1993).
439. Ferreira, C. R.; Fantini Jr, M.; Colodette, J. L.; Gomide, J. L. and Carvalho, A. M. M. L. – Avaliação tecnológica de clones de eucalipto. Parte 1: qualidade da madeira para produção de celulose kraft. *Scientia Forestalis* (70): 161–170 (2006).
440. Lennholm, H. and Iversen, T. – The effects of laboratory beating on cellulose structure. *Nordic Pulp and Paper Research Journal, 10* (2): 104–109 (1995).
441. Salmén, L. – Wood morphology and properties from molecular perspectives. *Annals of Forest Science,* (72):679–684 (2015).
442. Park, S. W. and Pinto, J. M. – Cinética da polpação kraft. Parte 1: a evolução de modelos empíricos para modelos mecanísticos. *In:* XXIII Congresso Anual de Celulose e Papel, ABTCP, São Paulo, 1990 pgs. 69–91.
443. Scallan, A. M. and Tigerström, A. C. – Swelling and elasticity of the cell walls of pulp fibres. *Jornal of Pulp and Paper Science, 18* (5): 188–193 (1992).
444. Sjöström, E. – Wood chemistry – fundamentals and applications, 2nd. Edition. Academic Press, San Diego, 1993 pg. 204–222.
445. Milichovsky, M. – A new concept of chemistry refining process. *Tappi Journal, 62* (10): 221–231 (1990).
446. Homas, L. H.; Kennedye, C. J.; Mayb, R. P.; Altanerf, C. M.; Apperleyg, D. C.; Wessh, T.J. and Jarvisi, M. C. – Structure of cellulose microfibrils in primary cell walls from collenchyma. *Plant Physiology, 161* (1): 465–476 (2013).
447. Genco, I.; Boufi, S.; Pèlach, M. A.; Alcalà, M.; Vilaseca, F. and Mutjéa, P. – Nanofibrillated cellulose as paper additive in eucalyptus pulps. *BioResources, 7* (4): 5167–5180 (2012).
448. Chinga-Carrasco, G. – Cellulose fibres, nanofibrils and microfibrils: The morphological sequence of MFC components from a plant physiology and fibre technology point of view. *Nanoscale Research Letter,. 6* 7 pgs. (2011).
449. Donaldson, L. – Cellulose microfibril aggregates and their size variation with cell wall type. *Wood Science and Technology,* (41): 443–460 (2007).
450. Sjösted A. – Preparation and characterization of nanoporous cellulose fibres and their use in new material concepts. Doctoral Thesis. KTH Royal Institute of Technology. Stockholm, 2014 65 pgs. https://www.diva-portal.org/smash/get/diva2:761478/FULLTEXT01.pdf
451. Zhang, M.; Hubbe, M. A.; Venditti, R. A. and Heitamann, J. A. – Can recycled kraft benefit from chemical addition before they are first dried? *Appita Journal, 55* (2): 135–144.
452. Smook, G. A. – Preparation of papermaking stock *In:* Handbook of Pulp and Paper Thechnologists – Chapter 13. Angus Wilde, 2002. pgs 190–204
453. Abdul Khalil, H. P. S.; Davoudpour, Y.; Nazrul Islama, Md; Mustaphaa, A.; K. Sudeshd, Rudi Dungani, R. and Jawaidb, M. – Production and modification of nanofibrillated cellulose using various mechanical processes: A review. *Carbohydrate Polymers 99*: 649–665 (2014)
454. Sjöström, E. – Production of microfibrillated cellulose by LC-refining. Master's Thesis, Abo Akademi, 2018 77 pgs https://www.doria.fi/handle/10024/165125
455. Winandy, J. E. and Rowell, R. M. – Chemistry of wood strength. *In:* Handbook of wood chemistry and wood. Chapter 11. CRC Press, Boca Raton, 2005 Ed. Rowell, R. M. – pgs. 303–347.

456. Gharehkhani, S.; Sadeghinezhada, E.; Kazi, S. N.; Yarmanda, H.; Badarudina, A.; Safaei, M. R. and Zubir, M. N. M. – Effect of pulp consistency during refining of pulp refining on fiber properties—A review. *Carbohydrate Polymers Journal, 115:* 785–803 (2015).
457. Burger, L. M. and Richter, H. G. – Anatomia da madeira. Editora Nobel, São Paulo, 1991 154 pgs.
458. Kahilahti, M. – Roundwood chipping for quality pulp. *World Pulp and Paper Technology,* 1995/1996: 57–58.459.
459. Salmén, L. – Micromechanical understanding of the cell-wall structure. *Comptes Rendus Biologies, 327* (9–10):873–880 (2004 https://www.sciencedirect.com/science/article/pii/S1631069104001726
460. Groom, L.; Motlt, L. and Shalert, S. – Mechanical properties of individual Southern pine fibers. Part I – determination and variability of stress-strain curves with respect to tree height and juvenility. *Wood na Fiber Science, 34* (1): 14–27 (2002) https://www.fs.usda.gov/treesearch/pubs/8010
461. Gorshkova, T.; Brutch, N.; Chabbert, B.; Deyholos, M.; Hayashi, T.; Lev-Yadun, S.; Mellerowicz, E. J.; Morvan, C.; Neutelings, G. and Pilate, G. – Plant fiber formation: state of the art, recent and expected progress, and open questions. *Critical Reviews in Plant Sciences, 31:* 201–228 (2012).
462. Brandberg, A. – Insights in paper and paperboard performance by fiber network micromechanics. KTH Royal Institute of Technology. Stockolm, 2019 20 pgs www.diva-portal.org/smash/get/diva2:1355441/FULLTEXT01
463. Grignon, J. and Scallan, A. M. – The effect of pH and neutral salts upon the swelling of cellulose gels. *Journal of Applied Polymer Science*, 25: 2829 (1990).
464. Franklin, E. C. – Selection strategies for eucalypt tree improvement – four generations of selection in *Eucalyptus grandis* demonstrate valuable methodology. *In:* Breeding Tropical Trees: Population Structure and Genetic Improvement Strategies in Clonal and Seedling Forestry. Ed. Gibson, G. L.; Griffin, A. R. and Matheson A. C. IUFRO, Pattaya, 1988 pgs. 197–209.
465. Mascarenhas, A. F. and Muralidharan, E. M. – Clonal forestry with tropical hardwoods. *In:* Clonal Forestry II – Conservation and Application. Ed. Ahuja, M. R. and Libby, W. J. Springer-Verlag, Berlin, 1993 Chapter 10: 167–187.
466. Berthold, J. and Salmén, L. – Effects of mechanical and chemical treatments on the pore-size distribution in wood pulps examined by inverse size-exclusion chromatography. *Journal of Pulp and Paper Science, 23* (6): 245–253 (1997).
467. Dinus, R. J. and Welt, T. – Tailoring fiber properties to paper manufacture: recent developments. *Tappi Journal, 80* (4): 127–139 (1997).
468. Higgins, H. G.; Young, J. de; Balodis, V.; Phillips, F. H. and Colley, J. – The density and structure of hardwoods in relation to paper surface characteristics and other properties. *In:* Process Engineering Handbook. Ed. Process Engineering Committee of the Engineering Division. Tappi Press, Atlanta, 1992, 2nd Edition Appendix: 77–81.
469. Luukko, K.; Laine, J. and Pere, J. – Chemical characterization of different mechanical pulp fines. *Appita Jorunal, 51* (2): 126–131 (1999).
470. Waterhouse, J. F. and Riipa, T. – Hardwoods from softwoods? *In:* III International Refining Conference and Exhibition. PIRA/IPST, Atlanta, 1995 Paper 12 24 pgs.
471. Read, J. I. – Dry strength additives. *In:* Pulp and Paper Manufacture – Volume 6. Ed. Hagemeyer, R.W., Manson, D.W. and Kocurek, M.J. The Joint Textbook Committee of the Paper Industry, Atlanta/Montreal, 1992: 86–98.
472. Aspler, J. S. – Interactions of ink and water with the paper surface in printing – a review. *Nordic Pulp and Paper Research Journal, 8* (1): 68–74 (1993).
473. Fürst, T. – A new instrument capable of measuring cross machine profiles of internal bond strength. *In:* Engineering/Process and Product Quality Conference & Trade Fair. TAPPI, Anaheim, 1999 Volume 1 pgs. 1–8.

474. Snowman, V. R.; Genco, J. M.; Cole, B. J. W.; Kwon, H. B. and Miller, W. J. – Bond strength of oxygen-delignified kraft pulps. *Tappi Journal, 82* (2): 103–109 (1999).
475. Clayton, D.; Easty, D.; Einspahr, D.; Lonsky, W.; Malcon, E.; McDonough, T.; Schroeder, L. and Thompson, N. – Part I- chemistry of alkaline pulping. *In:* Pulp and Paper Manufacture – Alkaline Pulping, Volume 5. Grace, T. M.; Leopold, B.; Malcon, E. W. and Kokureck, M. J. Ed. The Joint Textbook Committee of the Paper Industry, Atlanta/Montreal, 1989, 3th Edition, pgs. 1–128.
476. Uesaka, T.; Ferahi, M.; Hristopulos, D.; Deng N. and Moss C. – Factors controlling pressroom runnability of paper. *In:* The science of papermaking. XII Fundamentl Research Symppsium. Oxford, 2001 pgs. 1423–1440.
477. Oliveira, M. H. de – Wet web strength development of paper. Master's Thesis, McGill University, Montreal, 2007 118 pgs.
478. Jardim, C. – Variações na densidade básica da madeira versus impacto no processo produtivo. *In:* 1° Encontro de Operadores de Pátio de Madeira e 5° Encontro de Operadores de Linhas de Fibras. ABTCP. 2019.
479. Dasgupta, S. – Mechanism of paper tensile-strength development due to pulps beating. *Tappi Journal, 77* (6): 158–166 (1994).
480. Safadi, T. – Testing into the future. *World Pulp and Paper Technology,* 1994/1995: 184–187.
481. Bonham, J.S., Reich, M. and Faltas, R. – A new technique to measure paper roughness. *Appita Journal, 59* (4):302–305 (2006).
482. Xu, E. C. and Saborin, M. J. – Evaluation of APMP and BCTMP for market pulps from South American eucalyptus. *Tappi Journal, 82* (12): 75–82 (1999).
483. Bertolucci, F. L. G.; Demuner, B. J.; Garcia, S. L. R. and Ikemori, Y. K. – Increasing fiber yield and quality at Aracruz. *In:* Eucalypt Plantations: Improving Fibre Yield and Quality Conference. CRCTHF/IUFRO, Hobart, 1995 pgs. 31–34.
484. Clark, J. d'Á. – Some troughts on fiber classification and length. *In:* Process Engineering Handbook. Ed. Process Engineering Committee of the Engineering Division. Tappi Press, Atlanta, 1992, 2nd Edition Appendix: 74–76.
485. Xie, Y. Q.; Tong, Q. J. and Chen, Y. – Manufacture and properties of a novel ultra-low-density fiberboard from wood fibre. *BioResources, 6* (4): 4055–4066 (2011).
486. Perissoto, D. O.; Nascimento, E. A. and Morais, S. A. L. – Estudo dos extrativos da polpa kraft de eucalipto. Parte 2: Branqueamento dos extrativos e da polpa kraft. *O Papel, 61* (8): 69–72 (2000).
487. Page, D.H. – A note on the cell-wall structure of softwood tracheids. *Wood and Fiber, 7*: 246–248 (1976).
488. Salmén, N. L. and Olsson, A-M. – Interaction between hemicelluloses, lignin and cellulose: structure-property relationships. *Journal of Pulp and Paper Science, 24* (3): 99–103 (1998).
489. Gomide, J. L.;Colodette, J. L.; Oliveira, R. C. de; Girard, R. and Argyropoulos, D. S. – Fatores que afetam a branqueabilidade de polpas kraft de eucalyptus. Parte 2: Influência de parâmetros da Polpação. *O Papel, 60* (12): 61–70 (2000).
490. Gomide, J. L. and Almeida, J. M. – Carbohydrate and lignin degradation during continuous kraft pulping of Eucalyptus wood. *In:* Sixt Brazilian Symposium on the Chemistry of Lignins and Other Wood Components. FAENQUIL, Guaratinguetá, 1999 pgs. 45–51.
491. Al-Dajani, W. W. and Tschirner, U. W. – Pre-extraction of hemicelluloses and subsequent kraft pulping. Part I: alkaline extraction. *Tappi Journal, 7* (6): 3–8 (2008).
492. Kubelka, V.; Wizani, W.; Neubauer, G. and Kappel, J. – ENERBATCH extended delignification for TCF pulp. *In:* International Non-Chlorine Bleaching Conference. Pulp & Paper/ Emerging Technology Transfer, Amelia Island, 1994 Paper 3-2 23 pgs.
493. Demuner, B. J. Viana Doria, E. L.; Claudio-da-Silva Jr., E. and Manfredi, V. – As propriedades do papel e as características das fibras de eucalipto. *In:* XXIV Congresso Anual de Celulose e Papel. ABTCP, São Paulo, 1991 pgs. 621–641.
494. Cao, B.; Tschirner, U. and Ramaswamy, S. – Impact of pulp chemical composition on recycling. *Tappi Journal, 81* (12): 119–127 (1998).

495. Hanna, K. R., Fisher, J. J.; Krotm, M. J.; Goyal, G. C.; Packwood, R. E. and Ragauskas, A. J – Differences in bleaching and refining responses of displacement batch hardwood and softwood caused by pulping conditions and structure of residual lignin. *In:* International Pulp Bleaching Conference. TAPPI, Helsinki, 1998 Book 2. pgs. 323–328.
496. Backstrom, M. and Jensen, A. – Modified kraft pulping to high kappa numbers. *In:* Annual General Conference. APPITA, (cidade: 1999 Volume 1 pgs. 101–109.
497. Dahlman, O.; Sjöberg, J.; Jansson, U. B. and Larsson, P. O. – Effects of surface hardwood xylan on the quality of softwood pulps. *Nordic Pulp and Paper Research Journal, 18* (3): 310–315 (2003).
498. Saake, B.; Till, B. and Puls, J. – The effect of xylan adsorption on the properties of sufite and kraft pulps. *In:* International Symposium on Wood, Fibre and Pulping Chemistry. APPITA, Auckland, 2005 pgs 131–146.
499. Danielsson, S. and Lindström, M. E. – Influence of birch xylan adsorption during kraft cooking on softwood pulp strength. *Nordic Pulp and Paper Research Journal, 20* (4): 436 (2005).
500. Walker, E. – Effects of the uronic acid carboxyls on the sorption of 4-O-methylglucuronoarabinoxylans and their influence on papermaking properties of cellulose fibres. *Tappi, 48* (5): 298 (1965).
501. Colodette, J. L., Gomide, J. L., Girard, R., Jaaskelainen, A. S., Argyropoulos, D. S. – Influence of pulping conditions on hardwood pulp yield. Quality and bleachability. *In:* International Pulp Bleaching Conference, Halifax, 2000, pg 41–48 – Oral presentation.
502. El-Hosseiny, F. – Influence of the "Giertz effect" on development of short-span compression strength. *Tappi Journal, 81* (2): 177–180 (1996).
503. Moss, P. A. and Pere, J. – Microscopical study on the effects of partial removal of xylan on the swelling properties of birch Kraft pulp fibres. *Nordic Pulp and Paper Research Journal, 21* (1): 8–12 (2006).
504. Antes, R. and Joustino, O. P. – Fiber surface and paper technical properties of *Eucalyptus globulus* and *Eucalyptus nitens* – pulps after modified cooking and bleaching. *BioResources, 10* (1): 1599–1616 (2015).
505. Wang, X. – Improving the papermaking properties of kraft pulp by controlling hornification and internal fibrillation. Doctoral Thesis, Helsinki University of Technology, Espoo, 2006 88 pos http://citeseerx.ist.psu.edu/viewdoc/download?doi=10.1.1.575.9047&rep=rep1&type=pdf
506. Pere, J.; Pääkkönen, E., Ji, Y.; Retulainen, E. – Influence of the hemicellulose content on the fiber properties, strength, and formability of handsheets. *BioResources, 14* (1): 251–263 (2019).
507. Saucedo, J. J. and González, S. F. – Identificación y efecto de las hemiceluloses sobre las propriedades físicas de la celulosa y papel. *In:* II Congresso Latino-Americano de Celulose y Papel. AITIPE, Torremolinos, 1981 pgs. 143–151.
508. Hu, G.; Shiyu, F. and Liu, H. – Hemicellulose in pulp affects paper properties and printability. *Appita Journal, 66* (2) 139–141 (2013).
509. Hakanen, A. and Teder, A. – Modified kraft cooking with polysulfide: yield, viscosity, and physical properties. *Tappi Journal, 80* (7): 189 (1997).
510. Wågberg, E. and Lindström, M. E. – The hemicellulose composition of pulp fibres and their ability to endure mechanical treatment. *Tappi Journal, 6* (10): 19–24 (2007).
511. Genco, J. M. – Fundamental process in stock preparation and refining. *In:* Pulping Conference. TAPPI, Orlando, /1999 pgs. 57–95.
512. Resquin, F.; Barrichelo, L. E. G.; Silva Jr, F. G. da; Brito, J. O. and Sansigolo, C. A. – Wood quality for Kraft pulping of *Eucalyptus globulus* origins planted in Uruguay. *Scientia Forestalis* (72): 57 (2006).
513. Barbosa, B. M.; Colodette, J. L.; Muguet, M. C dos S.; Gomes, V. J.; e Rubens, C. de O. – Effects of xylan in eucalyptus pulp production. *CERNE, 22* (2): 207–213 (2016).
514. Antes, R. and Joutsimo, O. P. – Effect of modified cooking on fiber wall structure of *E. globulus* and *E. nitens*. *BioResources, 10* (2): 2195–2212 (2015).

515. Kersavage, P. C. – Moisture content effect on tensile properties of individual Douglas-fir latewood tracheids. *Wood and Fiber Science, 5* (2): 105–117 (1973).
516. Hall, M. J.; Hansen, N. W. and Rudra, A. B. – The effect of species, age and wood characteristics on eucalypt Kraft pulp quality. *Appita Journal, 26* (5): 348 (1973).
517. Whiteman, P. H.; Cameron, J. W. and Farrington, A. – Breeding trees for improved pulp and paper production. *Appita Journal,* 49 (1): 50–53 (1996).
518. Ferreira, M. and Santos, P. E. F. – Eucalyptus wood traits for species/provenances/plus trees and clones planted in Brazil – a review applied to genetic improvement for pulp production. *In:* Eucalypt Plantations: Improving Fibre Yield and Quality Conference. CRCTHF/IUFRO, Hobart, 1995 pgs. 258–260.
519. Banham, P. W.; Orme, K. and Russell, S. L. – Pulpwood qualities required for the cold soda pulping process. *In:* Eucalypt Plantations: Improving Fibre Yield and Quality Conference. CRCTHF/IUFRO, Hobart, 1995 pgs. 1–4.
520. Ferrari, G. and Mughini, G. – Variation of growth and wood quality traits in provenances of some *Eucalyptus* species in Italy. *In:* Eucalypt Plantations: Improving Fibre Yield and Quality Conference. CRCTHF/IUFRO, Hobart, 1995 pgs. 35–39.
521. Gonzaga, J. V. – Qualidade da madeira e da celulose kraft de treze espécies de *Eucalyptus*. Master's Thesis, Universidade Federal de Viçosa, Viçosa, 1983 119 p.
522. Bhat, K. M. and Benny, A. G. – Improving fibre yield and quality of eucalyptus in India: problems and prospects. *In:* Eucalypt Plantations: Improving Fibre Yield and Quality Conference. CRCTHF/IUFRO, Hobart, 1995 pgs. 255–257.
523. Chabbert, B.; Tollier, M. T.; Monties, B. – Biological variability in lignification of maize: Expression oft he brown midbrid bm2 mutation. *Journal of Science Food Agriculture, 64,* 455–460 (1994).
524. Lani, T. B. T; Iiyama, K.; Stone, B. – Lignin and hydroxycin-namic acids in walls of brown midrib mutants of sorghun, pearl millet and maize stems. *Journal of Science Food Agriculture, 406y* 174–178 (1996).
525. Thorstensson, E. M. G.; Buxton, D. R.; Chemey, J. H. Apparent inhibition to digestion by lignin in normal and brown midrib stenu. *Journal of Science Food Agriculture,* 59: 183–188 (1992).
526. Morrison, W H.; Akin, D. E.; Himmelsbach, D. S.; Gamble, G. R. – Investigation and esterand ether-linked phenolic constituents of cell wall types of normal and brown midrib pearl millet using chemical isolation, microspectrophotom-etryandl3CNMRspectroscopy. *Journal of Science Food Agriculture*, 63, 329–337 (1993).
527. Grenet, E.; Barry, R. – Microbial degradation of normal maize and bm3 maize in the rumen observed by scanning electron microscopy. *Journal of Science Food Agriculture,* 59y, 199–210 (1991).
528. Zhang, X.; Li, L. and Xu, F – Chemical Characteristics of Wood Cell Wall with an Emphasis on Ultrastructure: A Mini-Review. *Forests, 13* (3): 439–461 2022)
529. Claudio-da-Silva Jr, E. – The Flexibility of pulp fibers – a structural approach. *In:* International Paper Physics Conference. TAPPI/CPPA, Harwichport, 1983 pgs. 13–25.
530. Regmed Indústria Técnica de Precisão Ltda. www.regmed.com.br
531. Parhan, P. A. – Wood physical properties. *In:* Pulp and Paper Manufacture – Volume 1 – Properties of Fibrous Raw Materials and Their Preparation for Pulping. Ed. Kocureck, M. J. and Stevens, C. F. B. The Technical Committee of the Paper Industry, Atlanta/Montreal, 1983 pg. 46–54
532. Miranda, C. R. and Barrichelo, L. E. G. – Celulose de madeira de *E. citriodora*: influência do tamanho dos cavacos. *In:* XXIII Congresso Anual de Celulose e Papel. ABTCP, São Paulo: 1990 pgs. 01–34.
533. Evans, R.; Kibblewhite, R. P. and Stringer, S. – Kraft pulp fibre property prediction from wood properties in eleven radiata pine clones. *Appita Journal, 50* (1): 25 (1997)
534. Evans, R.; Kibblewhite, R. P. and Lausberg, M. – Relationships between wood and pulp properties of twenty-five 13 year old radiata pine trees. *Appita Journal, 52* (2): 133–139 (1999).

535. Oda, S.; Mello, E. J.; Menck, A. L. de M. and Costa, P. C. – Variação da densidade básica da madeira de clones de *E. grandis* e *E. saligna* em diferentes povoamentos, com seis anos de idade. *In:* VI Congresso Florestal Brasileiro. SBS/SBEF, Campos do Jordão, 1990 Volume 3: 701–705.
536. Rezende, M. A. de; Saglietti, J. R. C. and Marinez, J. C. – Estudo das variações de massa específica e retratibilidade da madeira do *Eucalyptus saligna*. *In:* VII Congresso Florestal Brasileiro. SBS/SBEF, Curitiba, 1993. Volume 2 pgs. 629–637.
537. Duffy, G. G. and Kibblewhite, R. P. – A new method of relating wood density, pulp quality and paper properties. *Appita Journal, 42* (3): 209–214 (1989).
538. Botrel, M. C. G.; da Silva, J. R. M.; Trugilho, P. F.; Rosado, S. C. da S. and Fernandes, B. R. – Ganho genético em propriedades físicas e mecânicas de clones de *Eucalyptus*. *Scientia Forestalis* (76): 13–19 (2007).
539. Dias, R. L. V. and Claudio-da-Silva Jr., E – Pulp and paper properties as influenced by wood density. *In:* 8th Fundamental Research Symposium. PIRA, Oxford, 1995 pgs. 7–35.
540. Moraes, M. L. T. – Variação genética da densidade básica da madeira em progênies de *Eucalyptus grandis* Hill ex-Maiden e suas relações com as características de crescimento. Master's Thesis, ESALQ/USP, Piracicaba, 1987 115 pgs.
541. Wehr, T. R. and Barrichelo, L. E. G. – Cozimento kraft com madeira de Eucalyptus grandis de diferentes densidades básicas e dimensões de cavacos. *In:* XXV Congresso Anual de Celulose e Papel. ABTCP, São Paulo, 1992 pgs. 161–177.
542. Shimoyama, V. R. de S. – Variações da densidade básica e características anatômicas e químicas da madeira em *Eucalyptus* spp. Master's Thesis, ESALQ/USP, Piracicaba, 1990 93 pgs.
543. Eskelinen, E.; Hu, S. H. and Marton, R. – Wood mechanics and mechanical pulping. *Appita, 36* (1): 32–38 (1992).
544. Demuner, B. J. and Bertolucci, F. de L. G. – Seleção florestal: uma nova abordagem a partir de estimativas de parâmetros genéticos e fenotípicos para características da madeira e polpa de eucalipto. *In:* XXVI Congresso Anual de Celulose e Papel. ABTCP, São Paulo, 1993 pgs. 411–423.
545. Malan, F. S. and Arbudthnot, A. L. – The inter-relationship between density and fibre properties of South African grown *Eucalyptus grandis*. *In:* Eucalypt Plantations: Improving Fibre Yield and Quality. CRCTHF/IUFRO, Hobart, 1995 pgs. 116–120.
546. Mohlin, U.-B e Miller, J. – Industrial refining – effects of refining conditions on fibre properties. *In:* III International Refining Conference and Exhibition. PIRA/IPST, Atlanta, 1995 Paper 4.
547. Zobel, B. J. and Buijteen, J. P. – Wood variation: its causes and control. Spring Series in Wood Science, Series Ed. Timell, T. E. Spring Verlag, Berlim, 1989 363 pgs.
548. Meadows, D. G. – An eye to the future: stock preparation. *Tappi Journal, 81* (2): 70–78 (1998).
549. Rezende, M. A. and Ferraz, E. S. B. – Densidade anual da madeira de *Eucalyptus grandis*. IPEF (30): 37–41 (1985).
550. Zhong, R. and Ye, Z-H. – Secondary cell walls: biosynthesis, patterned deposition and transcriptional regulation. *Plant and Cell Physiology, 56* (2): 195–214 (2015).
551. Barrichelo, L. E. G. and Brito, J. O. – Variabilidade longitudinal e radial da madeira de Eucalyptus grandis. *In:* XVII Congresso Anual de Celulose e Papel. ABTCP, São Paulo, 1984 pgs. 403–409.
552. Zobel, B. J. – Clonal forestry in the *Eucalyptus*. *In:* Clonal Forestry II – Conservation and Application. Ed. Ahuja, M. R. and Libby, W. J. Springer-Verlag, Berlin, 1993 Chapter 7 pgs. 139–148.
553. Fonseca, S. M. da; Fernandes, D. E.; Borges, J. F.; Martini, S. L. and Silveira, P. N. – Efeitos do espaçamento e da idade de corte sobre a produtividade e qualidade da madeira e o custo da celulose kraft de eucalipto. *In:* XXIX Congresso Anual de Celulose e Papel. ABTCP, São Paulo, 1996 pgs. 437–451.
554. Foelkel, C. E. B. – A madeira do eucalipto: da floresta ao digestor. *Boletim Informativo do IPEF, 6* (20): 1–25 (1978).

555. Dowens, G. M. – Wood properties of interest in plantation productivity. *In:* Sampling Plantation Eucalyptus for Wood and Fibres Properties. Chapter 1. Ed. Dewens, G. M. et al CSIRO Publishing, Collingwood, 1997 pgs. 1–8.
556. Corson, S. R. – Directions in New Zealand mechanical pulping research Part 1: development of research effort; research activity; influence of wood quality; and process influences. *Appita, 46* (2): 143–148 (1993).
557. Malan, F. S. – Variation, association and inheritance of juvenile wood properties of *Eucalyptus grandis* Hill ex-Maiden with special reference to the effect of rate of grown. *South African Forestry Journal* (157): 16–23 (1991).
558. Danforth, D. W. – Stock preparation: theory and practice. Southern *Pulp and Paper Manufacture, 32* (7): 52–53 (1969).
559. Malan, F. S. – Studies on the phenotypic variation in growth tree wood properties of South African grown *Eucalyptus grandis* (Hill ex-Maiden). Doctoral Thesis, University of Stellenbosch, Stellenbosch, 1984 https://scholar.sun.ac.za/handle/10019.1/64868
560. Shimoyama, V. R. S. and Barrichelo, L. E. G. – Influência das características anatômicas e químicas sobre a densidade básica da madeira de Eucalyptus. *In:* XXIV Congresso Anual de Celulose e Papel. ABTCP, São Paulo, 1991 pgs. 23–26.
561. Panshin, A. J. and Zeeuw, C. de – Textbook of Wood Technology, 4th Edition. McGraw Hill, New York, 1980. 722 pgs.
562. Laivans, G. V. and Scallan, A. M. – Removal of water from pulps by pressing, *Tappi Journal, 77*(3), 125 (1994).
563. Carrillo, I.; Aguayo, M. G.; Mendonça, S. V. R. T. and Elissetche, J. P. – Variations in wood anatomy and fiber biometry of *Eucalyptus globulus* genotypes with diferent wood density. *Wood Research. 60* (1): 1–10 (2015).
564. Barrichelo, L. E. G.; Brito, J. O.; Couto, M. T. Z. do e Campinhos Jr., E. – Densidade básica, teor de holocelulose e rendimento em celulose de madeira de *Eucalyptus grandis*. *Silvicultura, 8* (32): 802–808 (1983).
565. Trugilho, P. F.; Mendes, L. M.; Lima, J. T. and Silva, J. R. M. da – Uso de técnicas multivariadas na classificação de clones de eucalipto para produção de celulose e papel. *In:* XXX Congresso Anual de Celulose e Papel. ABTCP, São Paulo, 1997 pgs. 309–315.
566. Mohlin. U. B. – Cellulose fiber bonding. Part 3: The effect of beating and drying on inter fiber bonding. *Svensk Papperstiding, 78* (9): 338–341 (1975).
567. Silva, D. de J., Oliveira, R. C. de; Colodette, J. L. and Gomide, J. L – Impacto da qualidade da madeira na deslignificação, no branqueamento e nas propriedades físico-mecânicas da polpa kraft de eucalipto. *In:* XXIX Congresso Anual de Celulose e Papel. ABTCP, São Paulo 1996 pgs. 453–470.
568. Groot, B. de; Van der Kolk, J. C.; Van Dam, J. E. G. and Riet, K. V. – Papermaking characteristics of alkaline hemp-woodly-core pulps. *Tappi Journal, 82* (7): 107–111 (1999).
569. Hughes, M. – Wood and Wood Products – Cell wall and mass-volume relationships. Course CHEM-E2105 – Oral presentation. 2019 https://mycourses.aalto.fi/pluginfile.php/898005/mod_resource/content/1/2019_01_17_CHEM-E2105_Cell%20wall%20and%20mass-volume%20relationships.pdf
570. Kellogg, R. M. and Wangaard, F. F. – Variation in the cell wall density of wood. *Wood and Fiber Science,* 3: 1508–1542 (1969).
571. Salmén, L.; Olsson, A-M.; Stevanic, J. S.; Simonović, J. and Radotić, K. – Structural organization of the wood polymers in the wood fibre. *BioResources, 7* (1): 521–532 (2012).
572. Leask, R. A. – Introduction. In: Pulp and Paper Manufacture, Volume 2, Mechanical Pulping. R. A. Leask, Ed. Joint-Textbook Committee, Technical Section. CPPA/TAPPI.
573. Minor, J. L.; Atalla, R. H. and Harten, T. M. – Improving inter fibre bonding of recycled fibres. *Journal of Pulp and Paper Science, 19* (4): J152–155 (1993).
574. Tikka, P.; Tähkanen, H. and Kovasin, K. – Chip thickness versus kraft pulping performance: experimental by multiple hanging basket in batch digesters. *Tappi Journal, 76* (3): 131–136 (1993).

575. Berggren, R. – Cellulose degradation in pulp fibers studied as changes in molar mass distributions. Doctoral Thesis, Royal Institute of Technology, Stockholm, 2003 94 pgs https://www.diva-portal.org/smash/get/diva2:9304/FULLTEXT01.pdf
576. Fagerstedt, K. V.; Kukkola, E. M.; Koistinen, V. V. T.; Takahashi, J. and Marjamaa, K. – Cell wall lignin is polymerised by class III secretable plant peroxidases in Norway spruce. *Journal of Integrative Plant Biology, 52* (2): 186–194 (2010).
577. Du, X.; Gellerstedt, G. and Li, J. – Universal fractionation of lignin–carbohydrate complexes (LCCS) from lignocellulosic biomass: An example using spruce wood. *Plant Journal, 74* (2): 328–338 (2013).
578. Youssefian, S. and Rahbar, N. – Molecular origin of strength and stiffness in bamboo fibrils. *Scientific Reports, 5*: 1–13 (2015).
579. Niemelä, K.; Tamminen, T. and Ohra-aho, T. – Black liquor components as potential raw materials. TAPPSA (2008) http://www.tappsa.co.za/archive3/Journal_papers/BlackLiquorComponents/blackliquor components.html.
580. Tubino, M. and Mambrim filho, O. – Em busca da seletividade no branqueamento com peróxido de hidrogênio: estudos de parâmetros de ativação termodinâmicos e catalíticos. *O Papel, 60* (9): 41–44, 48–55 (1999).
581. He, L.; Terashima, N. – Formation and structure of lignin in monocotyledons. IV Deposition process and structural diversity of the lignin in the cell wall ofsugarcane and rice plant studied by ultraviolet microscopic spectroscopy. *Holzforschung, 45*: 191–198 (1991).
582. Neves, J. M. and Maistro, M. C. – Redutor hidrossulftio ou diotionito de sódio ($Na_2S_2O_4$) como reagente alternativo no uso em seqüência de branqueamento de pasta kraft. *O Papel, 60* (12): 71–79 (2000).
583. Meltzer, F. P. and Sepke, P.-W. – New ways to forecast the technological results of refining. *In:* III International Refining Conference and Exhibition. PIRA/IPST, Atlanta, 1995 Paper 2 26 pgs.
584. Derakhshandeh, B.; Kerekes, R.; Hatzikiriakos, S. and Bennington, C. – Rheology of pulp fibre suspensions: A critical review. *Chemical Engineering Science 66*: 3460–3470 (2011).
585. Stuart, S.-A. and Evans, R. – X-ray diffraction estimation of the microfibril angle variation in eucalypt wood. *Appita Journal, 48* (3): 197–200 (1995).
586. Eichhom, S. J. et al – Review: current international research into cellulosic fibres and composites. *Journal of Material Science, 36*: 2107–2131 (2001).
587. Adler, D. – A Study of the molecular mechanics of wood cell walls. Masterthesis, Massachusettes University of Technology, Cambridge, 2013 49 pgs https://dspace.mit.edu/bitstream/handle/1721.1/82339/861698451-MIT.pdf;sequence=2
588. Eklund, D. and Lindström, T. – Paper chemistry – An introduction. DT Paper Science Publications, Grankulla, Finland, 1991 305 pgs.
589. Martinez, P. C. and, S. W. – Modelos matemáticos de uma fibra celulósica em processo de refino em baixa consistência. *In:* V CIADICYP – Congreso Iberoamericano de Investigación en Celulosa y Papel, Guadalajara, 2006 12 pgs https://www.eucalyptus.com.br/artigos/2006_Modelagem+Fibras+Refinador.pdf
590. Rawat, N. and McDonough, T. J. – Effects of pulping conditions on the bleachability of hardwood kraft pulps: Effects of effective alkali charge in the pulping of birch and maple. *In:* Pulping Conference. TAPPI, Montreal, 1998.
591. Seth, R. S.; Kingsland, A. – The reinforcing properties of softwood kraft pulps. *Pulp Paper Canada, 91* (7): 273–279 (1990).
592. Retulainen, E.; Niskanen, K. and Nilsen, N. – Fibers and bonds. *In:* Papermaking Science and Technology. Volume 16 – Paper physics, Chapter 2. Ed. Gullichsen, J.; Paulapuro, H. and Niskanen, K. Finnish Paper Engineers' Association/Paperi ja Puu Oy Helsinki: 1998 pgs. 54–87.
593. Chauhan, A.; Kumari, A. and Ghosh, U. K. – Blending impact of softwood pulp with hardwood pulp on different paper properties. *Tappsa Journal,. 2*: 16–22 (2013).

594. Hubbe, M. A.; Venditti, R. V. and Rojas, O. J. – What happens to cellulosic fibers during papermaking and recycling? A Review. *BioResources, 2* (4): 739–788 (2007).
595. Martinez, P. C. and Park, S. W – Review of physical principles in low consistency refining. *O Papel, 73* (8): 65–72 (2012)
596. Eder, M.; Arnould, O.; Dunlop, J. W. C.; Hornatowska. J. and Salmén, L. – Experimental micromechanical characterization of wood cell walls *Wood Science and Technology, 47* :163–182 (2013).
597. Ratnieks, E. and Martins, M. – Refining pulp blends. *World Pulp and Paper Technology,* 1995/1996: 61–63.
598. Bovin, A. and Teder, A. – Some aspects on additivity. *In:* International Symposium on Paper and Pulp Characterization. EUCEPA, Ronneby, 1971
599. Maaß, M-C.; Saleh, S.; Militz, H. and Volkert, C. A. – The structural origins of wood cell wall toughness. *Advance Material*, 32 6 pgs (2020).
600. Stevens, W. V. – Principles of stock preparation and refining – refiners and refiner systems. In: Pulping Conference. TAPPI, Orlando, 1999, pgs. 97–102
601. Uesaka, T. and Moss, C. – Effects of fibre morphology on hygroexpansivity of paper: a micromechanics approach. *In:* 11st Fundamental Research Symposium. PIRA, Leatherhead, 1997 pgs. 663.
602. Leopold, B. – Effect of pulp processing on individual fiber strength. *Tappi, 49* (7): 315 (1966).
603. Bjurhager, I. – Mechanical behaviour of hardwoods – effects from cellular and cell wall structures. KTH Royal Institute of Technology, Stockolm, 2008 36 pgs. www.diva-portal.org/smash/get/diva2:14173/FULLTEXT01
604. Cave, I. D. and Hurt, L. – The anisotropic elasticity of the plant cell wall. *Wood Science and Technology, 2* (1): 268—278 (1968).
605. Seth, R. S.; Francis, D. W. and Bennington, C. P. J. – The effect of mechanical treatment during medium stock concentration fluidization on pulp properties. *Appita, 46* (1): 55–58, 60 (1993).
606. MacLeod, M. and Pelletier, L. J. – Basket cases: kraft pulps inside digesters. *Tappi Journal, 70* (11): 47 (1987)
607. De Grace, J. and Page, D. H. – The extensional behavior of commercial softwood bleached kraft pulps. *Tappi Journal, 59* (7): 98 (1976).
608. Costa, G. A. A.; Oliveira, E. D. de; Park, S. W. and Cardoso, M. – Overall heat transfer coefficients in a kraft black liquor industrial evaporation unit. Part I – simulation of multiple effect evaporation system. *Appita Journal, 60* (4): 321–326 (2007).
609. Boland, D. J.; Brooker, M. I. H.; Chippendale, G. M.; Hall, N.; Hyland, B. P. M.; Johnston, R. E.; Kleining, D. A. and Turner, J. D. – Trees of Australia. Nelson/CSIRO, Melbourne, 1984.
610. Lumiainen, J. – Refining – a key to upgrading the papermaking potential of recycled fibre. *Paper Technology,35* (9): 41–44 (1994).
611. Laine, J.; Lindström, T.; Bremberg, C. and Nordmark, G. – Studies on topochemical modification of cellulosic fibres. Part 5: Comparison of the effects of surface and bulk chemical modification and beating of pulp on paper properties. *Nordic Pulp and Paper Research Journal, 18* (3): 325–332 (2003).
612. Donaldson, L. – Microfibril angle: measurement, variation and relationships—A review. *IAWA Journal, 29* (4): 345–386 (2008).
613. Somerville, C.; Bauer, S.; Brininstool, G.; Facette, M.; Hamann, T.; Milne, J.; Osborne, E.; Paredez, A.; Persson, S.; Raab, T.; Vorwerk, S e Youngs, H. – Toward a systems approach to understanding plant cell walls. *Science, 306*: 2206–2211 (2004).
614. Liu, H.; Jixian Dong. J.; Guo, X.; Jiang, X.; Luo, C.; Xiaohui Tian, X.; Yang, R.; Zhang, L.; Bo Wang, B.; Yan, Y. – Refining characteristics of hardwood pulp using straight and curved bar plates. *Journal of Korea TAPPI, 51* (5): 45–60 (2019).
615. Li, M.; Johnston, R. E.; Xu, L.; Filonenko, Y. and Parker, I. – Characterization of hydrocycone-separated eucalypt fibre fractions. *Journal of Pulp and Paper Science, 25* (8): 299–304 (1999).
616. Nazhad, M. M. – Recycled fibre quality – a review. *Journal of Industrial and Engineering Chemistry, 11* (3): 314 (2005).

617. Brecht, W. and Klemm, K. H. – The mixture of structures in a mechanical pulp as a key to the knowledge of its technological properties. *Pulp and Paper Magazine Canada, 54* (1): 72–80 (1953).
618. Sharp, S. – A technical look at pulp & paper fiber. *FI Insights*. Fisher International. (2020) www.fisheri.com/blog/a-technical-look-at-pulp-paper-fiber-properties
619. Ruffo, R. and Malton, S. – Energy savings in stock preparation for recycled paper. *Appita Journal, 61* (4): 277–283 (2008).
620. Jang, H. F. – A theory for the transverse collapse of wood pulp fibres. *In:* 13th Fundamental Research Symposium, PIRA, Oxford, 2001, pg. 193.
621. Hill, J. – On line testing of mechanical pulp provides data for process control. *Pulp and Paper, 55* (7): 72–75 (1981).
622. Johansson, K.; Thuvander, F. and Germgård, U. – Single fibre fragmentation: a new measure of fibre strength loss during brown stock washing and oxygen delignification. *Appita Journal, 54* (3): 276–280 (2001).
623. Miletzky, F.; Erhard, K. and Alber, W. – Neue methoden zur chemischen und physikalischen charakterisierung von Zellstoffen. PTS Symposium, Dresden, 1995.
624. Area, C, – Updating the knowledge about the relationship between fibers characteristics and pulp properties. *In: 48th congresso Internacional da ABTCP, São Paulo,* 2012. Oral presentation) https://www.eucalyptus.com.br/artigos/2012_Fibers+and+Pulps.pdf
625. Mariani, S. A.; Marco Torres U.; Alicia Fernández, R. and Eduardo Morales – Efeito da formação de cerne em *Eucalyptus nitens* na polpação kraft. *O Papel, 65* (6): 55–59 (2004).
626. Smith, S. – The action of the beater in papermaking. *Paper Trade Journal, 106* (26): 47–48 (1922) [583, 1031].
627. Azevedo, C. A.; Sofia M. C.; Rebola, S. M. C.; Domingues, E. M.; Figueiredo, F. M. L. and Evtuguin, D. V. – Relationship between surface properties and fiber network parameters of eucalyptus kraft pulps and their absorption capacity *Surfaces,* 3: 265–281 (2020) https://www.researchgate.net/publication/342581602_Relationship_between_Surface_Properties_and_Fiber_Network_Parameters_of_Eucalyptus_Kraft_Pulps_and_Their_Absorption_Capacity
628. Dalpke, B. and Kerekes, R. J. (2005). The influence of fibre properties on the apparent yield stress of flocculated fibre suspensions. *Journal of Pulp and Paper Science, 31* (1): 39–43 (2005).
629. Brecht, W. – Modern concepts and practices of fibre refining. *In:* 11th Annual Pulp and Paper Conference. Kalamazoo, 1967.
630. Westman, L. – Iddling losses in the low-consistency refining of chemical pulp. *Svensk Papperstiding, 87* (3): 3–8 (1984).
631. Jagenberg, F. – Das holländergeschirr in briefen an einen papiermacher (1887) [178, 583, 1031].
632. Havelock, G. – Dynamic simulation of press drying. *Paper Technology, 31* (8) (1990).
633. Ellis, R. L. and Sedlachek, K. – Recycled vs. virgin fiber characteristics: a comparison. *Tappi Journal, 76* (2): 143–146 (1993).
634. Pulkkinen, I.; Kuitunen, S.; Alopaeus, V. – The most important eucalypt fiber properties for fiber network strength and structural property development. *In:* 5th ICEP International Colloquium on Eucalyptus Pulp. Porto Seguro, 2011 9 pgs www.eucalyptus.com.br/artigos/outros/49_Fibers_Characteristics.pdf
635. Dowens, G. M.; Hudson, I. L. and Raymond, C. A. – Variation in fibre dimensions in plantation eucalyptus. *In:* Sampling Plantation Eucalyptus for Wood and Fibre Properties. Ed. Dowens, G. M. and co-authors. CSRIO Publishing, Collingwood, 1997 Appendix 2 pgs. 100–109.
636. Mandlez, D.; Zangl-Jagiello, L.; Eckhart, R. and Bauer, W. – Softwood kraft pulp fines: application and impact on specific refining energy and strength properties. *Cellulose,* 27:10359–10367 (2020).
637. Paavilainen, L. and Luner, P. – Wet flexibility as a predictor of sheet properties. *Espra Research Reports, 84* (IX): 2 (1986).

638. Wardrop, A. B. – Fibre morphology and papermaking. *Tappi Journal, 52* (3): 396 (1969).
639. Seth, R. S. – The effect of fiber length and coarseness on the tensile strength of wet-webs: a statistical geometry explanation. *Tappi Journal, 78* (3): 99–102 (1995).
640. Gao, Y.; Li, K. and Wang, Z. – The influence of pulp furnish components on the property of supercalendered paper. *Pulp and Paper Canada, 108* (1): 44–49 (2007).
641. Tiikkaja, E. – The role of fibre length distribution measurement in refining. *In:* IV International Refining Conference. PIRA, Fiuggi, 1997 Paper 15 pgs. 239–246.
642. Pulkkinen, I.; Fiskari, J. and Alopaeus, V. – The use of fibre wall thickness data to predict handsheet properties of eucalypt pulp fibres. *Journal of Applied Science, 9* (22): 1 (2009)
643. Vainio, A. – Interfibre bonding and fibre segment activation in paper – observations on the phenomena and their influence on paper strength properties. Doctoral Thesis, Helsinki University of Technology, Helsink, 2007 84 pgs. https://aaltodoc.aalto.fi/handle/123456789/2958
644. Görres, J.; Amiri, R.; Wood, J. R. and Karnis, A. – The apparent density of sheets made from pulp blends. *Tappi Journal,* 79 (1): 179–187 (1996).
645. Carvalho, M. G.; Martins, A. A. and Figueiredo, M. M. – Fracionamento de pasta kraft de eucalipto: caracterização físico-química e desempenho papeleiro. *O Papel, 58* (7): 83–86 (1997).
646. Correa, A. de A. – A polpa química do eucalipto "urograndis" contra a da andiroba e do marupá: um estudo comparativo. *In:* XXX Congresso Anual de Celulose e Papel. ABTCP, São Paulo, 1997 pgs. 251–270.
647. Gaioto, M. R. – Influência da idade e do espaçamento na qualidade da madeira para a produção de papel e celulose. *In:* Seminário da Qualidade da Madeira. ESALQ/USP, Piracicaba, 1991
648. Moss, P. A.; Retulainen, E. and Paulapuro, H. and Aaltonen, P. – Taking a new look at pulp and paper: Applications of confocal laser scanning microscopy (CLSM) to pulp and paper research. *Paperi ja Puu – Paper and Timber, 75* (1–2): 74–79 (1993).
649. Jang, H. F.; Amiri, R.; Seth, R. S. and Karnis, A. – Fiber characterization using confocal microscopy-collapse behavior of mechanical pulp fibres. *Tappi Journal, 79* (4): 203–210 (1996).
650. Maloney, T. C. – On the pore structure and dewatering properties of the pulp fiber cell wall. Doctoral Thesis, Helsinki University of Technology, Esppo, 2000. 52 pgs https://www.researchgate.net/publication/34731327_On_the_pore_structure_and_dewatering_properties_of_the_pulp_fiber_cell_wall
651. Yu, Y.; Kettunen, H.; Hiltunen, E. and Niskanen, K. – Comparison of abaca and spruce as reinforcement. *In:* International Paper Phuysics Conference. TAPPI, Sao Diego, 1999 p. 161–169.
652. Sansígolo, C. A. – Efeitos da viscosidade nas propriedades da polpa de *Eucalyptus saligna*. Master's Thesis, ESALQ/USP, Piracicaba, 1988 66 pgs.
653. Karnis, A. – The mechanism of fiber development in mechanical pulping. *Journal of Pulp and Paper Science, 20* (10): 280–288 (1994).
654. Mansfield, S.D. and Kibblewhite, R.P. – Reinforcing potential of different eucalypt:softwood blends during separate and co-PFI mill refining. *Appita Journal, 53* (5): 385–392 (2000).
655. L'Anson, S. J.; Karademir, A.; Sampson, W. W. – Specific contact area and the tensile strength of paper. *Appita Journal, 59* (4): 297–301 (2006).
656. Yli-Viitala, P.; Jokinen, H.; Niinimäki e Ämmälä, A. – Hydrocyclone sand separation. *In:* 59th Annual Meeting. APPITA, Auckland, 2005 pg 39–43.
657. Bäckström, M.; Melander, E. and Brännall, E. – Study of the influence of charges on refinability of bleached softwood kraft pulp. *Nordic Pulp and Paper Research Journal, 28* (4): 588–595 (2013)
658. Ataíde, M. R. and Figueiredo, M. M. L. – Algumas considerações sobre a caracterização de fibras celulósicas. *Pasta e Papel* (4): 55–58 (1992).
659. Lumiainen, J. – Energy saving in low consistency refining. *Journal of Pulp and Paper Science, 19* (3): J125-J130 (1993).
660. Dillner, B e Tibbing, P. – Iso-thermal cooking to low kappa numbers facilitates TCF bleaching to full brightness. *In:* International Non-Chlorine Bleaching Conference. Pulp & Paper/Emerging Technology Transfer, Hilton Head, 1993 Paper 37 27 pgs.

661. Wallbäcks, L.; Edlund, U.; Lindgren, T. and Agnemo, R. – Multivariate characterization of pulp. Part 3: evaluation of physical and spectroscopic data on unbeaten and beaten kraft pulp samples. *Nordic Pulp and Paper Research Journal, 10* (2): 88–93 (1995).
662. Carlsson, J.; Persson, W.; Hellentin, P. and Malmqvist, L. – The propagation of light in paper: modeling and Monte Carlo simulations. *In:* International Paper Physics Conference. TAPPI, Niagara-on-the-lake, 1995 pgs. 83–86.
663. Laivins, G. V. and Scallan, A. M. – Removal of water from pulps by pressing. Part 1: inter- and intra-wall water. *Tappi Journal, 77* (3): 125–131 (1994).
664. Hietanen, S. and Ebeling, K. – A new hypothesis for the mechanics of refining. *Paperi ja Puu, 72* (2): 172–179 (1990).
665. Crotogino, R. H. and Gratton, M. F. – On-machine calendering – the process. *In:* TECH95 Theory & Practice of Papermaking Course. CPPA, Ottawa, 1995 Section F1 33 pgs.
666. Salvador, E.; Colodette, J. L.; Gomide, J. L. and Oliveira, R. C. de – Efeito da deslignificação com oxigênio nas propriedades físico-mecânicas de polpas kraft. *O Papel, 61* (2): 75–96 (2001).
667. Heikkurinen, A.; Levlin, J.-E. and Paulapuro, H. – Principles and methods in pulp characterization – basic fiber properties. *Papperi ja Puu, 73* (5): 411–417 (1991).
668. Lumiainen, J. – Specific surface load theory. *In:* III International Refining Conference and Exhibition. PIRA/IPST, Atlanta, 1995. Paper 5 15 pgs.
669. Kibblewhite, R. P. and Bawden, A. D. – Blends of extreme high and low coarseness radiata pine kraft pulps – fibre and handsheet properties. *Appita, 43* (3): 199–207 (1990).
670. Clarke, C. R. E.; Garbutt, D. C. F. and Pearce, J. – Growth and wood properties of preveances and trees of nine eucalypt especies. *Appita Journal, 50* (2): 121–130 (1997).
671. Lee, M-S.; Kim, C-H.; Lee J-Y.; Park, S-H. and Min, B-G. – Performance evaluation of bar plates without a draft angle to improve refining efficiency. *Journal of Korea TAPPI, 51* (5): 84–90 (2019).
672. Carpim, M. A.; Barrichelo, L. E. G.; Claudio-da-Silva Jr., E. and Dias, R. L. de V. – A influência do número de fibras por grama nas propriedades ópticas do papel. *In:* XX Congresso Anual de Celulose e Papel. ABTCP, São Paulo, 1987 pgs. 183–205.
673. Kibblewhite, R. P. – Effects of pulp drying and refining on softwood fibre wall structural organizations. *In:* 9th Fundamental Research Symposium, PIRA, Cambridge, 1989 pgs. 121–152.
674. Robertson, G.; Olson, J.; Allen, P.; Chan, B. and Seth, R. S. – Measurement of fiber length, coarseness, and shape with the fiber quality analyser. *Tappi Journal, 82* (10): 93–98 (1999).
675. Kibbewhite, R. P. and Bawden, A. D. – Fibre and fibre wall response to refining in softwood and hardwood kraft pulps. PAPDRO-New Zealand B. Report, 69, 1990.
676. Hortal, J. G. – Composicion quimica y estructura de la fibra. *In:* Constituyentes fibrosos de pastas y papeles. Ed. Escuela Técnica Superior de Ingenieros Industriales de Terrassa, 1988 Capitulo 2 pgs. 11–36.
677. Atack, D. – Towards a theory of refining mechanical pulping. *Appita, 34* (3): 223–227 (1980).
678. Rantanen, J.; Hitunen, E.; Nieminen, K.;Kerekers, R. and Paulapuro, H. – Construction or a wintl3 bar refinin34. *Tappi Journal, 10* (7): 45–51. (2011).
679. Ratnieks, E.; Massoquete, A.; Demuner, B.J.; Robinson, D. and DeFoe, R. – Ultra low intensity refining of eucalyptus pulp for papermaking. *In:* International Pulp Refinig Seminar. HUT, Helsinki (2007) 7 pgs
680. Lewis, D. W. and Danforth, J. – Stock preparation analysis. *Tappi Journal, 45* (3): 185–188 (1962).
681. Gottlieb, P. M.; Nutt, W. E.; Miller, S. R. and Macas, T. S. – Mill experience in high-consistency ozone bleaching of Southern pine pulp. *Tappi Journal, 77* (6): 117–124 (1994).
682. Schakford, L. D.; Miller, W. J. and Foelkel, C. E. B. – Commercial TCF bleach plant designed and effect of solids and thermal balance on operations. *In:* XXVII Congresso Anual de Celulose e Papel. ABTCP, São Paulo, 1994 pgs. 179–194.
683. Leskelä, M. – A model for the optical properties of paper. Part 1: theory. *Paperi ja Puu, 75* (9–10): 683–688 (1993).

684. Luce, J. E. – Transverse collapse of wood pulps fibers: fiber models, the physics and chemistry of wood pulp fibers. *In:* Tappi STAP number 8, 1970.
685. Mark, R. E. – Handbook of physical and mechanical testing of paper and paperboard. Marcel Deckker, Inc., New York, 1984 508 pgs.
686. Mayr, M.; Eckard, R.; Thaller, A. and Bauer, W. – Characterization of fines quality and that independent effect on sheet properties. *In:* 15th Fundamental Research Simposium, Oxford, 2017 pgs 299–322.
687. Liu, H.; Dong, J.; Jing, H.; Guo, X.; Duan, C.; Qi, K.; Yang, R.; Guo, H.; Wang, B, e Qiao, L. – Refining characteristics of isometric straight bar plates with different bar angles. *BioResources, 15* (4): 7844–7860 (2020).
688. Motamedian, H. R.; Halilovic, A. E. and Kulachenko, A. – Mechanisms of strength and stiffness improvement of paper after PFI refining with a focus on the effect of fines. *Cellulose 26*: 4099–4124 (2019).
689. Fischer, W. J.; Mayr, M.; Spirk, S.; Reishofer, D.; Jagiello, L. A.; Schmiedt, R.; Colson, J.; Zankel, A. and Bauer, W. – Pulp fines—characterization, sheet formation, and comparison to microfibrillated cellulose. *Polymers, 9*: 12 pgs (2017).
690. Comelato, J. S.; Ventorim, G.; Caraschi, J. C.; Santos, I. R. dos – Geração de finos no branqueamento de pasta kraft de eucalipto e seu efeito nas propriedades do papel. *Revista Árvore, 37* (1) (2013).
691. Bäckström, M.; Kolar, M.-C. and Htun, M. – Characterization of fines from unbleached kraft pulps and their impact on sheet properties, *Holzforschung, 62* (5): 546–552 (2008).
692. Lönnberg, L B.; Jäkärä, J.; Parèn, A. and Lundin, T. – Beating of chemical pulps from various wood raw materials. *In:* PTS Symposium (1999) https://www.researchgate.net/publication/284187811_Beating_of_chemical_pulps_made_from_various_wood_raw_materials
693. Retulainen, E. and Ebeling, K. – Fibre to fibre bonding and ways to characterize the bond strength. *Appita Journal, 46* (2): 282–288 (1993).
694. Kibblewhite, R. P. – The fibres of mechanical pulps – drying and latency effects. *Appita, 35* (3): 216–224 (1981).
695. Carvalho, M. G., Fernandes, J. R., Ferreira, L. A., Van Zeller, C. A., Ferreira, J. P. and Ferreira, P. J. T. – Efeito da secagem no potencial papeleiro de pastas kraft de eucalipto. *O Papel, 65*(10): 84–92 (2005).
696. Forsskåhl, I.; Tylli, H. and Olkkonen, C. – Participation of carbohydrate-derived chromophores in the yellowing of high-yield and TCF pulps. *Journal of Pulp and Paper Science, 26* (7): 245–249 (2000).
697. Park, J-M. – Fracture toughness, measurement, and applications related to paper break. *Journal of Korea TAPPI, 53* (1): 5–14 (2021).
698. O'Neil, P. L.; Karen, D. H. and Michell, A. J. – and Tijero, J. – Focused beam reflectant measurement as a tool to measure flocculation. *Tappi Journal, 1* (10): 14–20 (2002).
699. Htun, M. and Ruvo, A. D. – Implication of fines fraction for properties of bleached Kraft sheet. *Svensk Papperstidning, 81* (16): 507–510 (1978)
700. Wang, Q. Q.; Zhu, J. K.; Gleisner, R.; Kuster, T. A.; Baxa, U. and McNeil, S. E. – Morphological development of cellulose fibrils of a bleached eucalyptus pulp by mechanical fibrillation. *Cellulose, 19*: 1631–1643 (2012).
701. Ratnieks, E. and Martins, M. A. L. – Eucalyptus refining and white water quality. *In:* XXXV Congresso Anual. ABTCP, São Paulo, 1992 11 pgs.
702. Lindström, T. – Chemical factors affecting the behavior of fibres during papermaking. *Nordic Pulp and Paper Research Journal, 7* (4): 181–192 (1992).
703. Scallan, A. M. – The Effect of Acidic Groups on the Swelling of Pulps, *Tappi Journal, 66*(11), 73 (1983).
704. Hammar, L-Å; Bäckström, M. and Htun, M. – Efeitos da concentração de eletrólito e do pH na característica de refino de celuloses kraft não branqueadas. *O Papel, 63* (8): 79–86 (2003).
705. Kuitunen; S.; Pulkkinen, I.; Tarvo, V. and Alopaeus, V. – Modeling of fiber swelling. *In:* 5th ICEP – International Colloquium on Eucalyptus Pulp, Porto Seguro, 2011 9 pgs.

706. Kuitunen, S., and, Tarvo, V. – Modeling of fiber swelling. *In:* 5th International Colloquium on Eucalyptus Pulp. Porto Seguro, 2011 9 pgs.
707. Lobben, H. T. – The tensile stiffness of paper. Part 2: Activation studied by freeze drying. *Nordic Skogindustri, 30* (3): 43–48 (1976)
708. Lindström, T. and Carlsson, G. – The effect of carboxyl groups and their ionic form during drying aon the hornification of cellulose fibers. *Svensk Papperstiding,85* (15): 146–151 (1982).
709. Wang, X., Paulapuro, H. and Maloney, T. C. – Chemical pulp refining for optimum combination of dewatering and tensile strength. *Nordic Pulp and Paper Research Journal, 20* (4): 442–447 (2005)
710. Joris, G. – Optimized fillings for LC refiners. *In:* III International Refining Conference and Exhibition. PIRA/IPST, Atlanta, 1995 Paper 22 89 pgs.
711. Stone, J. E. and Scallan, A. M. – The effect of component removal upon the porous structure of the cell wall of wood. Part 2 – swelling in water and the fibre saturation point. *Tappi, 50* (10): 496–501 (1967).
712. Lindström, T. and Kolman, M. – The effect of pH and electrolyte concentration during beating and sheet forming on paper strength. *Svensk Paperstiding, 85* (15): 140 (1982).
713. Katz, S. and Scallan, A. M. – Ozone and caustic soda treatments of mechanical pulps. *Tappi Journal, 66* (1): 85–87 (1983).
714. Lukins, J. – A refiner control design. *In:* Papermakers Conference. TAPPI, Seattle, 1991 pgs. 523–532.
715. Scallan, A. M. and Grignon, J. – The efeect of cations on pulp and paper properties. *Svensk Papperstiding., 82* (2): 40–47 (1979)
716. Barzyk, D.; Page, D. H. and Ragauskas, A. – Acidic group topochemistry and fibre-to-fibre specific bond strength. *Journal of Pulp and Paper Science, 23* (2): 59–61 (1997).
717. Laine, J. – The effect of surface chemical composition and charge on the fibre and paper properties of unbleached and bleached kraft pulps.Doctoral Thesis, Helsinki University of Technology, Espoo, 1996.
718. Engstrand, P.; Sjögren, B.; Ölander, K. and Htun, M. – Significance of carboxylic groups for the physical properties of mechanical pulp fibres. *In:* 6th International Symposium on Wood and Pulping Chemistry. APPITA, Victoria, 1991 Volume 1 pg. 75.
719. Gellerstedt, G. and Lindfors, E-L. – On the structure and reactivity of residual lignin in kraft pulp fibers. *In:* International Pulp Bleaching Conference. P&PC, Stockholm, 1991 Volume 1 pg. 73.
720. Koljonen, K.; Österberg, M.; Kleen, M.; Fuhrmann, A. and Stenius, P. – Precipitation of lignin and extractives on kraft pulp: Effect on surface chemistry, surface morphology and paper strength. *Cellulose, 11*: 209–224 (2004).
721. Laivins, G. and Scallan, T. – Acidic versus alkaline beating of pulp. *Journal of Pulp and Paper Science, 26* (6): 228–233 (2000).
722. Babus'Haq, R. F.; O'Callaghan, P. W.; Al-Kader, M. and Nagle, L. – Development of an interative multimedia training package for stock refining in the paper and board industry. *In:* III International Refining Conference and Exhibition. PIRA/IPST, Atlanta, 1995 Paper 8 6 pgs.
723. Shen, W.; Yao, W.; Li, M. and Parker, I. – Characterization of eucalypt fibre surface using inverse gas chromatography and X-ray photoelectron spectroscopy. *Appita Journal, 51* (2): 147–151 (1998).
724. Daniel, A.; Neto, C.; Evtuguin, D. and Silvestre, A. – Hexenuronic acid contents of Eucalyptus globulus kraft pulps: variation with pulping conditions and effect onf ECF bleachability. *Tappi Journal, 2* (5): 3 (2003).
725. Torgnysdotter, A. and Wågberg, L. – Study of the joint strength between regenerated cellulose and its influence on the sheet strength. *Nordic Pulp and Paper Research Journal, 18* (4): 455–459 (2003).
726. Zhang, Y.; Sjögren, B.; Engstrand, P. and Htun, M. – Determination of charged groups in mechanical pulp fibres and their influence on pulp properties. *Journal of Wood Chemistry Technology, 14* (1): 83 (1994).

727. Tarvo, V.; Kuitunen, S.; Lehtimaa, T.; Tervola, P.; Räsänen, E.; Tamminen, T.; Aittamaa, J.; Vuorinen, T. and Henricson, K. – Modelling of chemical pulp bleaching. *Nordic Pulp and Paper Research Journal, 23* (1): 91–101 (2008).
728. Wågberg, L.; Bjorklund, M – On the mechanism behind wet strength development in papers containing wet strength resins. *Nordic Pulp and Paper Research Journal, 8* (8): 53 (1993).
729. Laine, J. and Stenius, P. – The surface chemistry of bleached kraft pulp fibres and its effects on final paper properties. *In:* Pulping Conference. TAPPI, Chicago: 1995 Book 1 pg. 1.
730. Isogai, A.; Kitaoka, C.; Onabe, F. – Effects of carboxyl groups in pulp on retention of alkylketene dimer. *Journal of Pulp and Paper Science, 23* (5): 215 (1997).
731. Chen, S-L.; Wang, S. and Lucia, L. A. – New insights into the fundamental nature of lignocellulosic fiber surface charge. *Journal of Colloid and Interface Science, 275* (2): 392–397 (2004).
732. Hirn, U. and Schennach, R – Fiber-fiber bond formation and failure: mechanisms and analytical techniques. *In:* 16th Fundamental Research Symposium, Oxford, 2017 pgs 839–863. https://bioresources.cnr.ncsu.edu/wp-content/uploads/2019/03/2017.2.839.pdf
733. Bäckström, M. and Hammar, L-A – The influence of the counter-ions to the charged groups on the refinability of the never-dried bleached pulps. *BioResources, 5*(4): 2751–2764 (2010).
734. Niemelä, K.; Alén, R. and Sjöström, E. – The formation of carboxylic acids during kraft and kraft-anthraquinone pulping of birch wood. *Hozforschung, 39*: 167–172 (1985).
735. Sjöström, E. – The origin of charge on cellulosic fibres. *Nordic Pulp and Paper Research Journal, 4* (4): 181 (1989).
736. Jansson, J. – The influence of pH on fiber and paper properties: Different pH levels during beating and sheet forming. Master's Thesis. Karlstad University, Karlstad, 2015 43 pgs www.diva-portal.org/smash/get/diva2:823180/FULLTEXT01
737. Choi, K. H.; Kim, A. R. and Cho, B. U. – Effects of alkali swelling and beating treatments on properties of kraft pulp fibers. *BioResources, 11* (2): 3769–3782 (2016).
738. Barzyk, D.; Page, D. H. and Ragauskas, A. – Carboxylic acid groups and fibre bonding. The fundamentals of papermaking materials, Volume 2 pg. 893 (1997).
739. Zhang, D.; Kim, D.; Allison, L.; Dang, Z. and Ragauskas, A. – The fate of fiber charge during peroxide and oxygen delignification. *In:* International Pulp Bleaching Conference. P&PC, Stockholm, 2005. pg. 93.
740. Katz, S.; Liebergott, N. and Scallan, A. M. – A mechanism for the alkali strengthenin of mechanical pulps. *Tappi Journal, 64* (7): 97–100 (1981).
741. Ruusumo, P.; Sacon, V. and Fardim, P. – Grupos aniônicos em fibras de polpa de eucalipto: perfil de concentração numa linha de branqueamento. *O Papel, 69* (10): 56–69 (2008).
742. Levlin, J. E. – Some differences in the beating behavior of softwood and hardwood kraft pulps. *In:* International Symposium on Fundamental Concepts of Refining. IPC, Appleton, 1980.
743. Nordström, B. and Hermansson, L. – Efect of softwood kraft fiber coarseness on formation and strength efciency in twin-wire roll forming. *Nordic Pulp & Paper Research Journal, 33* (2): 237–245 (2018).
744. Mohlin, U-B. – On the complexity of LC refining – changing consistency and flow rate in the Beloit DD refiner. *In:* Scientific and Technical Advances in Refining & Mechanical Pulping – Course Notes. PIRA: Stockholm, 2003 Paper 14.
745. Fiskari, J. – Oxalic acid formation in oxygen chemical bleaching. *In:* Pulping Conference. TAPPI, Orlando, 1999, volume 2, pgs. 7–10.
746. Bäckström, M.; Bolivar, S. and Paltakari, J – Effect of ionic form onn fibrillation and the development of fibre network strength during the refining of kraft pulps. *O Papel, 73* (7): 57–65 (2012).
747. Keckes, J.; Burgert, I.; Fruhmann, K.; Muller, M.; Kolling, K.; Hamilton, M.; Burghammer, M.; Roth, S.V.; Stanzl-Tschegg, S. and Fratzl, P. – Cell-wall recovery after irreversible deformation in wood. *Natural Materials, 2*: 810–814 (2003).
748. Vázquez, G.; López, M.; González, J. and Antorrena, G. – Adsorption of sodium on O_2-prebleached *Eucalyptus globulus* Kraft pulp. *Solutions, 85* (1): 1–3 (2002).

749. Torgnysdotter, A. and Wågberg, L. – Influence of electrostatic interactions on fibre/fibre joint and paper strength. *Nordic Pulp and Paper Research Journal, 19* (4): 440–447 (2004).
750. Evans, B. E. – The effect of process water on paper properties. *Paper Technology Ind., 22* (3): 99 (1981).
751. Zhao, C.; Zhang, H.; Zeng, X.; Li, H. and Sun, D, – Enhancing the inter-fiber bonding properties of cellulosic fibers by increasing different fiber charges. *Cellulose,* (23): 1617–1628 (2016).
752. Räsänen, E. – Modelling ion exchange and flow in pulp suspensions. Doctoral Thesis, Helsinki University of Technology, Espoo, 2003 62 pgs www.vttresearch.comsitesdefaultfilespdfpublications2003P495.pdf
753. Wang, F. and Hubbe, M. A. – Charge properties of fibers in the paper mill environment. 1. effect of electrical conductivity. *Journal of Pulp Paper Science, 28* (10): 347–353 (2002).
754. Paulapuro, H. – Wet pressing – present understanding and future chalenges. Translations of Fundamental Research Symposium, D.F.Baker Ed., BPBMA, London, 2001 639 pgs.
755. Brown, A. G. and Hills, W. E. – General introduction. *In:* Eucalyptus for Wood Production. Ed. Hillis, W. E. and Brown, A. G., CSIRO/Academic Press, London, 1984 Chapter 1 pg 3–5.
756. Scallan, A. M. and Carles, J. E. – Correlation of water retention value with fiber saturation point. *Svensk Papperstiding, 75* (17): 699–703 (1972).
757. Guo, X.; Dong, J.; Liu, H.; Duan, C.; Yang, R.; Kai Qi, K.; and Kong, L. – Refining mechanism of combined refining plates with different bar angles. *Journal of Korea TAPPI, 52* (2): 32–42 (2020).
758. Seth, R. S.; Barbe, M. C.; Willians, J. C. R. and Page, D. H. – The strength of wet webs. *Tappi, 65* (3): 135 (1982).
759. Vaz, A.; Silvy, J.; Gil, C. and Simões, R – Otimização na refinação de pastas químicas: minimização enegética e otimização conjugada de propriedades. *In:* 45th ABTCP International Pulp and Paper Congress and VII IberoAmerican Congress on Pulp and Paper Research. Sao Paulo., 2012 14 pgs.
760. Palacios, M. G. V. – Producción y caracterización de nanofibras de celulosa desde diferentes especies de eucalyptus para biomateriales en base de celulosa. Doctoral Thesis, Universidad de Concepción, Concepción, 2018 62 pgs http://repositorio.udec.cl/xmlui/handle/11594/3005
761. Stone, J. E.; Scallan, A. M. and Abrahamson, B. – Influence of beating on cell wall swelling and internal fibrillation. *Svensk Papperstiding, 71* (19): 687–694 (1968).
762. Maloney, T. C.; Laine, J. E. and Paulapuro, H. – Comments on the measurement of cell wall water. *Tappi Journal, 82* (9): 125–127 (1999).
763. Thode, E. F.; Bergomi Jr., J. G. and Unson, R. E. – The application of a centrifugal water retention test to pulp evaluation. *Tappi Journal, 43* (5): 505–511 (1960).
764. Hiltunen, E. – Papermaking properties of pulp. *In:* Papermaking Science and Technology. Volume 17 – Pulp and paper testing, Chapter 3, Ed. Gullichsen, J. and Paulapuro, H. Finnish Paper Engineers' Association/Paperi ja Puu Oy, Hesinki: 1999.
765. Lumiainen, J. – Comparison of the mode of operation between conical and disc refiners. *In:* IV International Refining Conference. PIRA, Fiuggi, 1997 Paper 14 pgs. 227–236.
766. Lee, J-Y.; Kim, C-H.; Kwon, S.; Park, H-H.; Yim, H-T.; Gu, H-G. and Min, B-G. – Study of mixed refining behaviors of softwood kraft pulps and hardwood kraft pulps using different bar fillings. *Journal of Korea TAPPI, 50*(5): 31–38 (2018).
767. Pantha, S. and Somboon, P. – Examination of separate and mixed refining methods on Softwood and hardwood pulps for linerboard production. *Kasetsart Journal* (Natural Science) 48: 540 – 547 (2014).
768. Srndovic, J. S. – Interactions between wood polymers in wood cell walls and cellulose/hemicellulose biocomposites. Doctoral Thesis. Chalmers University of Technology, Goteborg, 2011 98 pgs https://core.ac.uk/download/pdf/70588365.pdf
769. McIntoch, D. C. – The effect of refining on the structure of the fiber wall. *Tappi, 50* (10): 482–488 (1967).
770. Emerton, H. W. – Fundamentals of the beating process. Marshall Press, London, 1957 137 pgs.

771. Mayr, M.; Eckhart, R.; Winter, H. and Bauer, W. – A novel approach to determining the contribution of the fiber and fines fraction to the water retention value (WRV) of chemical and mechanical pulps. *Cellulose 24*: 3029–3036 (2017
772. Koskenhely K.; Ämmälä, A.; Jokinen, H. and Paulapuro, H. – Effect of refining intensity on pressure screen fractionated softwood kraft. *Nordic Pulp and Paper Research Journal, 20* (2): 169–175 (2005).
773. Bäckström, M.; Lindblom, A. T. and Wågberg, L. – Studies of the influence of deflocculants and flocculants on the refining efficiency on a pilot scale. *Nordic Pulp and Paper Research Journal, 24* (3): 319–326 (2006)
774. Campos, E. da S. – A influência do perfil transversal de umidade da folha na estabilidade dimensional do papel. *In:* XXVIII Congresso Anual de Celulose e Papel. ABTCP, São Paulo, 1995 pgs. 667–675.
775. Laivins, G. V. and Scallan, A. M. – The mechanism of hornification of wood pulps. *In:* 10th Fundamental Research Symposium. PIRA, Oxford, 1993 pgs. 1235–1260.
776. Jayme, G.; Ghoneim, A-F. and Krüger, H. – Verbesserte messung des wasser-zellstoffe. *Das Papier, 90* (12) (1958) [721].
777. Duarte, A. P., Martins, S., Abrantes, C., Ismael, M. I., Simões, R. and Figueiredo, J. A. – Aprimoramento das propriedades da polpa Kraft branqueada mediante a oxidação da celulose. *O Papel, 67* (10): 76–82 (2006).
778. Wang, X.; Maloney, T. C. and Paulapuro, H. – Improving the properties of never-dried chemical pulp by pressing before refining. *Nordic Pulp and Paper Research Journal, 21* (1): 135–139 (2006).
779. Giertz, H. W. – Contribution to the theory of tensile strength. *In:* Proceedings of EUCEPA/ Europe TAPPI Beating Conference. Venice, 1964 pg. 39.
780. Giertz, W. W. and Rodland, G – Elongation of segments – bonds in the secondary regime of the lead/elongation curve. *In:* Proceedings International Paper Physics Conference. CPPA, Montreal, 1979 pg. 129-
781. Mohlin, U-B.; Mohlin, U. and Puiseau, M. W. – Some aspects of using zero-span tensile index as a measure of fiber strength. *In:* International Paper Physics Conference. TAPPI, Victoria, 2003 pgs. 107–113
782. Duker, E.; Ankerfors, M.; Lindström, T. and Nordmark, G. G. – The use of CMC as a dry strength agent – the interplay between CMC attachment and drying. *Nordic Pulp and Paper Research Journal, 23* (1): 65–71 (2008).
783. Chen, Y.; Wan, J.; Dong, X. and Ma, Y. – Fiber properties of eucalyptus kraft pulp with different carboxyl group contents. *Cellulose,* 20:2839–2846 (2013).
784. Waterhouse, J. F. – The ultimate strength of paper. *In:* Design Criteria for Paper Performance. Ed. Kolseth, P; Fellers, C. and Salmen, L. STFI, Meddelande A 969, 1987.
785. Ingmanson, W. and Thode, E. – Factors contributing to the strength of paper – relative bonded area. *Tappi, 42* (1): 83–93 (1959).
786. Luner, P.; Karna, A. E. U. and Donofrio, C. P. – Studies in inter fiber bonding of paper – the use of optical bonded areas with high yield pulps. *Tappi, 46* (6): 409–414 (1961).
787. Tanaka, F. and Fukui, N. – The behavior of cellulose molecules in aqueous environments. *Cellulose, 11*: 33–38 (2004).
788. Retulainen, E. – The role of fibre bonding in paper properties. Doctoral Thesis, Helsinki University of Technology, Espoo, 1997 78 pgs https://ntrl.ntis.gov/NTRL/dashboard/searchResults/titleDetail/PB98102098.xhtml
789. Vainio, A., Paulapuro, H., Koljonen, K. and Laine, J. – The effect of drying stress and polyelectrolyte complexes on the strength properties of paper. *Journal of Pulp and Paper Science, 32* (1): 9–13 (2006).
790. Vainio, A. and Paulapuro, H. – Observations on inter fibre bonding and fibre segment activation based on the strength properties of laboratory sheets. *Nordic Pulp and Paper Research Journal, 20* (3): 340–344 (2005).
791. Moss, P. A. and Retulainen, E. – The effect of fines on fibre bonding: cross-sectional dimensions of TMP fibres at potential bonding sites. *Journal of Pulp and Paper Science, 23* (8): 382–388 (1997).

792. van den Akker, J. A. – Note on the theory of fiber-fiber bonding in paper. The influence on paper strength of drying by sublimation. *Tappi, 35* (1): 13–15 (1952).
793. Weisse, U. and Paulapuro, H. – Effect of drying and rewetting cycles on fibre swelling. *Journal of Pulp and Paper Science, 25* (5): 163–166 (1999).
794. Sykes, M.; Yang, V.; Blankenburg, J. and AbuBakr, S. – Biothechnology: working with nature to improve forest resources. *In:* International Environmental Conference. TAPPI, Nashville: 18-21/04/1999 Volume 2 pgs. 631–637.
795. Pelton, R. – A model of the external surface of wood pulp fibres. *Nordic Pulp and Paper Research Journal, 8* (1): 113–119 (1993).
796. Ratnieks, E.; Mora, E. and Martins, M. A. L. – Propriedades papeleiras de misturas de polpas: fibras de eucalipto, aparas destintadas e fibras longas. *In:* XXVII Congresso Anual. ABTCP, São Paulo (1994) 12 pgs
797. Rusu M, Mörseburg K, Gregersen Ø, Yamakawa A, Liukkonen S. – Relation between fibre flexibility and cross sectional properties. *Bioresources* 6(1):641–655 (2011).
798. Springer, A.; Nabors, L. A. and Bhatia, O. – The influence of fiber, sheet structual properties and chemical additives on wet pressing. *Tappi Journal, 74* (4):221 (1991).
799. Bergander, A. and Salmém, L. – The transverse elastic modulus of the native wood fibre wall. *Journal of Pulp and Paper Science, 26* (6): 234–238 (2000).
800. Brodeur, P. H. and Runge, T. M. – Compactability of a wet fibre mat using acoustic radiation pressure. *Journal of Pulp and Paper Science, 22* (8): J278–282 (1996).
801. Santos, A.; Amaral, M. E.; Vaz, A.; Anjos, O. and Simões, R. – Effect of *Eucalyptus globulus* wood density on papermaking potential. *Tappi Journal, 7* (5): 25–32 (2008).
802. Abitz, P. and Luner, P. – Applications of wet fiber flexibility (WFF) to papermaking: wet web strength and refining. *Espra Research Reports, 90* (IV): 109–110 (1989).
803. Bentley, R. G.; Hamilton, R. K. and Jack, J. S. – An optical method for monitoring pulp refining. *Journal of Pulp and Paper Science, 23* (10): 504–509 (1997).
804. Thode, E. F. and Ingmanson, W. L. – Factor contributing to the strength of a sheet paper. *Tappi, 42* (1): 74–83 (1959).
805. Claudio-da-Silva Jr., E.; Marton, R. and Granzows, S. – Effect of beating on wet-web properties. *Tappi Journal, 65* (11): 99–103 (1982).
806. Hartler, N. and Rennel, J. – Opacity in relation to strength properties of pulps. Part 2: light scattering. *Svensk Papperstiding, 72* (1): 9–13 (1969).
807. Kim, B. Y. – Influences of beating on photo degradation. *J.Tappik,* 25 (4): 27–33 (1993).
808. Mannström, B. – On the characterization and quality control of mechanical pulp. *Paperi ja Puu, 49* (4a): 137–146 (1967).
809. Kropholler, H. W.; Sampson, W. W. and Rosli, W. – Strength properties of paper not sensitive to tackle used for beating or refining. *In:* International Paper Physics Conference. CPPA/ TAPPI, Niagara-on-the-lake, 1995 pgs. 59–63.
810. Abitz, P.; Cresson, T; Brown, A. and Luner, P. – Role of web fiber flexibility in governing wet web properties – preliminary report. *Espra Research Reports, 85* (VI): 93 (1986).
811. Mohlin, U-B., Tubek-Lindblom, A., Luukko, Woo, Y. D. and Burman, A – Industrial refining of acacia. *Appita Journal, 59* (1): 53–57 (2006).
812. Görres, J.; Amiri, R.; Wood, J. R. and Karnis, A. – Mechanical pulp fines and sheet structure. *Journal of Pulp and Paper Science, 22* (12): 491–496 (1996).
813. Lawryshyn, Y. A. and Kuhn, D. C. S. – Large deflection analysis of wet fibre flexibility measurement techniques. *Journal of Pulp and Paper Science, 22* (11): 423–431 (1995).
814. Torgnysdotter, A. Wågberg, L. – Tailoring of fibre/fibre joints in order to avoid the negative impacts of drying on paper properties. *Nordic Pulp and Paper Research Journal, 21* (3): 411–418 (2006)
815. Shekhar, C. D. – Fine edged parallel and curved bar plates in refining system for pulp and paper industries. *IPPTA Journal, 24* (1): 115–119 (2012).
816. Shekhar, C. D. – Fine bar technology in refining system for pulp and paper industries. *IPPTA Journal, 22* (3): 109–112 (2010).

817. Arjas, A. – Effect of the evenness of the pulp bleaching stage on fibre bonding. *Paperi Ja Puu, 52* (12): 825–829 (1970).
818. Vainio, A. and Paulapuro, H. – The effect of wet pressing and drying on bonding and activation in paper. *Nordic Pulp and Paper Research Journal, 22* (4): 403–408 (2007).
819. Nanko, H. and Ohsawa, J. – Structure of fibre bond formation. *In:* Fundamentals of Papermaking. Transaction of 9th Fundamental Research Symposium. Pulp and Paper Fundamental Research Society, 786–830 (1989).
820. Ketola, H. and Andersson, T. – Dry strength additives. *In:* Papermaking Chemistry, Volume 4. Neimo, L. Ed., Papermaking Science Technology. Fapet Oy, 268–287, (1999);
821. Koljonen, K., Vainio, A., Hiltunen, E., Laine, J. and Stenius, P. – The effect of polyelectrolyte adsorption on the strength properties of paper made from mixtures of mechanical and chemical pulps. *In:* International Paper Coating Chemistry Symposium. PAPTAC, Montreal, 2003 pgs. 223–227.
822. Clark, J.d'A. – Fibrillation, free water and fiber bonding. *Tappi, 52* (2): 335 (1969).
823. van Eperen, R. H. – What do test result mean. *In:* Process and Product Quality Conference. TAPPI, Savannah, 1994 pgs. 45–50.
824. Allison, R. W.; Ellis, M. J. and Wrathall, S. H. – Interaction of mechanical and chemical treatments on pulp strength during kraft pulp bleaching. *Appita Journal, 51* (2): 107–113 (1998).
825. Gurnagul, N.; Page, D. H. and Paice, M. – The effect of cellulose degradation on the strength of wood pulp fibres. *Nordic Pulp and Paper Research Journal, 7* (3): 152 (1992).
826. Seth, R. S. and Chan, B. K. – Measuring fiber strength of papermaking pulps. *Tappi Journal, 82* (11): 115–120 (1999).
827. Mott, L.; Shaler, S.M. and Grrom, L.H. – A technique to measure strain distributions in single wood pulp fibers. *Wood Fiber Science, 28* (4): 429 (1999).
828. Duker, E. and Lindström, T. – On the mechanisms behind the ability of CMC to enhance paper strength. *Nordic Pulp and Paper Research Journal, 23* (1): 57–64 (2008).
829. Baker, C. – Introduction to control in the refining process. *In:* African Pulp and Paper Week. TAPPSA (2002) http://www.tappsa.co.za/archive2/APPW_2002/Title/Introduction_to_control_in_the_refining_process.html
830. Stoere, P.; Nazhad, M. and Kerekes, R. J. – An experimental study of the effect of refining on paper formation. *Tappi Journal, 84* (7): 52–58 (2001)
831. Andersson, S. R. – Network disruptions and turbulence in fibre suspensions. Doctoral Thesis. Chalmers Universit of Technology, Göteborg, 1998 [879].
832. Paul, T.; Duffy, G. and Chen, D. – New insight sinto the flow of pulp suspensions. Solutions, 1 (1):94 (2001)
833. Schmid, C.; Switzer, L. and Klingenberg, D. – Simulations of fiber flocculation effects of fiber properties and interfiber friction. *Journal of Rheology, 44* (4): 781–809 (2000).
834. Switzer, L. H. and Kingenberg, D. J. – Simulations of fiber floc dispersion in linear flow fields. *Nordic Pulp and Paper Research Journal, 18* (2): 141–144 (2003).
835. Kärenlampi, P. – Tensile strength of a mixture of two pulps. *Journal of Pulp and Paper Science, 21* (12): J432–436 (1995).
836. Steenberg, B. – A model of refining as a special case of milling. *In:* International Symposium on Fundamental Concepts of Refining. IPC, Appleton, 1980 pgs 107.
837. Koskenhely, K.; Ämmälä, A.; Jokinen, H. and Paulapuro, H. – Refining characteristics of softwood fibre fractions. *In:* 13th Fundamental Research Symposium – Advances in Paper Science and Technology Cambridge. 2005 pgs. 427–456.
838. Manfredi, V. and Claudio-da-Silva Jr, E. – Refining operating variables vs raw material. *In:* International Conference – Advances in Refining Technologies. PIRA, Birmingham, 1986 41 pgs.
839. Joy, E.; Robinson, D. and Mathew, J. – Deformation of fiber flocs in refining. *In:* Papermakers Conference. TAPPI, 2001 5 pgs.
840. Holik, H. – Mais rápidas, mais largas, melhores – progresso em máquinas de papel nos últimos 100 anos. *O Papel, 71* (8): 66–93. (2010)

841. Rihs, J. and Josephson, W. E. – Refining systems with flow recirculation. *In:* IV International Refining Conference. PIRA, Fiuggi, 1997 Paper 7 pgs. 117–124.
842. Koch, G – Raw material for pulp. *In:* Handbook of Pulp. Volume 1, Chapter 2, Ed. Sixta, H. WILEY-VCH Verlag GmbH &Co. KGaA, Weinheim, 2006 pgs. 28–68.
843. Statie, E.; Salcudean, M.; Gartshore, I. and Bibeau, E. – A computational study of particle separation in hydrocyclone. *Journal of Pulp and Paper Science, 28* (3): 84 (2002).
844. Wood, J. R., Grondin, M. and Karnis, A. – Characterization of mechanical pulp fines with a small hydrocyclone. Part I: the principle and nature of the separation. *Journal of Pulp and Paper Science, 17* (1): 1 (1991).
845. Sevilla, E. M. and Branion, R. M. R. – The fluid dynamics of hydrocyclones. *Journal of Pulp and Paper Science, 23* (2): 85 (1997).
846. Wood, J. R. and Karnis, A. – Distribution of fibre specific surface of papermaking pulps. *Pulp and Paper Canada, 80* (4): 73–79 (1979).
847. Jones, E. D.; Campbell, R. T. and Nelson, G. G. – Springwood-summerwood separation of southern pine pulp to improve paper qualities. *Tappi Journal, 49* (9): 410 (1966).
848. Kure, K. A.; Dahlqvist, G, Ekstrom, J., and Helle, T. – Hydrocyclone separation, and refect refining, of thick-walled mechanical pulp fibres. *Nordic Pulp and Paper Research Journal, 14* (2): 100–105 (1999).
849. Paavilainen, L. – The possibility of fractionating softwood sulfate pulp according to cell wall thickness. *Appita, 45* (5): 319–326 (1992).
850. Park, S.; Venditti, R.; Jammel, H.; Pawlak, J. – The effect of fibre properties on fibre fractionation using a hydrocyclone. *Journal of Pulp and Paper Science, 31*(3): 132–137 (2005).
851. Asikainen, S. – Applicability of fractionation of softwood and hardwood kraft pulp and utilization of the fractions. Doctoral Thesis. Aalto University School of Chemical Technology, Espoo, 2014 91 pgs.
852. Mansfield, S. D.; Swanson, D. J.; Roberts, N.; Olson, J. A. and Saddler, J. N. – Enhancing Douglas-fir pulp properties with a combination of enzyme treatments and fiber fractionation. *Tappi Journal, 82* (5): 152–158 (1999).
853. Scott, G. M. and Abubakr, S. M. – Fractionation of secondary fibre – a review. *Prog. Paper Recycling, 3* (3): 50–59 1994).
854. Eckert, W. F.; Afacan, A. and Masliyah, J. H. – Fractionation of wood pulps by continuous flotation. *Tappi Journal, 83* (8): 1–8 (2000).
855. Yu, C. J.; DeFoe, R. J. and Crossley, B. R. – Fractionation technology and its applications. *In:* Pulping Conference. TAPPI, Orlando, 1999 pgs. 451–466.
856. Olson, J. A. – Fibre length fractionation caused by pulp screening, slotted screen plates. *Journal of Pulp and Paper Science, 27* (8): 255–261 (2001).
857. Gooding, R.; Olson, J. and Roberts, N. – Parameters for assessing fibre fractionation and their application to screen rotor effects. *Tappi Journal, 3* (3): 16 (2004).
858. El-Sharkawy, K.; Rousu, P.; Haavisto, S. and Paulapuro, H. – Control of bagasse pulp quality by fractionation. *Appita Journal, 60* (5): 404–409 (2007).
859. Schakford, L. D.; Shin, N. H. and Wizani, W. – Cozimento avançado para o próximo século: RDH2000. *In:* II Seminário de Deslignificação. ABTCP, Ribeirão Preto, 1999 Paper 1 12 pgs.
860. Coelho, A. C. de M. and Souza, A. de – Papéis para sacaria. *O Papel, 61* (9): 64–68 (2000).
861. Galland, G. and Vernac, Y. – New attitudes towards increased consumption of secondary fibres in papermaking. *World Pulp and Paper Technology*, 1994/1995: 75, 76, 78–81.
862. Allison, R. W. – Effect of pulping and bleaching conditions on the strength of radiata pine kraft pulp. *Appita, 46* (6): 424–429, 444 (November 1993).
863. Sandeep, T.; Nirmal, S.; Om P. M.; Pratima, B.; Pramod K. B. – Enzymatic Refining of Chemical Pulp. IPPTA Journal, 20 (3): 129–132 2008 https://www.researchgate.net/publication/305527618_Enzymatic_Refining_of_Chemical_Pulp
864. Kibblewhite, R. P. Johnson, B. I. and Shelbourne, C. J. A. – Kraft pulp qualities of *Eucalyptus nitens*, *E globulus* and *E maidenii* at 8 and 11 years. *New Zealand Journal of Forestry Science, 30*(3): 447–457 (2000).

865. Almeida, M. D. de; Sevrini, G. I.; Leodoro, L. M., Faez, M. S., Soto, M. R., Kaneco, S. Y. – Tratamento mecânico de fibra curta de eucalipto com utilização de discos de refino com maior comprimento de corte. *O Papel, 66* (6): 80–87 (2006).
866. El-Sharkawy, K.; Haavisto, S.; Koskenhely, K. and Paulapuro, H. – Effect of fiber flocculation and filling design on refiner loadability and refining characteristics. *BioResources,* 3 (2):403–424 (2008).
867. Batchelor, W., Lundin, T. and Fardim, P. – A method to estimate fiber trapping in low-consistency refining. *Tappi Journal, 5* (8): 31–36 (2006).
868. Demuner, B. J.; Viana Doria, E. L.; Claudio-da-Silva Jr., E. and Manfredi, V. – Influência das características dos flocos sobre o refino de polpas químicas. *O Papel, 54* (2): 29–39 (1993).
869. Gess, J. M. – The refinability index. *In:* TAPPI Papermakers Conference. TAPPI, Chicago, 1985 pg: 279.
870. Lumiainen, J. – Behavior of fibres in mixed low consistency refining. *Paper Technology, 38* (6): 26–33 (1997).
871. Lumiainen, J. – Refining performances with separated and mixed pulps for fine-paper production. *Revue ATIP,* 51 (3): 99–107 (1997).
872. Joris, G. – Optimization of industrial refining unit through Fibrologic 4.0. *In:* IV International Refining Conference. PIRA, Fiuggi, 1997 Paper 17 pgs. 267–303.
873. Beghello, L. and Eklund, D. – Some mechanisms that govern fibre flocculation. *Nordic Pulp and Paper Research Journal, 12* (2): 119–123 (1997).
874. Beghello, L. and Eklund, D. – The influence of chemical environment on fibre flocculation. *Journal of Pulp and Paper Science, 25* (7): 246–250 (1999).
875. Lumiainen, J. – Refining of fine paper fibers – separate or mixed refining?. *In:* Papermakers Conference. TAPPI, Philadelphia, 1996 pgs. 175–186.
876. Li, L.; Collins, A. and Pelton, R. – A new analysis of filler effects on paper strength. *Journal of Pulp and Paper Science, 28* (8):267–273 (2002).
877. Stevens, W. V. – Principles of stock preparation and refining – refiners and refiner systems. *In:* Pulping Conference. TAPPI, Orlando, 1999, pgs. 97–102.
878. Kerekes, R. – Perspectives on fibre flocculation in papermaking. *In:* International Paper Physics Conference. TAPPI, Niagara-on-the-Lake, 1995. pgs. 23–31.
879. Lundin, T.; Lönnberg, B.; Soini, P. and Harju, K – Laboratory LC-refining of SBK pulps: effects of pulp consistency and dispersion. *In:* Scientific and Technical Advances in Refining & Mechanical Pulping – Course Notes. PIRA, Stockholm, 2003 Paper 5.
880. Bajpai, P. – Technology developments in refining. Pira International Ltd. Surrey, UK. 140 pgs (2005).
881. Björkman, U. – Break-up of suspended fibre networks. *Nordic Pulp and Paper Research Journal, 18* (1): 32–37 (2003).
882. Mason, S. G. – The motion of fibres in flowing liquids. Pulp *and Paper Magazine Canada, x* (51): 93–100 (1950) [879]
883. Albert, K. – Improved refining performance and lower operating cost through new technology. *In:* Scientific and Technical Advances in Refining & Mechanical Pulping – Course Notes. PIRA: Stockholm, 2003 Paper 4.
884. Strömberg, B. – Hardwood yield increases with Lo-solids® cooking and Lo-level® feed system at Papelera Guipuzcoana de Zicuñaga, Spain. *In:* Pulping Conference, TAPPI, *Quebec,* 1998 7 pgs https://www.eucalyptus.com.br/artigos/2002_LoSolids+Pulping.pdf
885. Kerekes, R. J. and Schell, C. J. – Characterization of fibre flocculation regimes by a crowding factor. *Journal of Pulp and Paper Science, 18* (1): J32–38 (1992)
886. Gabl, H. and Gorton-Heulgerth, A. – A new low consistency refiner yields improved fibre properties while reducing idle energy by up 40%. *In:* Scientific and Technical Advances in Refining & Mechanical Pulping – Course Notes. PIRA: Stockholm, 2003 Paper 8.
887. Palsanen, J. – Influence of properties of coating base paper on the sensitivity of LWC paper to breaking on the coater. Wochenblatt für Papierfabrikation, *107* (11/12): 432 (1969)

888. Sharkawya, K. E.; Haavistob, S.; Koskenhely, K. and Paulapuroa, H. – Effect of fiber flocculation and filling design on refiner loadability and refining characteristics. BioResources, 3 (2): 403-424 (2008).
889. Nordström, B. and Hermansson, L. – Effect of fiber length on formation and strength efficiency in twin-wire roll forming. *Nordic Pulp & Paper Research Journal, 32* (1):119–125 (2017).
890. Alfano, J. C.; Carter, P. W. and Whitten, J. E. – Use of scanning laser microscopy to investigate microparticle flocculation. *Journal of Pulp and Paper Science, 25* (6): 189–195 (1999).
891. Antku, J. – Maintaining minimum flow requirements of the twin-disc refiner. *Tappi Journal, 67* (10):112 (1984).
892. Laivins, G. V. and Scallan, A. M. – The influence of drying and beating on the swelling of fines. *Journal of Pulp and Paper Science, 22* (5): 178–184.1996).
893. Retulainen, E.; Luukko, K.; Fagerholm. K.; Pere, J.; Laine, J. and Paulapuro, H. – Papermaking quality of fines from different pulps – the effect of shape, size and chemical composition. *Appita Journal, 55* (6): 457–467 (2002)
894. Liitiä, T.; Maunu, S. L. and Hortling, B. – Solid-state NMR studies of residual lignin and its association with carbohydrates. *Journal of Pulp and Paper Science, 26* (9): 323–330 (2000).
895. Dillen, S. – Heterogeneity – An Important Parameter in Low Consistency Refining. *In:* International Symposium of Fundamental Concepts of Refining. IPC, Appleton: 16-18/09/1980 pgs. 331.
896. Eastwood, F. G. and Clarke, B. – Handsheet and pilot machine recycling degradation mechanism. *In:* Fibre Water Interactions in Papermaking Symposium. BPBIF, Oxford, 1978 Volume II: 835–848.
897. Seth, R. S. and Page, D. H. – The problem of using Page's equation to determine loss in shear strength of fiber-fiber bonds upon pulp drying. *Tappi Journal*, 79 (9): 206–210 (1996).
898. Joutsimo, O.; Wathén, R. and Robertsén, L. – Role of fiber deformations and damage from fiber strength to end user. *In:* 13th Fundamental Research Symposium- Advances in Paper Science and Technology. Cambridge. 2005 pgs. 591–611.
899. Arjas, A. – Constant versus variable rotational speed during the defining process and adapt-set system. *In:* International Fundamental Concepts of Refining. IPC, Appleton, 1980 pg. 263.
900. Ellis, M.; Allison, R.W. and Kibblewhite, R. P. – Effects of medium consistency mixing on kraft pulp fibres. *Appita Journal, 48* (5): 358 (1995).
901. Page, D. H. and Seth, R. S. – A note on the effect of fibre strength on the tensile strength of paper. *Tappi Journal, 71* (10): 182 (1988).
902. Casey, J. P. – Pulp and paper chemistry and chemical technology– Volume 2. John Willey and Sons, 1980.
903. MacLeod, J. M.; McPhee, F. J.; Kingsland, K. A.; Tristram, R. W.; O'Hagan, T. J.; Kowaiska, R. and e Thomas, B. C. – Pulp strength delivery along complete kraft mill fiber lines. *Tappi Journal, 78* (8): 153–160 (1995).
904. Wathen, R, – Studies on fiber strength and its effect on paper properties. Doctoral Thesis. Helsinki University of Technology, Helsinki, 2006 98 pgs https://citeseerx.ist.psu.edu/viewdoc/download?doi=10.1.1.554.2638&rep=rep1&type=pdf
905. Page, D. H.; Seth, R. S. and El-Hosseiny, F. – Strength and chemical composition of wood pulp fibres. *In:* Papermaking raw materials, 8th Fundamental Research Symposium, London (1985).
906. Page, D. H.; Seth, R. S.; Jordan, B. D. and Barbe, M. C. – Papermaking raw materials. *In:* 8th Fundamental Research Symposium. Mech. Eng. Publ. Ltd. London, p.183 (1985).
907. Thomson, S. L. and Gustafson, R. R. – Effects of kappa number variability on pulp properties. *Tappi Journal, 83* (1): 157–163 (2000).
908. Olson, J. A.; Robertson, A. G.; Finnigan, T. D. and Turner, R. R. H. – An analyzer for fibre shape and length. *Journal of Pulp and Paper Science, 21* (11): 367–373 (1995).
909. Eriksen, Ø.; Holmqvist, C. and Mohlin, U-B. – Theoretical outline of the cause for observed cavitation in a low-consistency refiner. *Nordic Pulp and Paper Research Journal, 23* (3): 315–320 (2008).

910. Mohlin, U.-B. – Low consistency beating – laboratory evaluation contra industrial experience. *In:* Current and Future Technologies in Refining. PIRA, Leatherhead, 1991 pgs 1–14.
911. Foelkel, C. – O processo de refinação da celulose. *Eucalyptus Newsletter n° 89 (Jul)* 2021 *206* pgs https://www.eucalyptus.com.br/news/pt_jul2021.pdf
912. Buchert, J.; Tenkanen, M. and Tamminen, T. – Characterization of carboxylic acids during kraft and Superbatch pulping. *Tappi Journal, 84* (4): 70 (2001).
913. Mohlin, U.-B e Miller, J. – Influence of industrial beating on fiber swelling and fiber shape. *In:* 4th International Conference on New Available Technologies – Current Tendencies. EUCEPA, Bologna, 1992 Volume 2 pgs. 274–283.
914. Jordan, B. D. and Nguyen, G. N. – Curvature, kink and curl. *Paperi ja Puu, 68* (4): 313 (1986).
915. Kibblewhite, R. P. and Brookes, D. – Factors which influence the wet web strength of commercial pulps. *Appita Jornal, 28* (4): 227–231 (1975).
916. Joutsimo, O.; Wathen, R. and Tamminen, T. – Effects of fiber deformations on pulp sheet properties and fiber strength. *Paperi jaa Puu, 87* (6):392 (2005).
917. Mohlin, U.-B. and Alfredsonn, C. – Fibre deformation and its implications in pulp characterization. *Nordic Pulp and Paper Research Journal, 4* (5): 172–179 (1990)
918. Nikolaeva, M. – Measurement and improvement of wet paper web strength. Master's Thesis, Lappeenranta University of Technology, Lappeenranta, 2010 78 pgs. https://citeseerx.ist.psu.edu/viewdoc/download?doi=10.1.1.956.5834&rep=rep1&type=pdf
919. Bennington, C. P. J. – Simultaneous chemical and mechanical treatment of pulp fibre and the resulting changes in pulp properties. *Nordic Pulp and Paper Research Journal, 21* (1): 44–48 (2006).
920. Allison, R. W. and Wrathall, S. H. – Effect of pulping and bleaching conditions on the strength of kraft pulp. Part 2: comparison of three commercial softwoods. *Appita, 47* (5): 369–374 (1994).
921. Marcoccia, B.; Sweeney, B. and och Lowe, R. – On the use of laboratory references for predicting and measuring the performance of bleach plant operations. Part 1: The effects of bleaching on pulp strength. *In:* International Pulp Bleaching Conference. TAPPI, Atlanta, 1996.
922. MacLeod, M. – The strength potential of unbleached Kraft pulps. *Pulp and Paper Canada, 88* (9): 103 (1987).
923. MacLeod, M. – New improved kraft pulp quality. *Paperi ja Puu, 72* (8): 780 (1990).
924. Lindau, J.; Theliander, H. and Sjöström, K. – The compression of pulp before and during brown stock washing – Its influence on fibre properties. *Nordic Pulp and Paper Research Journal, 23* (2): 195–201 (2008).
925. Allison, R. W.; Ellis, M. J., Kibblewhite, R. P. and Duffy, G. G. – Effect of mechanical processes on the strength of oxygen delignified kraft pulp. *In:* International Pulp Bleaching Conference. Helsinki, 1998 pgs. 159–166.
926. Gärd, J. – The influence of fibre curl on the shrinkage and strength properties of paper. Master's Thesis. Luleå University of Technology. Luleå, 2002. 42 pgs. http://www.diva-portal.org/smash/get/diva2:1015864/FULLTEXT01.pdf
927. Mohlin, U-B. and Salmén, L. – Which are the important fiber properties? *In:* International Conference. SPCI; Stockholm, 2005 pg 105
928. Mohlin, U. B.; Dahlbom, J. and Hornatowska, J. – Fiber deformation and sheet strength. *Tappi Journal, 79* (6): 105–111 (1996).
929. Gominho, J. and Pereira, H. – Effect of refining in the fibre structure and properties in unbleached eucalypt pulps. *In:* International Conference on Cellulose and Cellulose Derivatives: Physico-Chemical Aspects and Industrial Applications, 1998 pgs. 529–534.
930. Kimura, M.; Kimura, S.; Qi, Z. D.; Kuga, S. and Isogai. A. – Porous structure of never dried pulp fi bers analyzed by nitrogen adsorption method. *In:* 15th Fundamental Research Symposium, Cambridge, 2013 pgs 821–836
931. DeRuvo, A. and Htun, M. – Fundamental and practical aspects of paper making with recycled fibers. *In:* The Role of Fundamental Research in Papermaking. BP&BMA, London, 1983 Volume 1 pgs. 195–225.

932. Szwarcsztajn, E. and Przybysz, K. – Investigation on changes in the properties of recycled pulps fractions. *Cellulose Chemistry and Technology,* (10): 737–749 (1976).
933. Bovin, A.; Hartler, N. and Teder, A. – Changes in pulp quality due to repeated papermaking. *Paper Technology* (5): 169–172 (1973).
934. Attwood, D. – The solid waste problem – Pira Proposes a Recycling Centre. *Paper Technology and Industry, 319* (11): 2–4 (1990).
935. Escalar, X.; Egusquiza, E.; Farhat, M.; Avellan, F. and Coussirat, M. – Detection of cavitation in hydraulic turbines. *Mechanical Systems and Signal Processing,* 20: 983–1007 (2006).
936. Page, D. H. and Tydeman, P. A. – Transverse swelling and shrinkage of softwood tracheids. *Nature, 199* (4892): 471–472 (1963).
937. Kibblewhite, R. P. and Bailey, D. – Measurement of fibre cross-section dimensions using image processing. *Appita Journal, 41* (4): 297–303 (1988).
938. Seth, R. S. – The difference between never dried and dried chemical pulps. *Solutions, 1* (1): 95 (2001).
939. Howard, R. C. – The effects of recycling on paper quality. *Journal of Pulp and Paper Science, 16* (5): 143 (1990).
940. Kitayama, T.; Okayama, T. and Oye, R. – Changes of chemical pulp fibers during recycling. *Sen'i Gakkaishi* 43:486–494 (1987) https://www.jstage.jst.go.jp/article/fiber1944/43/9/43_9_486/_article/-char/en#citedby-w
941. Weise, U. – Hornification – mechanisms and terminology. *Paperi Puu, 80* (2):110–115 (1998)
942. Carlsson, G. and Lindström, T. – Hornification of cellulosic fibers during wet pressing. *Svensk Papperstiding, 87* (15): R119 (1984).
943. Stone, J. E. and Scallan, A. M. – Influence of drying on the pore structures of the cell wall. *In:* Consolidation of the Paper Web, Transactions of the 3rd Fundamental Research Symposium, BRBMA, Cambridge, 1965 pgs. 145–166.
944. Robertson, A. A. – Some observations on the effects of drying papermaking fibres. *Pulp and Paper Magazine Canada, 65* (3): T161–168 (1964).
945. Howard, R. C. and Bichard, W. – The basic effects of recycling on pulp properties. *Journal of Pulp and Paper Science, 18* (4): 151–159 (1992).
946. Arroyo, L. M. – Optimización energética de una fábrica de papel de nueva planta. *El Papel* (46): 35–39 (1995).
947. McKee, R. C. – Effect on repulping on sheet properties and fiber characteristics. *Paper Trade Journal, 155* (2): 34 (1971).
948. Smook, G. A. – Hornification (keywod). *In:* Handbook of Pulp & Paper Terminology. Angus Wilde Publications, Vancouver, 1990 Chapter 11 pg 135.
949. Lumiainen, J. – Refining secondary fibres. *In:* III International Refining Conference and Exhibition. PIRA/IPST, Atlanta, 1995 Paper 16 11 pgs.
950. Kiyohara, R. S. and Patiño, O. A. – Estudo laboratorial das características fibrosas em função dos números de reciclagens. *O Papel, 61* (8): 39–42 (2000).
951. Ölander, K.; Htun, M. and Grén, U. – Specific surface area – an important property of mechanical pulps. *Journal of Pulp and Paper Science, 20* (11): 338–342 (1994).
952. Suurnäkki, A.; Heijnesson, A.; Buchert, J.; Viikari, L. and Westermark, U. – Chemical characterization of the surface layers of unbleached pine and birch kraft pulp fibres. *Journal of Pulp and Paper Science, 22* (2): J43–47 (1996).
953. Sjöholm, E., Gustafsson, K., Norman, E., Reitberger, T. and Colmsjö, A. – Fibre strength in relation to molecular weight distribution of hardwood kraft pulp: degradation by gamma irradiation, oxygen/alkali or alkali. *Nordic Pulp and Paper Research Journal, 5* (4): 326–332 (2000).
954. Hjort, A. – Effects on pulp and paper properties from iso thermal cooking (ITC) and black liquor impregnation in a continuous digester. *In:* International Non-Chlorine Bleaching Conference. Pulp & Paper/Emerging Technology Transfer, Amelia Island, 1994 Paper 3-1 12 pgs.
955. Rämark, H. and Kettunen, A. – Cocción de pastas softwood y hardwood – recientes experiencias. *El Papel* (71): 47–51 (1998).

956. Cardoso, G. da S.; Lopes, J. L.; dos Santos, M. R. and Lopes, R. J. F. – Tratamento enzimático sobre as fibras recicladas de papelão ondulado. *O Papel, 79* (7): 80–86 (2018).
957. Leopold, B.; Thorpe, J. L. and Marabito, J. – Effect of pulping on strength properties of dry and wet pulp fibers. *Espra Research Reports, 45* (IV): 20 (1967).
958. Tam Doo, P. A. and Kereke, R. – Flexibility of wet pulp fibred. *Pulp and Paper Canada, 83* (2): 46–50 (1982).
959. Irvine, G. M.; Wallis, A. F. A. and Wearne, R. H. – Application of a Monte Carlo procedure to the kraft pulping of eucalyptus from Eden forest resource. *Appita, 44* (6): 394–398, 409 (1991).
960. Ryynänen, H.; Nelson, P. J. and Chin, C. W. J. – Ozone bleaching of eucalypt kraft pulps. *Appita, 48* (6): 440–444 (1995).
961. Suckling, I. D., Allison, R. W., Campion, S. H., McGrouther, K. G., McDonald, A. G. – Cellulose degradation during conventional and modified kraft pulping. *Journal of Pulp and Paper Science, 27* (11):336–341 (2001)
962. Carvalho, M. G.; Martins, A. A. and Figueiredo, M. M. – Efeito de algumas variáveis de cozimento nas características de pastas kraft de *E.globulus*. *In:* XVI Encontro Nacional. Tecnicelpa, Covilhã, 1998 pgs. 113–119.
963. Shin, N. H. and Stromberg, B. – Impact of cooking conditions on physical strength of eucalyptus pulp. *In:* 2° Colóquio Internsvionsl Sobre Celulose de Eucalipto. ATCP. Concepción, 2005 9 pgs.
964. Rosli, W. U.; Mazlan, I. and Law, K. N. – Effects of kraft pulping variables on pulp and paper properties of acacia mangiumkraft pulp. *Cellulose Chenistry and Technology, 43* (1): 9–15 (2009).
965. Colodette, J. L.; Gomide, J. L.; Robles, Y. A. M.; Almeida, J. M.; Brito, A. C. H. de; Mehlman, S. K. and Argyropolus, D. S. – Fatores que afetam a branqueabilidade de polpas kraft de Eucalyptus por seqüências ECF, Z-ECF e TCF. 1: Influência do processo de polpação. *O Papel, 60* (05): 46–55 (1999).
966. Malinen, R.; Rantanen, T.; Rautonen, R. and Toikkanen, L. – TCF bleaching to high brightness – bleaching sequences and pulp properties. *In:* International Pulp Bleaching Conference. TAPPI/CPPA/SPCI/EUCEPA, Vancouver, 1994 pgs. 187–194.
967. Bengtsson, G.; Brunsvik, J.-J. and Mansson, E. – Herstellung von ECF und TCF sulfatzellstoff, qualiätas und umweltkonsequenzen. *Das Papier,* (10AA): 11–19 (1994).
968. Blechschmidt, J; Strunz, A.-M. and Baumgarten, H.-L. – Refining behavior of ECF and TCF pulps. *In:* First EcoPaperTech. FPPRI, Helsinki, 1995 pgs. 245–257.
969. Zhu, J. Y.; Chai, X. S. and Dhasmana, B. – Formation of volatile organic compounds (VOCs) during pulping. *Journal of Pulp and Paper Science, 25* (7): 256–263 (1999).
970. Gomide, J. L. and Fantuzzi Neto, H. – Aspectos fundamentais da polpação kraft de madeira de Eucalyptus. *O Papel, 61* (3): 62–68 (2000).
971. Wallis, A. F. A.; Wearne, R. H. and Wright, P. J. – Analytical characteristics of plantation eucalyptus woods relating to kraft pulp yields. *Appita Journal, 49* (6): 427–432 (1996).
972. Kube, P. D. and Raymond, C. A. – Prediction of whole-tree basic density and pulp yield using wood core samples in *Eucalyptus nitens*. *Appita Journal, 55* (1): 43 (2002).
973. Michell, A. J. – Pulpwood quality estimation by near-infrared spectroscopic measurements on eucalypt woods. *Appita Journal, 48* (6): 425 (1995).
974. Schimleck, L. R. and French, J. – Application of NIR spectroscopy to clonal *Eucalyptus globulus* samples covering a narrow range of pulp yield. *Appita Journal, 55* (2): 149 (2002).
975. Crawford, I. A., Prentice, F. J. and Turner, C. H. – Variation in pulping quality within two trees of *E. delegatensis*. *Appita Journal, 25* (5):353 (1972).
976. Pereira, H. – Variability in the chemical composition of plantation eucalypts (*Eucalyptus globulus labill*). *Wood Fiber Science, 20* (1): 82 (1988).
977. Danby, R. – Pulp to paper. *In:* TECH95 Theory and Practice of Papermaking Course. CPPA, Ottawa, 1995 Section D2 5 pgs.
978. Hubbe, M. A. – Bonding between cellulosic fibers in the absence and presence of dry-strength agents – a review. *BioResources, 1* (2): 281–318 (2016).

979. Yu, L., Zhan, H., Qian, B., Yue, B., He, W. – Study on modified kraft cooking of *Pinus taeda* and low pollution bleaching of the pulp. *Chung-kuo Tsao Chih, 18* (2):8–12 (1999)
980. Manfredi, V.; Gomide, J. L.; Carneiro, C. J. G.; Salvador, E. and Fantuzzi Neto, H. – Effects of delignification strategies on production and quality of eucalypt kraft pulp. *In:* 7th Brazilian Symposium on the chemistry of Lignins and Other Wood Components. UFV/SIF, Belo Horizonte, 2001 pgs. 119–127.
981. Redko, B. V. P. – Vasos de folhosas: *Eucalyptus deglupta, Eucalyptus urophylla, Gmelina arborea. In:* XVI Congresso Anual de Celulose e Papel. ABTCP, São Paulo, 1983 pgs 1169–1194.
982. Cohen, W. E. – Pulp and paper from Australian eucalypts. *Unasylva, 2* (6) – 8 pgs www.fao.org3x5347ex5347e05.htm
983. Glowacki, J. J. – Refiners: new designs achieved higher efficiency/capacity. *Pulp and Paper 70,* (1): 113–114 (1996).
984. Burke, T. M.; Skålén, B. and Talley, M. – Mill experience in refiner control through on-line measurement of stock drainage. *In:* III International Refining Conference and Exhibition. PIRA/IPST, Atlanta, 1995 Paper 20 10 pgs.
985. Tozzi, E. J.; Lavenson, D. M.; McCarthy, M. J. and Powell, R. L. – Effect of fiber length, flow rate, and concentration on velocity profiles of cellulosic fiber suspensions. *Acta Mechanica, 224*: 2301–2310 (2013) https://link.springer.com/article/10.1007/s00707-013-0922-2
986. Derakhshandeh, B.; Hatzikiakos, S. G. and Bennington, C. P. J. – The apparent yield stress of pulp fibre suspensions. Journal of Rheology, 54 (5): 1137–1154 (2001).
987. Martinez, M.; Buckley, K.; Jivan, S.; Lindström, A.; Thiruvengadaswamy, R.; Olson, J. A.; Ruth, T. J. and Kerekes, R. J. – Characterizing the mobility of papermaking fibres during sedimentation. *In:* 12th Fundamental Research Symposium. PIRA, Oxford, 2001 pgs. 225–254.
988. Arjas, A.; Aario, M. and Rythi, N. – Influence of the residence time distribution on the beating result of a Mill-size conical refiner. *Paperi ja Puu, 52* (10): 639–649 (1970)
989. Soszynski, R.H. and Kerekes, R. – Elastic interlocking of nylon fibers suspended in liquid. Part 1: nature of cohesion among fibers. *Nordic Pulp and Paper Research Journal, 3* (4): 172–179 (1988).
990. Meyer, R. and Wahren, R. – On the elastic properties of three-dimensional fibre networks. *Svensk Papperstidning, 57* (10): 432–436 (1964).
991. Zhao, R. N. and Kerekes, R. J. – The effect of suspending liquid viscosity on fiber flocculation. *Tappi Journal, 76* (2): 183–188 (1993).
992. Dodson, C. T. J. – Fiber crowding, fiber contacts, and fiber flocculation. *Tappi Journal, 79* (9): 211–216 (1996)
993. Huber, P.; Roux, J. C.; Mauret, E.; Belgacem, N. and Pierre, C – Suspension crowding for a general fibre-length distribution: application to flocculation of mixtures of short and long papermaking fibres. *Journal of Pulp and Paper Science, 29* (3): 77–85 (2003).
994. Andersson, S. R.; Ringnér, J. and Rasmuson, A. – The network strength of non-flocculated fibre suspensions. *Nordic Pulp and Paper Research Journal, 14* (1): 61–70 (1999).
995. Parker, J. D. – The sheet forming process. *In:* STAP, 9. TAPPI, Atlanta, 1972.
996. El-Sharkawy, K.; Koskenhely, K. and Paulapuro, H. – Tailoring softwood kraft pulp properties by fractionation and refining. *Tappi Journal, 7* (11): 15–22 (2008).
997. Alexandre, G.; Mello, B. H. de; Oliveira, V. C. and Alexandre, L. A. C. – Opwg. *In:* 48th ABTCP International Pulp and Paper Congress & 1st Ibero-American Conference on the Bio-Economy – BEM, São Paulo. 2015 8 pgs
998. Byrd, M.; Pawlak, H.; Jameel, H. and Venditti, R. – The effect of operating variables on the fractionating characteristics of a hydrocyclone. *In:* Fall Conference and Trade Fair. TAPPI, San Diego, 2002.
999. Negro, C.; Fuente, E.; Blanco, A. and Tijero, J. – Effect of chemical flocculation mechanisms on rheology of fibre pulp suspensions. *Nordic Pulp and Paper Research Journal, 21* (3): 336–341 (2006).

1000. Gurnagul, N. and Page, D. H. – The difference between dry and rewetted zero-span tensile strength. *Tappi Journal, 72* (12): 164–167 (1989).
1001. Retulainen, E.; Moss, P. and Nieminen, K. – Effect of calendering and wetting on paper properties. *Journal of Pulp and Paper Science, 23* (1): 34–39 (1997).
1002. Mohlin, U.-B. – Properties of TMP fractions and their importance for the quality of printing papers. Part 2: the influence of particles properties and particle size distribution on pulp properties. *Svensk Papperstiding, 83* (18): 513–519 (1980).
1003. Page, D. H. and MacLeod, M. – Fibre strength and its impact on tear strength, *Tappi Journal, 71* (1): 172 (1988).
1004. Ushakov, A. V.; Alashkevich, Y. D.; Kozhukhov, V. A. and Marchenko, R. A. – Effect of pulp consistency during refining on physical and mechanical characteristics of handsheets. *Journal of Physics: Conference Series, 2094*, (5): 052035 (2021) https://www.researchgate.net/publication/356625009_Effect_of_pulp_consistency_during_refining_on_physical_and_mechanical_characteristics_of_handsheets
1005. Vikharev, S. – Research of a fibrous layer at refining in the refiners. *In:* IOP Conferences Series: Earth and Environmental Science, 316 (2019) 7 pgs https://www.researchgate.net/publication/336007321_Research_of_a_fibrous_layer_at_refining_in_the_refiners
1006. Przybysz, P.; Dubowik, M.; Kucner, M. A.; Przybysz, K. and Buzała, K. P. – Contribution of hydrogen bonds to paper strength properties, *PLOS One, 11* (5):1–10 (2016) https://www.researchgate.net/publication/303552314_Contribution_of_Hydrogen_Bonds_to_Paper_Strength_Properties
1007. Soszynski, R. M. – Relative bonded area – a different approach. Nordic *Pulp and Paper Research Journal, 10* (2): 150 (1995).
1008. Torgnysdotter, A.; Kulachenko, A.; Gradin, P. and Wågberg, L. – Fibre/fibre crosses: finite element modelling and comparison with experiment. *Journal of Composite Materials, 41* (13): 1603–1618 (2007)
1009. Batchelor, W.; Kibblewhite, R. P. and He, J. – A new method for measuring RBA applied to the Page's Equation for the tensile strength of paper. *Appita Journal, 61* (4): 302–306 (2008).
1010. Batchelor, W. J. and He, J. – A new method for determining the relative bonded area. *Tappi Journal, 4* (6): 23 (2005).
1011. Batchelor, W. J. and Kibblewhite, R. P. – Calculation of relative bonded area and scattering coefficient from sheet density and fibre shape. Holzforschung, 60 (3): 253 (2006).
1012. Wågberg, L. and Annergren, G. – Physicochemical characterization of papermaking fibres. *In:* 16th Fundamental Research Symposium. Cambridge, 1997 pgs. 1–82 https://bioresources.cnr.ncsu.edu/wp-content/uploads/2020/03/1997.1.1.pdf
1013. Hirn, U.; Schennach, R.; Ganser, C.; Magnusson, M.; Teichert, C. and Östlund, S. – The area in molecular contact in fiber-fiber bonds.*In:* 15th Fundamental Research Symposium: Advances in Paper Research, Cambridge. 2013 pgs 201–226.
1014. Hirn, U. and Schennach, R. – Comprehensive analysis of individual pulp fiber bonds quantifies the mechanisms of fiber bonding in paper. *Scientific Reports, 5* (10503):1–9 (2015).
1015. Martinez, P.C. and Park, S. W. – Ações das forças em processo de refino em baixa consistência. *In:* 40th Congresso Anual. ABTCP: São Paulo, 2007 12 pgs.
1016. Anand, S. H.; Manigandan, P.; Kumar, S. and Jeffrey, J. E. C. – Design and manufacturing of disc refiner. *International Journal of Emerging Technology in Computer Science and Electronics, 21* (3): 79–82 (2016)
1017. Banks, W. A. – Design considerations and engineering characteristics of disc refiners. *Paper Technology,* 8 (4): 363–369 (1967).
1018. Stationwala, M. I.; Atack, D. and Karnis, A. – Distribution and motion of pulp fibres on refiner bar surface. *Journal of Pulp and Paper Science, 18* (4): 131–137 (1992).
1019. Atalla, R. H. and Wahren, D. On the energy requirement in refining. *Tappi, 63* (6): 121 (1980).
1020. Martinez, M. – The energy expended on pulp fibres during low consistency refiners. Doctoral Thesis, The University of British Columbia, Vancouver, 1995 106 pgs.

1021. Batchelor, W. J. and Ouellet, D. – Estimating forces on fibers in refining. *In:* IV International Refining Conference. PIRA, Fiuggi, 1997. Paper 2 pgs. 47–60.
1022. Demler, C. L. – Fundamentals of papermill refining for chemical, mechanical and secondary fibers. *In:* TECH95 Theory & Practice of Papermaking Course. CPPA;. Ottawa, 1995 Section A2 23 pgs.
1023. Guest, D. and Weston, J. – Fiber-water interactions: modified by recycling. *In:* Recycling Paper From Fiber to Finished Product. 2nd Edition, Ed. Coleman, M. J. Tappi Press, Atlanta, 1990 Volume 2: 169–175.
1024. Hartman, R. R. – Mechanical treatment of pulp fibers for property development. Doctoral Thesis, Institute of Paper Chemistry, Atlanta, 1984 131 pgs https://smartech.gatech.edu/handle/1853/5510
1025. Shallhorn, P. and Karnis, A. – The tear and tensile strength of mechanical pulps. *In:* CPPA. Proceedings, International Mechanical Pulping Conference, 1979 pgs. 25–36
1026. Kerekes, R. J. – Characterizing the refining action – linking the process to the refining result. *In:* Refining and Mechanical Pulping Conference. PIRA, Barcelona, 2005 Paper 1.
1027. Martinez, M. and Kerekes, R. J. – Forces on fibres in low-consistency refining. *Tappi Journal, 77* (12): 119–123 (1994).
1028. Nissan, A. H. – Lectures on fiber science in paper. Edited by W.C. Walker, Pulp and Paper Technology Series. Joint Textbook Committee of the Paper Industry. N°4. p27
1029. Senger, J.; Siadat, A. and Ouellet, D. – Measurement of normal and shear forces during refining using a piezoelectric force sensor. *Journal of Pulp and Paper Schience, 30* (9): 247–251 (2004).
1030. Senger, J. J. – The forces on pulp fibres during refining. Master Theais. Department of Mechanical Engineering. University of British Columbia, Vancouver, 1998.
1031. Roux, J-C.; Bloch, J-F. and Nortier, P. – A kinetic model for pulp refining, including the angular parameters of the equipment. *Appita Journal, 60* (1): 29–34 (2007).
1032. Kerekes, R. J. – Energy and forces in refining. *Journal of Pulp and Paper Science, 30* (1): 10–15 (2010).
1033. Kerekes, R. J. – Force-based characterization of refining intensity. *Nordic Pulp and Paper Research Journal, 26* (1): 14–20 (2011).
1034. Nugroho, D. D. P. – Low consistency refining of mixtures of softwood and hardwood bleached kraft pulp: effects of refining power. Master's Thesis. Asian Institute of Technology. Pathum Thani, 2012 66 pgs www.turbulence-initiated.sites.olt.ubc.ca/files/2013/02/LCRefining_Mixed-Softwood-Hardwood_Dimas-Thesis_.pdf
1035. Martinez, M.; Batchelor, W. J.; Kerekes, R. J. e Oucllet, D. – Forces on fibres in low-consistency refining: normal force. *Journal of Pulp and Paper Science, 23* (1): J11–J18 (1997).
1036. Batchelor, W. J.; Martinez, M.; Kerekes, R. J. and Ouellet, D. – Forces on fibres in low-consistency refining: shear force. *Journal of Pulp and Paper Science, 23* (1): J40–J45 (1997).
1037. Kerekes, R.J. and Olson, J.A. – Perspectives on fibre length reduction in refining. *In:* International Conference Scientific and Technical Advances in Refining & Mechanical Pulping. PIRA, Stockholm, 2003 Paper 2.
1038. Ouellet, D.; Bennington, C. P. J. and Potkins, D. – Wood comminution and material flow in a laboratory chip refiner. *Journal of Pulp and Paper Science, 21* (12): 415–42 (1995).
1039. Joris, G. – Gestion optimale d'une unite industrielle de raffinage. *Revue ATIP, 61* (2): 6–25 (2007).
1040. Kerekes, R. J. – Characterizing refining action in PFI mills, *Tappi Journal, 4* (3): 9–13 (2005).
1041. Kang, T, – Role of external fibrillation in pulp and paper properties. Doctoral Thesis, Helsinki University of Technology, Espoo, 2007 43 pgs www.citeseerx.ist.psu.edu/viewdoc/download?doi=10.1.1.575.6375&rep=rep1&type=pdf
1042. Kang, T. and Paulapuro, H. – Effect of external fibrillation on paper strengt. *Pulp and Paper Canada, 107* (7/8): 51–54 (2006).
1043. Kibblewhite, R. P. – Effect of beating on fiber morphology and fiber surface structure. *Appita, 26* (3): 196–202 (1972)

1044. Luukko, K.; Kemppainen-Kajola, P. and Paulapuro, H. – Characterization of mechanical pulp fines by image analysis. *Appita Journal,* 50 (5): 387–392 (1997).
1045. Page, D. H. and De Grace, J. H. – Delamination of fiber walls by beating and refining. *Tappi Journal, 50* (10): 489–493 (1967).
1046. Vishtal, A. and Retulainen, E. – Boosting the extensibility potential of fibre networks: A review. *BioResources, 9* (4): 7951–8001 (2014.
1047. Mason, S. G. – Some factors involved in the flocculation of pulp suspension and the formation of paper. *Pulp and Paper Magazine Canada, 52* (51): 94–98 (1950).
1048. Hubbe, M. A. and Heitmann, J. A. – Review of factors affecting the rekease if water from cellulosic fibers during paper manufacture. *BioResources, 2* (3): 500–535 (2007)
1049. Molin, U. and Daniel, G. – Effects of refining on the fibre structure of kraft pulps as revealed by FE-SEM and TEM: influence of alkaline degradation. *Holzforschung, 58* (3): 226–232 (2004).
1050. Batchelor, W.; Elahimehr, A.; Martinez, M. and Olson, J. – A new representation for low consistency refining data. *In:* 16th Fundamental Research Symposium. Advances in Pulp and Paper Research. Oxford. 2017 pgs. 195–207.
1051. Ebeling, K. – A critical review of current theories for the refining of chemical pulps. *In:* International Symposium of Fundamental Concepts of Refining. IPC, Appleton, 1980 pgs. 1–33.
1052. Gassman, H. – Influences on the screening and fractionating processes. *World Pulp and Paper Technology,* 1998/9: 40–41.
1053. Law, K. N.; Valade, J. L. and Yang, K. C. – Fibre development in thermomechanical pulping: comparison between black spruce and jack pine. *Journal of Pulp and Paper Science, 24* (2): 73–76 (1998).
1054. Robertson, A. A. and Mason, S. G. – Specific surface of cellulose by the liquid permeability method. *Pulp and Paper Canada, 50* (12): 103–110 (1949).
1055. Senger, J. and Ouellet, D. – Factors affecting the shear forces in high-consistency refining. *Journal of Pulp and Paper Science, 28* (11): 364–368 (2002).
1056. Kappel, L.; Hirn, U.; Bauer, W. and Schennach, R. – A novel method for the determination of bonded area of individual fiber-fiber bonds. *Nordic Pulp and Paper Research Journal, 24* (2): 199–205 (2009).
1057. Hietanen, S. and Ebeling, K. – Fundamental aspects of the refining process. *Paperi ja Puu, 72* (2): 158–170 (1990)
1058. El-Hosseiny, R. and Yan, J. F. – Analysis of Canadian standard freeness. Part II: practical implications. *Pulp & Paper Canada, 81* (6): 67–70 (1980
1059. Kure, K. A.; Saborin, M. J.; Dahlqvist, G. and Helle, T. – Adjusting refining intensity by changing refiner plate design and rotational speed-effects on structural fibre properties. *Journal of Pulp and Paper Science, 26* (10): 346–352 (2000).
1060. Johansson, M. H. and Samuelson, O. – Alkaline destruction of birch xylan in the light of recent investigations of its structure. *Svensk Papperstiding, 80*(16), 519–524 (1977).
1061. Dekker, J. – New insights in beating leading to innovative beating techniques. *In:* Scientific and Technical Advances in Refining & Mechanical Pulping – Course Notes. PIRA: Stockholm, 2003 Paper 10.
1062. Brogdon, B. N.; Mancosky, D. G. and Lucia, L. A. – New insight into modification during chlorine dioxide bleaching sequences. Part III: modifications in (EO). *In:* Fall Conference. TAPPI, 2003.
1063. Nasab, N. R. – Understanding of no-load power in low consistency refining. Doctoral Thesis, The University of British Columbia, Vancouver, 2013. 106 pgs www.open.library.ubc.ca/media/download/pdf/24/1.0074101BARRA1
1064. Andersson, S.; Serimaa, R.; Paakkari, T.; Saranpää, P. and Pesonen, E. – Crystallinity of wood and the size of cellulose crystallites in Norway spruce (*Picea abies*). *Journal of Wood Science, 49* (6):531–537 (2003).
1065. Sepke, P.-W. and Meltzer, F. P. – Mahlung von faserstoffer für LWC – papiere. *Das Papier, 47* (3): 123–127 (1993).

1066. Meltzer, F. P. and Rauntnbach, R. – Neue möglichkeiten zur vorherbestimmung des technologischer mahlergebnisses. *Das Papier, 47* (3): 578–583 (1994).
1067. Sepke, P.-W.; Meltzer, F. P. and Musselmann, W. – Influencing specific energy in low consistency refining – a new approach. *In:* III International Refining Conference and Exhibition. PIRA/PST, Atlanta, 1995 Paper 21.dillen
1068. Crow, H. – Energy saving in the refining of shor fibre pulps. *In:* Papermakers Conference. TAPPI, Philadelphia, 1996 pg. 9.
1069. Joris, G. – Théorie mathématique du raffinage de la pâte à papier en basse concentration. *Revue ATIP, 40* (10): 507–526 (1986).
1070. Krotscheck, A, W. – Pulp screening, cleaning, and fractionation. *In:* Handbook of Pulp. Volume 1, Chapter 6, Ed,Sixta, H. WILEY-VCH Verlag GmbH &Co. KGaA, Weinheim, 2006 pgs. 561–608.
1071. Croney, C.; Ouellet, D.; Kerekes, R. J. – Characterizing refining intensity for tensile strength development. *In:* 5th International Conference on Scientific and Technical Advances in Refining. PIRA, Viena, 1999.
1072. Leider, P. J. and Nissan, A. H. – Understanding the disk refiner: the mechanical treatment of fibers. *Tappi, 60* (10): 85–90 (1977).
1073. Kerekes, R. – Characterization of pulp refiners by a C-factor. *Nordic Pulp and Paper Research Journal, 05* (1): 3–8 (1990).
1074. Lumiainen, J. – A new approach to the critical factors effecting refining intensity result in low-consistency refining. *In:* EUCEPA Meeting. EUCEPA, Stockholm, 1990 pg. 310.
1075. Lumiainen, J. – New theory can improve practice. *Pulp and Paper International, 32* (8): 46–47, 54 (1990).
1076. Kerekes, R. J.; Clara, M.; Dharni, S. and Martinez, M. – Application of the C-factor to characterize pulp refiners. *Journal of Pulp and Paper Science, 19* (3): J125–130 (1993).
1077. Kerekes, R. J.; Ouellet, D. and Martinez, M. – New perspectives on refining intensity. *In:* III International Refining Conference and Exhibition. PIRA/IPST, Atlanta, 1995 Paper 3 23 pgs.
1078. Musselman, R.; Letarte, D. and Simard, R. – Trial stage low consistency refining o TMP for energy savings and quality enhancement. *In:* IV International Refining Conference. PIRA, Fiuggi, 1997 Paper 9 pgs. 141–147.
1079. Charuel, R. C.; Roux, J. C.; Agostini, F. de e Roussele, M. – Influence de la géometrie des lames sur le raffinage des pátes chimiques. *Revue ATIP, 42* (4): 153–160 (1988).
1080. Lumiainen, J. – Refining intensity at low consistency – critical factors. *Paper Technology, 32* (11): 22–26 (1991).
1081. Roux, J-C.; Franc, N.; Duffy, G. G. and Fabry, B. – Shear factor: a new way to characterize fiber suspension shear. *Tappi Journal, 84* (8): 46–63 (2001).
1082. Kerekes, R. and Meltzer, F. – The influence of bar width on bar forces and fibre shortening in low consistency pulp refining. *Nordic Pulp & Paper Research Journal* 33(2): 220–225 (2018).
1083. Olender, D.; Wild, P.; Byrnes, P.; Ouellet, D. and Sabourin, M. – Forces on bars in high-consistency mill-scale refiners: Effect of consistency. *Nordic Pulp and Paper Research Journal, 23* (2): 218–223 (2008).
1084. Berna, J. E. R.; Martinez, M. and Olson, J. – Power–gap relationships in low consistency refining. *Nordic Pulp and Paper Research Journal, 34* (1): 36–45 (2019).
1085. Demuner, B. J. – Alternatives to improve eucalypt kraft pulp refining. https://www.eucalyptus.com.br/artigos/outros/2001_Refining_7thBSCL_final.pdf
1086. Barbosa, F. E. V. – Análise das variáveis operacionais de um processo de produção de papel visando maior eficiência da operação. Dissertação de Mestrado, Universidade Estadual de Campinas, Campinas. 2015 145 pgs. http://repositorio.unicamp.br/bitstream/REPOSIP/265981/1/Barbosa_FranciscoElpidioViana_M.pdf
1087. Kondora G. and Asendrych, D – Flow modelling in a disc refinier. *In:* 14th International Conference on Fluid Flow Technologies, Budapest, 2009 8 pgs https://www.researchgate.net/publication/265563544_Flow_Modelling_in_a_Disc_Refiner

1088. Giertz, H. W. – A new way to look at the beating process. *NorskeSkog 18* (7): 239–248 (1964).
1089. Lundin, T.; Wurlitzer, F.; Park, S. W. and Fardim, P. – Análise energética em refinação de baixa consistência de madeira de coníferas. *O Papel, 70* (10): 41–60 (2009).
1090. Kurdin, J. A. – High consistency refining of chemical (pulp) fibres. *In:* Refining Chemical Pulps Seminar. Doschi Association, Appleton, 1986 pg 15.
1091. Lumiainen, J. – Refining of chemical pulp. *In:* Papermaking Science and Technology. Volume 8 – Papermaking Part 1 – Stock preparation and wet end, Chapter 4. Ed. Gullichsen, J. and Paulapuro, H. Finnish Paper Engineers' Association/Paperi ja Puu Oy, Hesinki: 2000 pg. 86.
1092. Roux, J. C. and Bloch, J. F. – Lubrication theory explains the modification of fiber properties in the refining process. AIP Conference Proceedings 1558, 1095 (2013) https://aip.scitation.org/doi/10.1063/1.4825697
1093. Min, B. G.; Lee, J. Y.; Kim, C. H.; Park, S. H.; Lee, M. S.; Gu, H. G. and Lee, C. Y. – New technology for developing a lightweight refiner plate for hardwood kraft pulp fibers. *BioResources, 15* (4): 9128–9142 (2020).
1094. Christopher, L.; Demler, C. L.; Beder-Miller, S. and e Aldridge, R. – Refining case study: using improved refiner control to ncrease pulp quality and paper machine performance. *In:* Paper Conference, New Orleans, 2012 39 pgs
1095. Rajendra D., Deshpande, R. D. – Energy considerations in stock preparation refining modified by recycling. Master's Thesis. Western Michigan University, Kalamazoo, 1988 164 pgs
1096. Baker, C. F. – Optimization of paper mill refining systems. *In:* 3rd International Refining Conference. PIRA/IPST, Atlanta, 1995 Paper 13.
1097. Rihs, J. – Low consistency refining – theory vs practice. *In:* III International Refining Conference and Exhibition. PIRA/IPST, Atlanta, 1995 Paper 10 15 pgs.
1098. Leblanc, J. and Light, J. – The performance of low and medium intensity low consistency refiner plates. *In:* 79th Annual Meeting. CPPA, Montreal, 1993 pg. 359.
1099. Radoslavova, D.; Roux, J. C. and Silvy, J. – The beating of pulp considered as a hydrodynamic process. *In:* III International Refining Conference and Exhibition. PIRA/IPST, Atlanta, 1995 Paper 6 32 pgs.
1100. Crook, J. – The dream revisited – an update on Brazilian refining technology. *In:* III International Refining Conference and Exhibition. PIRA/IPST, Atlanta: 20-22/03/1995 Paper 14 12 pgs.
1101. Luengo, J.; Segura, C. and González, R. – Effect of refining intensity on properties of eucalyptus fibers – Santa Fe pulp – CMPC. *In:* 8th International Colloquium on Eucalyptus Pulp, Concepción, 2012 28 slides
1102. Lumiainen, J. – Is the lowest refining intensity the best in the consistency refining of hardwood pulps? *In:* Papermakers Conference, Atlanta, 1994. 12 pgs.
1103. Bennett, R. D.; Turt, V.; Genco, J. M.; Gansson, S. and Lonnberg, B. K. – Effect of specific edge loading on freeness drop coefficient. *In:* Further Advances in the Forest Products Industries symposium. AIChE, Symposium Series, 93 (315), 146 pgs. 1997.
1104. Baker, C. F. – Various options for the control of refining process. *In:* IV International Refining Conference. PIRA, Fiuggi: 18-20/03/1997 Paper 16 pgs. 249–264.
1105. Baker, C. F. – Critical review of refiner theory. *In:* III International Refining Conference and Exhibition. PIRA/IPST, Atlanta: 20-22/03/1995 Paper 7 22 pgs.
1106. Ratnieks, E. and Demler, C. – O refino da polpa de eucalipto. *O Papel, 54*(8):22–25 (1993).
1107. Welch, L. V. and Kerekes, R. J. – Characterization of the PFI mill by t
1108. Berger, T. H. – The impact of plate material on pulp quality over time. *In:* IV International Refining Conference. PIRA, Fiuggi: 18-20/03/1997 Paper 18 pgs. 307–315.
1109. Mayade, T. L. – Statistical theory of chemical pulp refining: an innovative combined approach. *Appita Journal, 50* (3): 237–244 (1997).
1110. Li, S.; He, H. Z.; Yu, G. F. and Wei, Y. S. – Study on thickening-refining with a low-consistency disk refiner. *China Pulp and Paper, 13* (2): 29–34 (1994.
1111. Baker, C. F. – Optimisation of refining in fine and speciality paper mill systems. *In:* Papermakers Conference. TAPPI, Philadelphia: 24-27/03/1996 pgs. 215–224.

1112. Levlin, J. E. – Characterisation of Beating Result. *In:* International Symposium of Fundamental Concepts of Refining. IPC, Appleton: 16-18/09/1980 pgs. 131.
1113. Baker, C. F. – The refining of nonwood fibres. *In*: IV International Refining Conference. PIRA, Fiuggi: 18-20/03/1997 Paper 10 pgs. 151–180.492. Prasad, D. Y.; Jameel, H. and Gratzl, J. – Extended delignification of hardwood with AQ/Polysulfide. Tappi Journal, 78 (9): 151 (1995).
1114. Pycraft, C. – The optimization of energy consumption and strength during refining. *Paper Southern Africa, 6*: 6–8, 10, 14 (1986).
1115. Manfredi, V.; Vilela, C. B. and Claudio-da-Silva Jr. E. – Efeito das variáveis operacionais de refino na evolução das propriedades da polpa refinada. *In:* Congresso Anual de Celulose e Papel. ABTCP, São Paulo, 1986 pgs. 189–207.
1116. Manfredi, V. – Optimizing eucalyptus pulp refining. *In:* International Papermaking & Environment Conference. Ed. Yang, S.; Ni, Y e Liu, Z. Tianjin University of Science and Technology, Tianjin – China: 12-14/05/2004 Book A pgs. 41–50.
1117. Lundin, T.; Batchelor, W. and Fardim, P. – Fiber trapping in low-consistency refining: new parameters to describe the refining process. *Tappi Journal, 7* (7): 15–21 (July 2008).
1118. Shekhar, C. D. – Fine edged parallel and curved bar plates in refining system for pulp and paper industries. *IPPTA Journal, 24* (1): 115–119 (2012).
1119. Ratnieks, E. and Mora, E. – How the dryness of pulp influences the stock preparation. *In:* XXVI Congresso Anual de Celulose e Papel, ABTCP, São Paulo, 1993 pgs. 715–731.
1120. Page, D. H. – The mechanism of strength development of dried pulps by beating. *Svensk Papperstiding, 88* (3): R30–35 (1985).
1121. Page, D. H. – The beating of chemical pulps – the action and the effects. In Fundamentals of Papermaking, Trans. of the IXth Fund. Res. Symp. Cambridge, 1989, pgs. 1–38.
1122. Lindbrant, O. and Mohlin, U-B. – Changes in fiber structure due to refining as revealed by S.E.M. *In:* International Symposium of Fundamental Concepts of Refining. IPC, Apleton: 16-18/09/1980 pgs. 61-xxx.
1123. Swerin, A. and Ödberg, L. – Flocculation and floc strength in suspensions flocculated by retention aids. Nordic *Pulp and Paper Reserarch Jorunal, 8* (1): 141–146 (1993).
1124. Tam Doo, P. A. and Kerekes, R. J. – The effect of beating and low amplitude flexing on pulp fibre flexibility. *Journal of Pulp and Paper Science, 15* (11): 36–42 (1989).
1125. Sha, J.; Nikbakht, A.; Wang, C.; Zhang, H. and Olson, J. – The effect of consistency and freeness on the yield stress of chemical pulp fibre suspensions. *BioResources, 10*(3), 4287–4299 (2015).
1126. Koskenhely, K.; Nieminen, K. and Paulapuro, H. – Edge form profile of refiner filling bars and its impact on softwood fibre shortening. *Paperi ja Puu – Paper and Timber, 89* (4): 1–6 (2007).
1127. Lundin, T.; Batchelor, W. and Fardim, P. – Effect of bar edge conditions on fibre trapping in low consistency refining. *In:* 62nd Appita Annual Conference and Exhibition, Rotorua, 2008 pgs. 199–2006 https://search.informit.org/doi/book/10.3316/informit.APP62
1128. Hietanen, S. and Ebeling, K – Control of heterogeneity of the refining action. *In:* International Conference of New Technologies in Refining: Advances in Refining Technologies. PIRA, Birminghan, 1986 Volume I pgs. 1–27
1129. Fox, T. S.; Brodkey, R. S. and Nissan, A. H. – High-speed photography of stock transport in a disk refiner. *Tappi Journal, 62* (3): 55–58 (1979)
1130. Fox, T. S. – Inside a disc refiner. *In:* International Symposium of Fundamental Concepts of Refining. IPC, Appleton, 1980 pgs. 281–313.
1131. Jiang, Z.-H, van Lierop, B. and Berry, R. – Chlorine dioxide bleaching in the presence of an aldehyde. US Patent 6,235,154 B1 (issued May 22, 2001)
1132. Bordin, R.; Roux, J.-C. and Bloch, J.-F. – Global description of refiner plate wear in low consistency beating. *Nordic Pulp and Paper Research Journal, 22* (4): 529–534 (2007).
1133. Baker, C. F. – Specific edge load theory – application and limitations. *In:* Current and Future Technologies of Refining Conference. PIRA, 1991 Volume 2 pg. 18.

1134. Backlund, H.-O. – Measurement of shear force, temperature profiles and fibre development in mill-scale TMP refiners. Licentiate Thesis, Mid Sweeden University. Sundsvall, 2004.
1135. Parola, M. J. and Linna, H. I. – Runnability and the unstable stress fields in paper web. *Graphics Arts, 29* (3): 9 (2000).
1136. Baker, C. – The refining of non-wood fibres. *In:* African Pulp and Paper Week. TAPPSA (2004) http://www.tappsa.co.za/archive2/APPW_004/Title2004/The_refining_of_non-wood.html
1137. Kibblewhite, R. P. – Refining requirements of softwood and eucalypt market pulps and blends. *Appita Journal, 47* (5): 375–379, 401 (1994).
1138. Saltarelli, M. O. – Stock preparation process requirements. *Tappi, 57* (7): 83–85 (1974).
1139. Sallada, O. – Eucalyptus for specialty papers. *In:* New Pulps for the Paper Industry. Ed. L. Haas. Miller Freeman, Brussels, (1979).
1140. Soza, J. and Gonzalez, R. – Estudio de refinaci6n de mesclas de celulosa fibra corta fibra larga em refinador piloto Escher Wyss. www.eucalyptus.com.bricep03190Soza.text.pdf
1141. Chauhan, V. S.; Kumar, N.; Kumar, M.; Chakrabarti, S. K. and Thapar, S. K. – Effect of separate and mixed refining of hardwood and softwood pulps on paper properties. *Journal of Korea TAPPI, 43* (4): 1–10 (2011).
1142. Panthai, S. and Somboon, P. – Examination of separate and mixed refining methods on softwood and hardwood pulps for linerboard production. *Kasetsart Journal – Natural Science, 48*: 540–547 (2014).
1143. Brindley, C. L. and Klibbewhite, R. P. – Comparison of refining response of eucalypt and a mixed hardwood pulp and their blends with softwood. *Appita, 49* (1): 37–49 (1996).
1144. Wallace, B. W. – Dryer section energy use reduced up to 15 % with computer control. *In:* Processes Control for Pulp and Paper Mills. Ed. Smith, K. E. – Miller Freedman Publications, 1983 Chapter 25 pgs. 98–100.
1145. Kerekes, R. J. and Tam Doo, P. A. – The effect of beating and low-amplitude flexing on pulp fibre flexibility. *Journal of Pulp and Paper Science, 15* (1): 36 (1989).
1146. Ghatak, H. R. – A comminution model for laboratory beating. *Appita Journal, 58* (3): 127–128 (2005)
1147. Lindström, T.; Ljunggren, S.; Ruvo, A. de e Söremark, C. – Dissolution of carbohydrates and lignin during beating of kraft pulps. *Svensk Papperstiding, 81* (12): 397–402 (1978).
1148. Bajpai, P., Mishra, S. P., Mishra, O. P., Kumar, S. and Bajpai, P. K. – Use of enzymes for reduction in refining energy – laboratory studies. *Tappi Journal, 5* (11): 25–32 (2006).
1149. Roux, J-C. and Joris, G. – How load is distributed in an industrial disc refiner. *In:* International Refining Conference. PIRA, Barcelona, 2005 Paper 13.
1150. Yasumura, P. K. and Park, S. W. -Novos aspectos sobre ações de refino do moinho PFI e refinadores industriais de disco em fibras. *In:* 45th ABTCP International Pulp and Paper Congress e VII IberoAmerican Congress on Pulp and Paper Research. São Paulo, 2012 8 pgs
1151. Khokhar, G. M. – Numerical simulation of the flow in a disc refiner. Master's Thesis. Royal Institute of Technology, Stockholm, 2011 59 pgs https://www.lcrl.ppc.ubc.ca/files/2013/02/2011-Khokhar-MSc-Thesis.pdf
1152. Rihs, J. – Refining recycled fibers for brown and white grades. *In:* Papermakers Conference. TAPPI, Nashville, 1992 pgs. 239–245.
1153. Pashinskii, V.; Pedotova, B.; Komisarchuk, G; Maslakov, V. and Voronova, G. – Beating of hardwood pulp in a multi-disk refiner. Sub. Tr. Ukr. Nauch.-Issled. Inst. Tsellyul. – Bumazh. Prom. (18): 5–10. (1975) according to *ABIPST, 68* (3): 342 reference 39426(AS).
1154. Heymer, J. O.; Olson, J. A. and Kerekes, R. J. – The role of multiple loading cycles in pulp refiners. 5 pgs. (2012) www.fibrelab-mech.sites.olt.ubc.ca/files/2012/07/The-Role-of-Multiple-Loading-Cycles-in-Pulp-Refiners.pdf
1155. Ouellet, D.; Bennington, C. P. J.; Senger, J. J.; Borisoff, J. F. and Martiskainen, J. M. – Measurement of pulp residence time in a high-consistency refiner. *Journal of Pulp and Paper Science, 22* (8): J301–305 (1996).

1156. Bordin, R.; Toux, J-C. and Block, J-F. – No-load characterization of a low-consistency disc refiner for na efetive application of refining theories. Engineering, Pulping and Environmental Conference, TAPPI, TAPPI Engineering, Pulping and Environmental Conference, Jacksonville, 2007 33 pgs.
1157. Hietanen, S. and Ebeling, K. – Heterogeneity in refining action: effects on fiber and paper structure. *In:* International Paper Physics Conference. TAPPI/CPPA, Cape Cod, 1983 pgs. 27–39.
1158. Atack, D.; Stationwala, M. I.; Huusari, E.; Ahlqvist, P.; Fontebasso, J. and Perkola, M. – High-speed photography of pulp flow patterns in a 5 MW pressurized refiner. *Paperi ja Puu, 71* (6): 689 (1989)
1159. Hietanen, S. – The role of fiber flocculation in chemical pulp refining. *Paperi ja Puu, 73* (3): 249–259 (1991).
1160. Page, D. H., Kosky, J. and Booth, D. – Some initial observations on the action of the beater. BP&BIRA Bulletin, 28: 15–21 (1962).
1161. Waterhouse, J. F. – Characterizing pulps for paper-machine runnability. *In:* Engineering/ Process and Product Quality Conference & Trade Fair. TAPPI, Anaheim, 1999 Volume 3 pgs. 1195–1204.
1162. Liu, H.; Dong, J.; Jing, H.; Guo X.; Duan, C.; Qi, K.; Yang, R.; Gio, H.; Wang, B. and Qiao, L. – Correlation between bar angle and characterization parameters of the isometric straight bar plate. *BioResources 15* (4), 7844–7860 (2020).
1163. Eriksen, O.; Holmqvist, C. and Mohlin, U-B. – Fibre floc drainage – a possible cause for substantial pressure peaks in low-consistency refiners. *Nordic Pulp and Paper Research Journal, 23* (3): 321–326 (2008).
1164. Almeida, L. F. – Nova tecnologia para refinação em baixa consistência. *In:* 45th ABTCP International Pulp and Paper Congress and VII Ibero American Congress on Pulp and Paper Research. São Paulo, 2012 9 pgs.
1165. Wittberg, L. P.; Björkman, M.; Khokhar, G.; Mohlin, U-B. and Dahlkild, A. – Flow conditions in the grooves of a low-consistency refiner. *Nordic Pulp & Paper Research Journal, 27* (2): 173–183 (2012).
1166. Fox, T. S.; Brodkey, R. S. and Nissan, A. H. – Inside a disk refiner. *Tappi Journal, 65* (7): 80–83 (1982).
1167. Groome, E. J. and Gerhardt, T. D. – Fiber retention time in a disk refiner. *In:* International Symposium on Fundamental Concepts of Refining. IPC, Appleton, 1980. pgs. 21–29.
1168. Bordin R.; Roux, J.-C. and Bloch, J.-F. – New technique for measuring clearance in low-consistency refiners. *Appita Journal, 61* (1): 71–77 (2008).
1169. Gupta, P. R. – Effects of variables in beating and refining. *In:* Pulping Conference. TAPPI, Dallas, 1980 pgs. 233–240
1170. Arjas, A. – Influence of residence time distribution on pulp properties. *In:* International Symposium of Fundamental Concepts of Refining. IPC, Appleton, 1980 pg 61
1171. Kim, A-R.; Choi, K-H. and Cho, B.-U. – Changes in fiber characteristics by low concentration sodium hydroxide swelling and beating. *Journal of Korea TAPPI. 46* (3): 65–72 (2014).
1172. Olejnik, K. – Effect of the free swelling of refined cellulose fibres on the mechanical properties of paper. *Fibres & Textiles in Eastern Europe, 1* (90): 113–116 (2012).
1173. Lahtinen, P.; Liukkonen, S.; Pere, J.; Sneck, A. and Kangas, H. – A comparative study of fibrillated fibers from different mechanical and chemical Pulps. *BioResources, 9* (2): 2115–2127 (2014).
1174. Ferreira, J. P.; Matos, S. and Figueiredo, M. M. – Size characterization of fibres and fines in hardwood. *Particle & Particle System Characterization, 16* (1):20–24 (1990).
1175. Gallay, W. – Some aspects of the theory of the beating process. *In:* Fundamentals of Papermaking Symposium. PIRA Cambridge, 1961 pgs. 377–387.
1176. Talaeipoor, M. and Imani, R. – Effects of debonding agents and refining on the properties of deinked pulp. *Tappi Journal, 7* (7): 12–14 (2008).

1177. Johansson, O. – A new optimization approach can lead to significant improvement of the refining efficiency. *In:* Scientific and Technical Advances in Refining & Mechanical Pulping – Course Notes. PIRA: Stockholm: 25-26/03/2003 Paper 13.
1178. Kang, T. and Paulapuro, H. – Characterization of chemical pulp fines. *TAPPI Journal 5* (2):25–28 (2006).
1179. Lindqvist, H.; Salminen, K.; Kataja, J.; Retulainen, E.; Fardim, P. and Sundberg, A. – The effect of fines on dewatering, wet and dry web properties. *In:* Paper Conference, Covington, 2011 pgs 887–894 https://www.tappi.org/content/events/11papercon/documents/285.315%20docx.pdf
1180. van Stiphout, J. M. J. – A preliminary study of the refining action on cellulose fibers. *Tappi, 47* (2): 189–191 (1964).
1181. Reed, R. W. – The colloidal and physical phenomena relating to freeness and stock drainage. Technical Association of Pulp and Paper Industry, New York, 1942.
1182. Reiner, F. J. M. – Mecanismos de ligação entre as fibras e a importância do refino. *O Papel, 55* (1): 29–32 (1994).
1183. Lumiainen, J. – Post-refining of mechanical pulps. *In:* IV International Refining Conference. PIRA, Fiuggi, 1997 Paper 8 pgs. 127–137.
1184. Steffens, D. – Ultrasonic technology – measurements of paper orientation and elastic properties. *In:* XXVII Congresso Anual de Celulose e Papel. ABTCP, São Paulo, 1994 pgs. 661–681.
1185. Cowan, W. F. – High-shear laboratory beating and fiber-quality testing offer new insinghts into pulp evaluation. *Tappi Journal, 78* (3): 133–137 (1995).
1186. Niskanen, K. – Strength and fracture of paper. *In:* 10th Fundamental Research Symposium – Products of Papermaking. PIRA, Oxford, 1993 Volume 2 pgs. 641–725.
1187. Basile, F. C. – Energy considerations in stock preparation refining. *Tappi Journal, 97* (4): 84–88 (1984).
1188. Zhang, G.; Hiltunen, E.; Kettunen, H.; Nisjanen, K.; Laine, J. E. and Paulapuro, H. – Effects of wet strain on fracture properties of paper in comparison to refining. *Nordic Pulp and Paper Research Journal, 17* (1): 45 (2002).
1189. Lindström, T., Laine, J. and Bremberg, C. – Fundamental aspects of wet seb consolidation and paper strength. *In:* Progress in Paper Physics Seminar. ESPR, SUNY-ESF, TAPPI, Skaneatles, 2002 pg. 10.
1190. Goosen, D. R., Olson, J. A. and Kerekes, R. J. – The role of heterogeneity in compression refining. *Journal of Pulp and Paper Science, 33* (2): 110–114 (2007).
1191. Ghatak, H. R. – Analysis of laboratory beating by fiber settling time frequency distributions. *Tappi Journal, 2* (11): 3–6 (2003)
1192. Vainio A. K. and Paulapuro, H – Interfiber bonding and fiber segment activation in paper. *Bioresources, 2* (3):442–458 (2007).
1193. Ueno, P. Y. – A drenagem na mesa plana hoje. *In:* XXIV Congresso Anual de Celulose e Papel. ABTCP, São Paulo, 1991 pgs. 525–563.
1194. Heitmann, J. A. – Pulp properties. *In:* Pulp and Paper Manufacture – Mill Control & Control Systems: Quality & Testing, Environmental, Corrosion, Electrical. Ed.: Kouris, M e Kocurek, M. J. The Textbook, Committee of the Paper Industry, Atlanta/Montreal, 1992, pgs. 85–98.
1195. Naik, B. H.; Nishi K. B. and Ray, A. K. – Statistical evaluation of the effect of refining on fibre charge of bagasse pulp. *IPPTA Journal, 21* (2): 141–146 (2009).
1196. Marton, J. and Marton, T. – Some new principles to optimize rosing sizing. *Tappi Journal, 65* (11):105–109 (1982).
1197. French, J. and Maddern, K. N. – A mini pulp evaluation procedure. *Appita Journal, 47* (1): 38–44 (1994).
1198. Frölander, U. and Hartler, H. – Determination of the bulk fibre stiffness from the network shear modulus. *Svensk Papperstiding, 73* (2): 21 (1970).

1199. Curto, J.; Simões, R. and Silvy, J. – The influence of bleaching, beating and drying in the wet fibre flexibility of *Pinus pinaster* kraft pulp. *In:* 12th International Symposium on Wood and Pulping Chemistry. Madison, 2003 Volume 3 pgs. 287–289.
1200. Paavilainen, L. – Bonding potential of softwood sulphate pulp fibres. *Paperi ja Puu, 76* (3): 162 (1994).
1201. Grén, U. and Ljungqvist, K. – Compressibility and permeability of chemical pulps. Some notes on the effect of beating. *Cellulose Chemistry Technology, 17* (9): 515 (1983).
1202. Sampson, W. W. and Kropholler, H. W. – Batch-drainage curves for pulp characterization. Part 1: experimental. *Tappi Journal, 78* (12): 145–151 (1995).
1203. Brown, Jr, J. C. – Determination of the exposed specific surface of pulp fibers from air permeability measurements. *Tappi Journal, 33* (3): 130–137 (1950).
1204. Hunt, N. – Review of PIRA non-wood fibre conference. *In:* African Pulp and Paper Week. TAPPSA (2002) http://www.tappsa.co.za/archive2/APPW_2002/Title/ReviewofPIRAnonwoodfibreconference.html
1205. Scallan, A. M. and Borch, J. – An interpretation of paper reflectance based upon morphology: Part I. initial consideration. *Tappi, 55* (4): 583 (1972).
1206. McKinney, R. – A review of stickie control methods, including the role of surface phenomena in control. *In:* Paper and Board Division Conference. Gatwick, 1989 Paper 4.
1207. Agarwal, S.B.; Genco, J. M.; Cole, B. J. W. and Miller, W. – Kinetics of oxygen delignification. *Journal of Pulp and Paper Science, 25* (10): 361–366 (1999).
1208. Kibblewhite, R. P. – Interrelations between pulp refining treatments, fibre and fines quality, and pulp freeness. *Paperi Puu, 57* (8): 519–526 (1975).
1209. Patel, M. and Trivedi, R. – Variations in strength and bonding properties of fines from filler, fiber and their aggregates. *Tappi Journal, 77* (3): 185–191 (1994).
1210. Coutts, R. S. P. and Ridikas, V. – Refined wood fibre-cement products. *Appita, 35* (5): 395–400 (1982).
1211. Kibblewhite, R. P. – Fractures and dislocations in the walls of kraft and bisulphite pulp fibers *Cellulose Chemistry and Technology, 10* (2): 497–503 (1976)
1212. Gruber, E – Analytical characterization on pulps. *In:* Handbook of Pulp. Volume 2, Section IV Ed, Sixta, H. WILEY-VCH Verlag GmbH &Co. KGaA, Weinheim, 2006 pgs. 1211–1290.
1213. Mohlin, U.-B. and Wennberg, K. – Some aspects of the interaction between mechanical and chemical pulps. *Tappi Journal, 67* (1): 90–93 (1984).
1214. Marton, R. and Alexander, S. – Effect of beating and wet-pressing on single fiber and sheet properties. *Espra Research Reports, 43* (II): 17 (1966).
1215. Kure, K. A. and Dahlqvist, G. – Development of structural fibre properties in high intensity refining. *Pulp and Paper Canada, 99* (7): 59–63 (1998).
1216. Lonnberg, B. – Energy balance in low consistency (LC) refining. *In:* Refining and Mechanical Pulping conference. PIRA, Barcelona, 2005 Paper 7.
1217. Turt, V.; Genco, J. M. and Co A. – Effect of refining on fiber properties. *In:* Engineering Conference. TAPPI, San Francisco, 1994 pg. 273–280.
1218. Steenberg, B.; Sandgreen, B. and Wahren, D. – Studies on pulp crill. Part 1: suspended fibrils in paper pulp fines. *Svensk Paperstiding, 63* (12): 395–397 (1960).
1219. Szwarcsztajn, E. and Przybysz, K. – Investigations of water retention value of beaten cellulose fibers. *Cellulose Chemistry and Technology, 9:* 597–607 (1975).
1220. Joris, G. and Roux, J-C. – The dynamic freeness tester. *In:* Scientific and Technical Advances in Refining & Mechanical Pulping – Course Notes. PIRA: Stockholm, 2003 Paper 6.
1221. Byrd, V. L.; Horn, R. A. and Fahey, D. J. –How refining of vessel elements affects offset printing papers. *Paper Trade Journal, 151* (46): 55–59 (1967).
1222. Venkatesh, V. and Edwards, L. L. – Prediction of mechanical pulp handsheet and drainage properties for mill control and design. *Svensk Papperstiding, 4* (13): 422–430 (1977).
1223. Wultsch, F e Flucher, W. – Small Escher-Wyss refiner as a standard testing instrument for modern stock preparation plants. *Das Papier, 12* (13/14): 334–342 (1958).

1224. Brecht, W. and Siewert, W. H. – Zur theoretish technischen beurteilung des mahlprozesses. *Das Papier, 20* (1): S4–14. (1966).
1225. Brecht, W. – A method for the comparative evaluation of bar-equipped beating devices. *Tappi, 50* (8): 40–44 (1967).
1226. Kartowaara, I. –The average number of impacts experienced by a fiber in a disk refiner: comment. *Tappi, 61* (8): 90 (1978).
1227. Bian, L.; Liu, S.; Pu, Y.; Liu, C. and Yuan, L. – Appraisal of a disk refiner performance. *Guowai Zaozshi, 16* (1): 28–31 (1997) according to ABIPST, 68 (2): 123 reference 1409(AS).
1228. Formento, J. C.; Maximino, M. G., Adelli, A. M. and Taleb, M. C. – Selective refining actions on repulped long-fiber kraft paper. *Tappi Journal, 2* (9): 1–9 (2003).
1229. Fan, X.; Jeffrey, D. J. and Ouellet, D. – A stochastic model for the residence time of pulp in a single-disc chip refiner. *Journal of Pulp and Paper Science, 20* (11): 343–349 (1994).
1230. Manfredi, V. – Evaluation of refining strategies for combined use of softwood and eucalypt pulps in papermaking. *In:* Pan Pacific Conference – Advances in Pulp & Paper Sciences and Technologies. KTAPPI, Seoul, 2006. Book 1 pgs. 37–42.
1231. Karlström, A.; Eriksson, K.; Sikter, D. and Gustavsson, M. – Refining models for control purposes. *Nordic Pulp and Paper Research Journal, 23* (1): 129–138 (2008).
1232. Tonoli, G. H. D.; Teixeira, E. M.; Corrêa, A. C.; Marconcini, J. M.; Caixeta, L. A.; Silva, M. A. P. da e Mattoso, L. H. C. – Cellulose micro/nanofibres from *eucalyptus* kraft pulp: Preparation and properties. *Carbohydrate Polymers, 89* (1): 80–88 (2012).
1233. Eibinger, K; Eichinnger, R. and Bauer, W. – Property development of virgin and recycled fibres treated with compression refining in a PFI mill, *In:* International Refining Conference, PIRA, 2005 12 pgs https://www.researchgate.net/publication/267694160_Property_development_of_virgin_and_recycled_fibres_treat
1234. Grzegorz, K. and Dariusz, A. – Flow modelling in a low consistency disc refiner, *Nordic Pulp & Paper Research Journal* 28 (1), 119–130 (2013).
1235. Khlebnikov, A. A.; Pashinnskii, V. F.; Gonharov, V. N. and Smirnova, E. A. – Analysis of the forces involved in the operation of a conical refiner. *Tr. Leningr.,* 22: 129–136 (1969) [1039, 1163].
1236. Goncharov, V. N.; Smirnova, E. A. and Shemyakin, E. V. – Method for the determination of stresses between refiner blades. Bumazh. Prom. (27): 134–138 [1163]
1237. Goncharov, V. N. – Force factors in a disk refiner and their effect on the beating process. *Bumazh. Prom.,* (5): 12–14 [1163]
1238. Koskenhely K.; Nieminen, K.; Hiltunen, E. and Paulapuro, H. – Comparison of plate and conical fillings in refining of bleached softwood and hardwood pulps. *Paperi ja Puu, 87* (7): 458–463 (2005).
1239. Olejnik, K. – The effect of specific refining intensity on paper breaking length. *Paper Technology, 52* (2): 8–11 (2011).
1240. Batchelor, W.J. – Effects of flocculation and floc trapping on fibre treatment in low consistency refining. *Journal of Pulp and Paper Science. 27* (7): 249–252 (2001).
1241. Stephansen, E. – Contribution to the understanding of the mechanism of the beating process. *Norsk Skogindustri, 21* (8): 266 (1967).
1242. Odabas, N.; Henniges, U.; Potthast, A. and Rosenau, T. – Cellulosic fines: properties and effects. *Progress in Materials Science, 83*: 574–594 (2006).
1243. Colodette, J. L.; Ghosh, A. K.; Dhasmana, B.; Singh, U. P.; Gomide, J. L. and Singh, R. P. – Bleaching processes for market grade TCF pulps. *In:* International Non-Chlorine Bleaching Conference. Pulp & Paper/Emerging Technology Transfer, Amelia Island, 1994 Paper 7-1 22 pgs.
1244. Milichovsky, M. – Scientific Papers, Institute of Chemical Technology. Pardubice 51: 115 (1988) [445].
1245. El-Sharkawy, K. – Different approaches to tailoring chemical pulp fibres. Doctoral Thesis. Helsinki University of Technology, Espoo, 2008 64 pgs http://lib.tkk.fi/Diss/2008/isbn9789512296224/isbn9789512296224.pdf

1246. Steenberg, B. – Review of the effect of mechanical treatment of fibres. *Svensk Paperstidning, 66* (22): 933–939 (1963)
1247. Wang, Z.; Dong, S.; Gartshore, I. and Salcudean, M. – Fiber separation in hydrocyclones. *In:* Fall Conference and Trade Fair. TAPPI, Chicago, 2002 pgs. 1–7.
1248. Heymer, J. O. – Measurement of heterogeneity in low consistency pulp refining by comminution modeling. Doctoral Thesis. The University of British Columbia, Vancouver, 2009 107 pgs. http://turbulence-initiated.sites.olt.ubc.ca/files/2013/01/2009-Heymer-PhD-thesis.pdf
1249. Mohlin, U-B. – Refining intensity and gap clearance. *In:* 9th International Refining Conference. No14, PIRA, Vienna, 2006.
1250. Loijas, M. – Factors affecting the axial force in low-consistency refining. Licentiate Degree. Tampere University of Applied Sciences. Tampere. 2010 53 pgs https://www.theseus.fi/bitstream/handle/10024/16798/Loijas_Marko.pdf?sequence=2&isAllowed=y
1251. Vaz, A.; Simões, R. and Silvy, J. – Refining rheological response of chemical pulp fibre suspensions. *In:* 10th World Congress of Chemical Engineering, EFCE, Barcelona, 2017 Abstract https://ubibliorum.ubi.pt/handle/104 00.6/9627.
1252. Nordman, L.; Levlin, J.-E.; Makkonen, T. and Jokisalo, H. – Conditions in na LC refiner as observed by physical measurements. *In:* International Symposium on Fundamental Concepts of Refining. IPC, Appleton, 1980 pgs. 121–130.
1253. Koskenhely, K. – Effect of selected filling and pulp suspension variables in improving the performance of low consistency refining. Doctoral Thesis, Helsinki, Espoo, 2007 xx pgs www.core.ac.uk/download/pdf/80701885.pdf
1254. Wagle, D. G.; Lee, C. W. and Brodley, R. S. – Further comments on a visual study of pulp floc dispersion mechanisms. *Tappi, 71* (9): 137–141 (1988).
1255. Peralta, C.; Piçarra, E.; Santos, A.; Anjos, O. and Simões, R. – Effect of specific edge load on *Eucalyptus globulus* paper properties. *In:* 2° International Conference on Environmentally – Compatible Forest Products. Proceedings of Ecowwod – 2006 pgs: 499–505 https://www.researchgate.net/publication/259822897_Effect_of_specified_edge_load_on_Eucalyptus_globulus_paper_properties
1256. Cowan, W. F. – Wet pulp characterization by means of specific surface and compressibility. *Pulp and Paper Magazine Canada, 71* (9): 63–66 (1970).
1257. Przybysz, P.; Dubowik, M.; Małachowska, E., Kucner, M., Gajadhur, M. and Przybysz, K. – The effect of the refining intensity on the progress of internal fibrillation and shortening of cellulose fibers. *BioResources, 15* (1): 1482–1499 (2020)
1258. Nunes, T. F. G. – Produção, caracterização e aplicação de nanofibras de Celulose. Master's Thesis Universidade de Coimbra, Coimbra. 2014 81 pgs. www.eg.uc.pt/bitstream/10316/30004/3/Produ%C3%A7%C3%A3o,%20caracteriza%C3%A7%C3%A3o%20e%20aplica%C3%A7P%C3%A3%20nanofibras%20de%20celulose
1259. Reid, D. W.; Billmark, G. G. and Sutton, C. J. – Bleaching with peroxide to reduce dioxin formation in a mill with limited chlorine dioxide generator capacity. *Tappi Journal, 74* (7): 99–103 (1991).
1260. Conn, A.B. and Batchelor, W.J. – Conversion of an Instron to mechanical testing of single fibers. APPITA Annual Conference and Exhibition, Brisbane, 1998 pgs. 43–548
1261. Lipshitz, H.; Bridger, M. and Derman, G. – On the relationship between topography and gloss. *Tappi Journal, 73* (10): 237 (1990).
1262. Young, J. H. – Fiber preparation and approach flow. *In:* Pulp and Paper Chemistry and Chemical Technology – Volume III. Ed. Casey, J.P., John Wiley & Sons, New York, 1980.
1263. Lumiainen, J. – The specific surface load, impact length, number of impacts, refiner geometry and fiber characteristics are key factors in the refining system design. *In: SPCI-92.* Aticelpa, Bologna, 1992.
1264. Rangamannar, G. and Silveri, L. – Diskpersion – an effective secondary fiber treatament process for high quality deinking pulp. *In:* Pulping Conference. TAPPI, Seatle, 1989 pgs. 381–390.
1265. Rangamannar, G. and Silveri, L. – Effective secondary fiber treatment process for high quality deinked pulp. *Tappi Journal, 73* (7): 188–191 (1990).

1266. Moore, G. K.; Cathie, K.; Crow, H. and Smith, J. – Developing mixed office waste as a furnish component for printings and writings. *In:* III International Refining Conference and Exhibition. PIRA/IPST, Atlanta, 1995 Paper 15 11 pgs.
1267. McKenzie, A. W. and Tercelli, F. – Tearing resistance of crumbed chemical pulps. *Appita, 46* (1): 59–60 (1993).
1268. Gabl, H. – Pilot scale optimisation of refining conditions for fractionated recycled furnishes. *In:* Refining and Mechanical Pulping. PIRA, Barcelona: 2005 Paper 12.
1269. Paulapuro, H. – Papermaking – part 1: stock preparation and wet end. *In:* Papermaking Science and Technology. Ed. Gullichsen, J. and Paulapuro, H. Finnish Paper Engineers' Association/Paperi ja Puu Oy, Hesinki: 2000 Volume 8 461 pgs.
1270. Egenes, T. H. and Helle, T. – Water removal from pulp suspensions in screw press. *In:* XXIII Congresso Anual de Celulose e Papel. ABTCP, São Paulo, 1990 pgs. 113–128.
1271. Page, D. H. – The axial compression of fibres – a newly discovered beating action. *Pulp and Paper Magazine Canada, 67* (1): 2–12 (1966
1272. Belle, J. Kleemann, S.; Odermatt, J. and Olbrich, A. – A new method showing the impact of pulp refining on fiber-fiber interactions in wet webs. *Nordic Pulp & Paper Research Journal, 31* (2): 205–212 (2016).
1273. Strazdins, E. – Chemistry and application of rosin size. *In:* The sizing of paper. Ed. Raynolds, W. F. TAPPI Press, Atlanta, 156 p 2nd Ed. 1989.
1274. Atack, D. – Advances in beating and refining. *In:* Transactions of the Oxford Symposium, British Pulp Paper and Board Industry Federation, London, 1978 pg. 261–297.
1275. Bhardwaj, N. K. – Refining of bamboo long fiber fraction pulp effects on wet web and dry strength properties of paper. *Cellulose Chemistry and Technology, 53* (1–2): 113–120 (2019).
1276. Hou, Q.; Yang, B.; Liu, W.; Liu, H.; Hong, Y. and Zhang, R. – Co-refining of wheat straw pulp and hardwood kraft pulp, *Carbohydrate Polymers, 86 (1),* 255–259 (2011).
1277. Hietanen, S. – Development of a fundamentally new refining method. Part 2: operating conditions of the new refiner. *Paperi ja Puu, 72* (10): 952–960 (1990).
1278. Câmara, A. L. da e Brandão, P. R. G. – Avaliação da adsorção de piche (pitch) no processo de fabricação do papel por talco brasileiro. *O Papel, 65* (4): 67–78 (2004).
1279. Sundström, L.; Brolin, A. and Hartler, N. – Fibrillation and its importance for the properties of mechanical pulp fiber sheets. *Nordic Pulp and Paper Research Journal, 8* (4): 379–383 (1993).
1280. Yasumura, P. K.; D'Almeida, M. L. O. and Park, S. W. – Multivariate statistical evaluation of physical properties of pulps refinied in a PFI mil. O PapeL, 73 (3): 59–65 (2012)
1281. El-Hosseiny, F. – The effect of sheet densification on the shape of its stress-strain curve. *Journal of Pulp and Paper Science, 20* (12): J366–370 (1994).
1282. Seth, R. S.; Robertson, A. G.; Mai, Y.-W. and Hoffmann, J. D. – Plane stress fracture toughness of paper. *Tappi Journal, 76* (2): 109–116 (1993).
1283. Parham, R. A. – Ultra-Structure and Chemistry. *In:* Pulp and Paper manufacture – Volume I – Properties of Fibrous Raw Materials and their Preparation for Pulping. E.: Kocurek, M. J. and Stevens, C. F. B. The Joint Textbook Committee of the Paper Industry, Atlanta/Montreal, 1983 pgs. 35–45.
1284. Retulainen, E. – Strength properties of mechanical and chemical pulp blends. *Paperi ja Puu, 74* (5): 419–426 (1992).
1285. Atack, D.; Stationwala, M. I. and Karmis, A. – What happens in refining. *Pulp and Paper Canada, 85* (12): 119–124 (1984)
1286. Goncharov, V. N. – Stostoyanie i perspektivy sovershnstvovaniya oborudovaniya dlya razmola voloknistykh materialov. *Tsellyul. Bum. Karton, 7/8*: 20–24 (1995) according to ABIPST, 67 (9): 1337 reference 12645(AS)
1287. Tousignant, D. and Madgett, H. H. – Results of drainage measurements and related controls on a newsprint Forurdninier. *Pulp and Paper Canada, 64* (10): 427 (1963).
1288. Ohsawa, J.; Wakai, M.; Komatsu, Y.; Yoneda, Y.; Nagasawa, T. – Prevention of vessel picking trouble in tropical hardwood pulps – II. Vessel separation and high consistency bleating. *Journal of Japanese Wood Research Society, 30* (9): 742–749 (1984).

1289. Sari, A.; Agneta, F.; Kariniemi Merja, K. and Särkilahti Airi, S, – Evaluation of vessel picking tendency in printing. *O Papel, 73* (1): 79–85 (2012)
1290. Mukoyoshi, S.; Ohtake, T.; Ohsawa J. – Mechanism of vessel separation with hydrocyclone I. Vessel separation with centri-cleaner. *Japan Tappi Journal, 40* (11) 55–63 (1986).
1291. Rissato, T. C. – Influência do refino de baixa intensidade no consumo energético e nas propriedades do papel. Monografia Latu-sensu. Universidade Federal de Viçosa, Viçosa, 2010 71 pgs
1292. Ratnieks, E., Demler, C. L. – Estamos obtendo o melhor na refinação de polpas de eucalipto? *In:* 24° Congresso Anual. ABTCP, São Paulo, 1991 19 pgs.
1293. Miles, K. B. and May, W. D. – The flow of pulp in chip refiners. *Journal of Pulp and Paper Science, 16* (2): 63–71 (1990),
1294. Miles, K. B. – Refining intensity and pulp quality in high-consistency refining. *Paperi ja Puu, 72* (5): 508–514 (1990).
1295. Levlin, J. E. – The response of different fibre raw materials to refining. *In:* Advances in Refining Technologies Conference. PIRA, Birmingha, 1986.
1296. Miles, K. B. – A simplified method for calculating the residence time and refining intensity in a chip refiner. *Paperi ja Puu, 73* (9): 852–857 (1991).
1297. Swerin, A.; Odberg, L. and Lindström, T. – Deswelling of hardwood kraft pulp fibers by cationic polymers. The effect on wet pressing and sheet properties. *Nordic Pulp and Paper Research Journal, 5* (4): 188–196 (1990).
1298. Mantar, E.; Co, A. and Genco, J. M. – Drainage characteristics of pulp slurries under dynamic conditions. *Jornal of Pulp and Paper Sciency, 21* (2): 44–50 (1995).
1299. Hubbe, M. A. and Panczik, M – Dewatering of refined bleached hardwood kraft pule by gravity vacuum and centrifugation with applied pressure. *O Papel, 77* (10): 74–87 (2017).
1300. Britt, K. W. and Unbehend, J. E. – New methods for monitoring retention. *Tappi Journal, 59* (2): 67–70 (1976).
1301. Sayegh, N. N. and Gonzalez, T. O. – Compressibility of fibre mats during drainage. *Journal of Pulp and Paper Science, 21* (7): 255–261 (1995).
1302. Sepke, P.-W. – The effect of pulp fiber properties on the tearing work of the paper. *Tappi Journal, 79* (4):211–216 (1996).
1303. Sutton, P. – Pulp refiner trial trips. *Tappi Journal, 82* (9): 67 (1999).
1304. Dharni, S. – Estimation of the number of impacts impacted to the pulp in a disc refiner. Master's Thesis, Dept. of Chemical Engineering, Pulp and Paper Centre, University of British Columbia, Vancouver, 1989.
1305. Preddy, B – Current advances in papermaking. *World Pulp and Paper Technology, 1998* (9): 113 (1998).
1306. Belle, K. and Odermatt, L. – Initial wet web strength of paper *Cellulose,* 23:2249–2272 (2016).
1307. Olejnik, K. – Impact of pulp consistency on refining process conducted under constant intensity determined by SEL and SEC factors. *BioResources, 8*(3): 3212–3230 (2013).
1308. Claudio-da-Silva Jr., E. – Chemical pulp beating related to fiber structure. Doctoral Thesis, College of Environmental Science and Forestry, Syracuse, 1991 295 pgs.
1309. Kibblewhite, R. P. – Refining energy demand, freeness and strength of separate and co-refined softwood and eucalypt market pulps and blends. *In:* 47th General Conference. APPITA, Rotorua, 1993 Volume 1: 105–112.
1310. Araújo, J. A. de; Araújo, M. L. C. de; Dvorack, A. and Dvorack, M. E. S. – Schopper-Riegler eletrônico: mecanismos para avaliação da drenabilidade dinâmica, consistência e retenção de cargas e fibras. *In:* XXXI Congresso Anual de Celulose e Papel. ABTCP, São Paulo, 1998 pgs. 283–297.
1311. Wallbäcks, L. – Characterisation of chemical pulps using spectroscopy, fibre dimensions and multivariate data analysis. *In:* International Symposium on Wood and Pulping Chemistry. Beijing, 1993 pgs. 802
1312. Peng, Y. X. and Li, Z. Z. – Beating and papermaking characteristics of eucalyptus chemical pulps. *China Pulp and Paper, 15* (5): 35–40 (1995).

1313. Kerekes, R. J. and Senger, J. J. – Characterizing refining action in low-consistency refiners by forces on fibres. *Journal of Pulp and Paper Science, 32* (1): 1–8 (2006).
1314. van den Akker, J. A. – Energy considerations in the beating of pulp. *In:* Trans. Fund. Res. Symposium. Ed. F. Bolam, BPBMA, Cambridge, 1957 pgs. 435–446.
1315. Bordin, R.; Roux, J-C. and Bloch, J-F. – No-load power evolution during low-consistency pulp beating. *Nordic Pulp and paper Research Journal, 23* (1): 34–38 (2008).
1316. Amero, B. A. – Refining optimization single-pass refiners versus multiple-pass laboratory beaters. *In:* International Symposium on Fundamental Concepts of Refining. IPST, Appleton, 1980 pg. 179
1317. Liu, H.; Dong, J.; Qi, K.; Guo, X.; Yan, Y, Qiao, L.; Duan, C. and Zhao, Z. – The effect of pulp properties on the consumption in low consistency refining. *Journal of Korean Wood Science and Technology, 48* (6): 869–877 (2020)
1318. Kerekes, R. J. – A theoretical analysis of recirculation in pulp refiners. *Tappi Journal, 13* (4):29–32 (2014).
1319. Kral, P.; Klimek, P. and Mishra, P. K. – Effect of refiner plate pattern design on refined fibre size distribution – a time series study. *Drvna Industrua, 66* (1): 63–67 (2015).
1320. Liu, H.; Dong, J.; Guo, X.; Wang, B.; Lijie, Q.; Chuanwu Duan, C.; Kai Qi, K.; Kong, L. – No-load power of disc refiner in low consistency refining. *Journal of Korea TAPPI, 52*(2): 87–96 (2020).
1321. Oslon, J.; Allison, B.; Friesen, A. and Petrs, C. – Fibre fractionation for high porosity sack paper. *Tappi Journal, 84* (6): 1-xx (2001).
1322. Panula-Ontto, S. – Fractionation of unbleached softwood draft pulp with wedge wire screen and hydrocyclone. Lic.Techn Thesis, Department of Forest Products Technology, Helsinki University of Technology, Helsinki, 2002 83 pgs.
1323. Sloane, C. M. – Kraft pulp processing – pressure screen fractionation. *Appita Journal, 53* (4): 220–226 (2000).
1324. Vomhoff, H. and Grundström, K.-J. – Fractionation of a bleached softwood pulp and separate refining of the earlywood and latewood-enriched fractions. *Das Papier* (2): 37–41.
1325. Mayovsky, J. – Fractionation of OCC. How can it help you. *In:* Recycling Symposium. TAPPI, 1998 pgs. 407–416.
1326. Lapierre, L.; Pitre, D. and Bouchard, J. – Bleaching of deinked recycled pulp: benefits of fibre fractionation. *Pulp and Paper Canada, 102* (2): 43–46 (2001).
1327. Braaten, K. R. and Wakelin, R. F. – Fiber length fractionation of TMP using pressure screens. *Tappi Journal, 82* (6): 129–135 (1999).
1328. Hill, H. S.; Edwards, J. and Beath, L. R. – Curlated pulp – a new approach to pulp processing. *Tappi, 33* (1):36 (1950
1329. Herbert, W.; Marsh, P. G. and Harbron, D. L. – New kinks in refining with the chemifiner. *Paper Trade Journal, 42* (1967)
1330. Pfarr, A. – Holländer und deren kraftverbrauch. *Wochenblatgt für Papierfabrikation, 35* (37): 3032 (1907)
1331. Kirchner, E. –Das Papier, 4 Teil Ganzstoffe. Günther. 1906.
1332. Meltzer, F. – Key figures for the assessment of the refining process. *In:* Refining Technology. C. Baker, Ed. Pira International, Leatherhead. Chapter 2, pg. 35 (2000).
1333. McKenzie, A. W. and Prosser, N. A. – The beating action of a PFI Mill. *Appita Journal, 34* (4): 239–297 (1981).
1334. Hietaniemi, J. and Gullichsen, J. – Flow properties of medium-consistency fibre suspensions. *Journal of Pulp and Paper Science, 22* (12): J469–474 (1996).
1335. Marcoccia, B. S. and Poulin, T. M. – A lo-solids pulping update. *In:* XXIX Congresso Anual de Celulose e Papel. ABTCP, São Paulo, 1996 pgs. 203–219.
1336. Arjas, A. – Residence time distribution in conical refiners and refining systems. *In:* International symposium of Fundamental Concepts of Refining. Appleton, 1980 pgs. 8–20.
1337. Björkman, U. – Stress generation and transmission in suspended fibre networks. *Nordic Pulp and Paper Research Journal, 18* (1): 38–43 (2003).

1338. Lindström, T. – Der einfluss chemischer factoren auf faserquellung und papierfestigkeit. *Das Papier, 34* (12): 561 (1980)
1339. Olejnik, K.; Skalski, B.; Stanislawska, A.; Wysocka-Robak, A. – Swelling properties and generation of cellulose fines originating from bleached kraft pulp refined under different operating conditions. *Cellulose 24*: 3955–3967 (2017).
1340. Claudio-da-Silva Jr., E. – Efeitos da moagem nas propriedades de fibras. Separata del boletin "Investigación y Técnica del Papel", 75 (1983).
1341. Brecht, W. and Siewert, W. – The development of small refiner into a universal laboratory beating machine. *Das Papier, 18* (10): 638–645 (1964).
1342. Mohlin, U-B. – Industrial refining of unbleached kraft pulps – the effect of pH and refining intensity, *In:* Technology Summit Proceedings. TAPPI, 2002 https://imisrise.tappi.org/TAPPI/Products/ts/ts0226.aspx
1343. Anderson, J. E. – Evaluation of the lasentec laser microscope for monitoring refining. *In:* Papermakers Conference. TAPPI, Philadelphia, 1996 pgs. 225–235.
1344. Paula, I. de; Guerta, A. C. and Martinez, R. M. – Comparisson of eucalypt, pine and sugarcane cellulose fibers used for paper production. *Revista Árvore 43* (4) (2019) https://www.scielo.br/j/rarv/a/WVYbtPRkDgXj3WCXdBjm48g/?lang=en
1345. Roux, J. C.; Chu, J. P.; Joris, G. and Caucal, G. – Théorie hydro-mécaque du fonctionnement d'um raffrineur à baisse concentration. *Revenue ATIP, 53* (4–5): 100 (1999).
1346. Roux, J-C.; Joris, G. and Caucal, G. – Angular parameters beyond specific edge load. *In:* 6th International Refining Conference. PIRA, Toronto, 2001.
1347. Roux, J-C. and Joris, G. – Specific edge load. *Japanese Journal of Paper Technology, 10* (2003) [1309]
1348. Roux, J. C. – Geometrical approach to crossing point theory. Pulp Refining Process, Volume I, Chapter 5 pgs. 245–295.
1349. Roux, J. C. and Joris, G. – Optimizing refining for economic and qualitative performance. Some advantages and some forecasts of development. IP91 – ATIP Congress, 1991.
1350. Alexandre, G.; Gomes, G. A.; Rodrigues, S.; Ferreira, L.; Nascimento, W. and Alexandre, L. A. C. – Influência da temperatura de alimentação da massa na refinação de fibras curtas não branqueadas de eucalipto. *In:* 48th ABTCP International Pulp and Paper Congress & 1st Ibero-American Conference on the Bio-Economy – BEM. São Paulo. 2015 9 pgs.
1351. Kerekes, R. J. – Pulp floc behavior in entry flow to constrictions. *Tappi Journal, 66* (1): 88–91 (1983).
1352. Sampson, W.W. and Wilde, R. – An investigation into the pilot scale refining of blended papermaking furnishes. *Appita Journal, 54* (6): 547–551 (2001).
1353. Ouellet, D. – Characterizing refining intensity for tensile strength development. *In:* 5th International Paper and Board Industry Conference. PIRA, Vienna, 1999 17 pgs.
1354. Ellery, M. – Selection of refiner control scheme requires total stock system review. *In:* Process Control for Pulp and Paper Mills. Ed. Smith, K. Miller Freeman Publications, San Francisco, 1983 Chapter 21 pgs. 89–93.
1355. Poppel, E. and Turtureanu, C. – Cercetari privind evaluarea severiatii proceselor de macinare. *Celul. Hirtie, 45* (1): 3–12 (1996).
1356. Meltzer, F. P. – Reducing specific energy in refining with a new key figure and innovative fillings materials. *In:* Papermakers Conference. TAPPI, Philadelphia, 1996 pgs. 187–194.
1357. Ettefagh, H. H. – Application of low consistency refining of pulp in producing multi-ply folding boxboards. Master's Thesis. Toronto University, Toroto, 2016 119 pgs.
1358. Kerekes, R. J. – Perspectives on high and low consistency refining in mechanical pulping. *BioResources, 10* (4): 8795–8811 (2015).
1359. Harju, K. and Hietanummi, J. – Low and high consistency refining of reiforcement pulp. *In:* Scientific and Technical Advances in Refining & Mechanical Pulping – Course Notes. PIRA: Stockholm: 25-26/03/2003 Paper 11.
1360. Marcoccia, B. S. – Lo-Solids pulping: theorical background and mill scale experiences. Kamyr Inc., *Status Report 130* (1994).

1361. Espenmiller, H. P. – The theory and practice of refining. *Southern Pulp and Paper Manufacturing, 32* (4): 50–57 (1969).
1362. Shi, Y.; Belosinschi, D.; Brouillette, F.; Belfkira, A. and Chabot, B. – The properties of phosphorylated kraft fibers. *BioResources 10* (3): 4375–4390 (2015).
1363. Naito, T.; Vsuda, M. and Kapoya, J. – Fatigue phenomena of single fibers under repeated torsional stress. *In:* International Paper Physics Conference. TAPPI, 1981. pgs. 197–201.
1364. Wild, P.M. and Provan, J.W. – Fatigue damage accumulation and failure mechanisms in single wood-pulp fibres subjected to cyclic loading. *In:* Sixth International Fatigue Conference. Berlin, 1996 pg. 1597
1365. Prairie, B.; Wild, P.; Byrnes, P.; Olender, D.; Francis, W. and Ouellet, D. – Forces during bar-passing events in low-consistency refining: Distributions and relationships to specific edge load. *Journal of Pulp and Paper Science, 34* (1): 1–8 (2008).
1366. Alaskevich, Y. D. – Study of the pressure in the liquid layer between knives of a beater. Bum. Promst, 46 (10): 16–17 (1971) [1077].
1367. Kibblewhite, R. P. and Brookes, D. – Fibre, beating and papermaking properties of kraft pulps froam New Zealand beech (*Nothofagus*) species. *General N.Z. Journal of Forestry Science, 7* (3): 425–44 (1977).
1368. Szwarcsztajn, E. and Przybysz, K. – External fibrilation of beaten cellulose fibres. *Cellulose Chemistry and Tecnhology* (6): 223–238 (1972).
1369. Dietmann, P. and Roux, J-C. – A study of disc refiner running in no-load conditions. *Cellulose Chemistry Technology, 39* (5–6): 459–471 (2005).
1370. Batchelor, W.; Krizinger, J.; Bauer, W.; Kuntzsch, T. and Meinl, G. – Improved characterization of changes in fibre coross section during sheet forming and drying using optical fibre analyzer data and a serial sectioning technique. *In:* COST Action E54 Fine Structure of Papermaking Fibres. Cost Action, Brussels, 2011 pgs 191–196.
1371. Kerekes, R. J.; Soszynski, R. M. and Tam Doo, P. A. – The flocculation of pulp fibres. *In:* Papermaking Raw Materials, Transactions 8th Fundamental Research Symposium, Ed. Ponton, V., Mech. Eng. Publ. Ltd., Oxford: 1985 pgs. 265–310.
1372. Rihs, J.; Albert, K. and Josephson, W. – Optimal refining of bleached tropical hardwood kraft for uncoated paper. *In:* 51st Annual General Conference. APPITA, Melbourne, 1997 Paper 4B21 pgs. 627–634.
1373. Lumianen, J. – Comparison of the mode of operation of conical and disc refiners. *In:* CALAR Symposium. PTS-IZP, Heidenan, 1995 [871]
1374. Soini, P. and Partanen, M. – The effect of recirculation on pulp properties in refining of bleached hardwood and softwood pulps with a conical refiner. *In:* Preparing for the Next Millennium Conference. TAPPI, Atlanta, 1999 Volume 2 pgs. 711–727.
1375. Bonfiglio, F.; Curbelo, V.; Santana, E. and Doldán, J. – Grinding of a PFI mill: a comparison between two PFI mills by evaluation of Eucalyptus pulp physical properties. *In:* 6th International Colloquium on Eucalyptus Pulp. Colonia del Sacramiento, 2013 10 pgs. www.researchgate.net/publication/308902583_Grinding_of_a_PFI_Mill_a_comparison_between_two_PFI_mills_by_evaluation_of_Eucalyptus_pulp_physical_properties
1376. Wang, X.; Maloney, T. and Paulapuro, H. – Fibre fibrillation and its impact on sheet properties. *Paperi ja Puu – Paper and Timber, 89* (3):148–151 (2007).
1377. Somboon, P. – Quality developments in pulp fibers using conventional papermaking laboratory beaters and an industrial pilot refiner. *Kasetsart Journal (Natural Science), 45*: 883–890 (2011)
1378. Keays, J. L.; McDowell, P. F. and Hatton, J. V. – Single-point procedure for PFI-mill evaluation of softwood kraft pulps. *Tappi Journal, 60* (6): 81–83 (1977).
1379. Demler, C. L. Low consistency refiner mechanics & troubleshooting. *In:* PaperCon, 2013 Oral presentation. https://www.eucalyptus.com.br/artigos/2013Low+Consistency+Refiners_Mechanics.pdf
1380. Dölle, K. and Bardhyl Bajrami, B. – Beating of eucalyptus pulp fibers under neutral and alkaline conditions – A Valley beater study. *Journal of Engineering Research and Reports, 20* (8): 86–96 (2021).

1381. Mohlin, U-B. – Refining of unbleached kraft pulps – the effect of pH and refining intensity. *In:* Tappi Paper Summit. TAPPI, Atlanta, 2002
1382. Manfredi, V. – Analisando curvas de refino de polpa kraft de eucalipto: efeitos do pH e da consistência de refino. *Eucalyptus Newsletter* (89): 17–28 (2021) https://www.eucalyptus.com.br/news/pt_jul2021.pdf
1383. Lundberg, R. and De Ruvo, A. – The influence of defibration and beating conditions on the paper making potential of recycled paper. *Svensk Papperstiding, 81* (12): 383–386 (1978).
1384. Mangini, L. F. K. and Andrade, A. A. – Ação de diferentes tipos de discos de refinação sobre as características das fibras químicas. *In:* 43th Congresso e Exposição Internacional de Celulose e Papel. ABTCP, São Paulo, 2010 11 pgs.
1385. Cai, H.; Yuan, Z.; Tong, G.; Zhang, X. and Zhang, H. – Comparison of two bar edge lengths of refining plates on the properties of American old corrugated container pulp during low consistency refining. *BioResources, 15* (1):347–359 (2020).
1386. Steel, C. L. – Evaluation of cell-wall modifying enzymes for improved refining of pulp from two eucalyptus species. Master's Thesis. University of the Free State, Bloemfontein, 2010 145 pgs https://scholar.ufs.ac.za/xmlui/handle/11660/5576
1387. Joy, E.; Rintamaki, J.; Weckroth, R. and Tuomelai, P. – Ultra-low intensity refining of short fibered pulps. *In:* African Pulp and Paper Week, 2004 7 pgs https://www.eucalyptus.com.br/artigos/2004_Ultra-low+Intensity+Refining.pdf
1388. Sanjay, C.; Akansha, S.; Rajeev, S.; Shankarshan, S. and Narendra, A. – Optimization of fibre properties using single pass refining vs recirculation – A Case Study. *IPPTA, 27* (04): 58–66 (2015).
1389. Radoslavova, D.; Roux, J. C. and Silvy, J. – Energy efficiency analysis in refining. *In:* IV International Refining Conference. PIRA, Fiuggi, 1997 Paper 3 pgs. 63–72.
1390. Crook, J. – The practical refiner. *In:* IV International Refining Conference. PIRA, Fiuggi, 1997 Paper 11 pgs. 185–189.
1391. Metelski, L. – Análise da ação enzimática na produção de papel visando a economia de energia. Trabalho de conclusão de curso, Universidade Tecnológica Federal do Paraná, Ponta Grossa, 2018 53 pgs. www.repositorio.utfpr.edu.br:8080/jspui/bitstream/1/16466/1/PG_COENQ_2018_2_18
1392. Elahimehr, A. – Low consistency refining of mechanical pulp: the relationship between plate pattern, operating variables and pulp properties. Doctoral Thesis, University of British Columbia, Vancouver, 2014 120 pgs. https://central.bac-lac.gc.ca/.item?id=TC-BVAU-50745&op=pdf&app=Library&oclc_number=1032967384
1393. Cisneros, H. A.; Williams, G. J. Hatton, J. V. – Fibre surface characteristics of hardwood refiner pulps. *Journal of Pulp and Paper Science, 21* (5): 178–184 (1995).
1394. Liu, H.; Dong, J. X.; Guo, X. Y.; Luo, C.; Tian, X. H.; Jiang, X. J.; Wang, S.; Yang, R. F.; Duan, C. W.; Wang, B.; Qi, K. – Study on fiber cutting performance of isometric straight bar plate with different bar angle. *Journal of Korea TAPPI, 51* (5): 16–26. (2019).
1395. Ryti, N. and Arjas, A. – Influence of residence time distribution of the flowing stock on the beating effect of a beating machine, Part II. *Paperi ja Puu, 51* (1): 69–84 (1969).
1396. Ryti, N. and Arjas, A. – Influence of residence time distribution of the flowing stock on the beating effect of a beating machine, Part III. *Paperi ja Puu, 51* (2): 163–168 (1969).
1397. Bedretchuk, J. P. – Estudo de caso: aplicação de um controle de energia específica em refinadores a disco. Monografia de Especialização. Universidade Tecnológica Federal do Paraná, Curitiba. 2016 51 pgs http://repositorio.roca.utfpr.edu.br/jspui/bitstream/1/13164/1/CT_CEAUT_2015_16.pdf
1398. Noe, P.; Joris, G. and Roux, J. C. – Pulp refining process. Volume I. Bonpertuis [710]
1399. Marton, R.; Crosby, C.; Brown, A.; Bachorik, R. and Shook, P. – High-consistency refining. *Espra Research Reports, 44* (III): 18–19 (1967) according to *ABIPST, 67* (5): 728 reference 6663(BS).
1400. Buzała K. P.; Przybysz, P.; Kalinowska, H.; Derkowska, M. – Effect of cellulases and xylanases on refining process and kraft pulp properties. *PLoS ONE 11* (8): e0161575 (2016) https://doi.org/10.1371/journal.pone.0161575

1401. da Silva, J. C.; de Oliveira, R. C.; Larisse Ribas Batalha, L. R. B and Manfredi, M – Combination of enzymatic, mechanical and ultrasonic treatments for improvement of the properties of secondary pulps. *Cerne, 19,* (4): 653–660 (2013) https://www.scielo.br/j/cerne/a/TZF4bFXjshfxhVqCHJrVj3L/?lang=en
1402. Singh, R.; Bhardwaj, N. K. and Choudhury, B. – Cellulase-assisted refining optimization for saving electrical energy demand and pulp quality evaluation. *Journal of Scientific & Industrial Research, 74* (8): 471–475 (2015).
1403. Kerekes, C. and McDonald, J. D. – Fiber-based characterization of pulp refining. *Tappi Journal, 21* (9): 497–503 (2022)
1404. Zhang, Q., Xu, M., Xing, L., Dang, C., Han, X., and Pu, J. – Enzymatic assisted ultrasonic pretreatment's effect on poplar pulp properties. *BioResources, 12* (3): 6832–6843 (2017).
1405. Liu, H.; Dong, J.; Luo, C.; Duan, C., Guo. X.; Qi, K.; Qiao, L.; Zhao, Z. – Accuracy of the different calculation methods of specific edge load. *Journal of Korea TAPPI, 52* (5): 45–54 (2020) http://www.ktappi.kr/xml/26604/26604.pdf
1406. Berglund, J., Mikkelsen, D., Flanagan, B.M. et al. – Wood hemicelluloses exert distinct biomechanical contributions to cellulose fibrillar networks. *Nature Commun*ications, 11, 4692 (2020) 15 pgs. https://doi.org/10.1038/s41467-020-18390-z
1407. Manfredi, M.; Oliveira, R. C. de; Siçba, J. C. da and, R. – Ultrasonic treatment of secondary fibers to improve paper properties. *Nordic Pulp and Paper Research Journal, 28* (2):297 (2013).
1408. Foelkel, C. – Importância da qualidade das madeiras e fibras celulósicas. *Eucalyptus Newsletter n° 85* (Fev) 2021 963 pgs https://www.eucalyptus.com.br/news/pt_fev2021.pdf
1409. Park, J-M. – Application of fracture toughness understanding to improve papermaking runnability. *Journal of Korea TAPPI, 54* (1): 3–9 (2022).
1410. Liu, H.; Dong, J.; Jing, H.; Xiya Guo, X. and Qiao, L. – Characterization of the parameters for the refining Intensity in terms of performance. *Journal of Korea TAPPI, 51* (2): 26–39 (2019);
1411. Kim, J-S. and Cho, B-U – Effects of refining degree, pulp mixing ratio and PAE-based strength agents on initial wet-web strength of HwBKP—SwBKP sheet. *Journal of Korea TAPPI, 52* (1): 110–120 (2020).
1412. Lee, J-Y.; Kim, C-H.; Park, S-H. and Sol Kwon, S. – Effects of refining intensities on physical properties of Interleaving paper for Stainless Steel. *Journal of* Korea *TAPPI, 51* (6): 45–51 (2019).
1413. Helmuth, G.; Gorton-Hulgerth, A. and Takeshita, Y. – A new low consistency refiner improved fibre properties while reducing idle energy by up to 40%. *In:* Japan Tappi Annual Meeting. Japan Tappi, Hiroshima, Paper C-12 (2003).
1414. Rennel, J. – Opacity in relation to strength properties of pulps III. Light scattering coefficient of sheets of model fibres. *Tappi, 52* (10): 1943–1947 (1969).

Index

© The Author(s), under exclusive license to Springer Nature Switzerland AG 2024
V. Manfredi, *Eucalyptus Kraft Pulp Refining*,
https://doi.org/10.1007/978-3-031-47285-5

GPSR Compliance
The European Union's (EU) General Product Safety Regulation (GPSR) is a set of rules that requires consumer products to be safe and our obligations to ensure this.

If you have any concerns about our products, you can contact us on

ProductSafety@springernature.com

In case Publisher is established outside the EU, the EU authorized representative is:

Springer Nature Customer Service Center GmbH
Europaplatz 3
69115 Heidelberg, Germany

www.ingramcontent.com/pod-product-compliance
Ingram Content Group UK Ltd.
Pitfield, Milton Keynes, MK11 3LW, UK
UKHW022214051026
15624UKWH00001B/155
9783031472848